RATE CONSTANT CALCULATION FOR THERMAL REACTIONS

RATE CONSTANT CALCULATION FOR THERMAL REACTIONS

METHODS AND APPLICATIONS

Edited by

Herbert DaCosta
Maohong Fan

WILEY

A JOHN WILEY & SONS, INC., PUBLICATION

Published by John Wiley & Sons, Inc., Hoboken, New Jersey
Published simultaneously in Canada

Library of Congress Cataloging-in-Publication Data:

Rate constant estimation for thermal reactions : methods and applications / edited by Herbert DaCosta, Maohong Fan.
p. cm.
Includes bibliographical references and index.
ISBN 978-0-470-58230-5
1. Thermochemistry–Mathematics. 2. Chemical kinetics–Effect of temperature on–Mathematics. 3. Numerical calculations. I. DaCosta, Herbert. II. Fan, Maohong.
QD504.R38 2012
541'.36–dc23

2011041445

Printed in the United States of America

10 9 8 7 6 5 4 3 2 1

CONTENTS

PREFACE

In the past 30 years, the computational chemistry field has experienced an exponential growth. This growth has been enabled by tremendous improvements in computer hardware, theoretical methods, and numerical methods to integrate the theoretical methods into computer software. Applications of computational chemistry are now abundant in diversified areas, including nanotechnology, drug design, materials design, molecular design, tribology, lubricants, coal chemistry, petroleum chemistry, biomass chemistry, combustion, and catalysis.

The recent developments in computational chemistry have also enabled a large qualitative leap in the field of computational kinetics, thus yielding significant contributions to the chemical and engineering literatures. Despite all these progresses, to our knowledge, a book describing the modern methods used by scientists and engineers in order to predict rate constants has not as yet been published. This book addresses this need, as it was designed to serve as a major reference for prediction of rate constants of thermal reactions. Some successful examples along with the highlights of certain computational methods currently used in the literature are presented in this book. Therefore, it will be a useful tool for academic and industrial chemists and engineers working in the areas of chemical kinetics and reaction engineering.

The first five chapters (Part I) present an overview of some methods that have been used in the recent literature to calculate rate constants and the associated case studies. The main topics covered in this part include thermochemistry and kinetics, computational chemistry and kinetics, quantum instanton, kinetic calculations in liquid solutions, and new applications of density functional theory in kinetic calculations. The remaining five chapters (Part II) are focused on applications even though methodologies are discussed. The topics in the second part include the kinetics of molecules relevant to combustion processes, intermolecular electron transfer reactivity of organic compounds, lignin model compounds, and coal model compounds in addition to free radical polymerization.

This book is also part of Wiley's special celebration activities in marking the International Year of Chemistry in 2011.

We would like to thank the whole Wiley team, in particular our Senior Acquisitions Editor, Mrs. Anita Lekhwani, for her vision, persistence, and support throughout the whole editing process, as well as Ms. Becky Amos and Ms. Catherine Odal, for their many helps.

We would also like to thank the outstanding body of researchers who contributed their time, knowledge, and expertise to the publication of this book.

Happy International Year of Chemistry!

HERBERT DACOSTA
MAOHONG FAN

CONTRIBUTORS

Elisabet Ahlberg, Department of Chemistry, Electrochemistry, University of Gothenburg, Gothenburg, Sweden

Ariana Beste, Computer Science and Mathematics Division, Oak Ridge National Laboratory, Oak Ridge, TN, USA

A.C. Buchanan, III, Chemical Sciences Division, Oak Ridge National Laboratory, Oak Ridge, TN, USA

Alexander Burcat, Faculty of Aerospace Engineering, Technion: Israel Institute of Technology, Haifa, Israel

Michael Busch, Department of Chemistry, Electrochemistry, University of Gothenburg, Gothenburg, Sweden

Michelle L. Coote, ARC Centre of Excellence in Free-Radical Chemistry and Biotechnology, Research School of Chemistry, Australian National University, Canberra, Australia

Fernando P. Cossío, Departamento de Química Orgánica I, Universidad del País Vasco-Euskal Herriko Unibertsitatea, San Sebastián-Donostia, Spain

Faina Dubnikova, Institute of Chemistry, The Hebrew University of Jerusalem, Jerusalem, Israel

Elke Goos, Institute of Combustion Technology, German Aerospace Center (DLR), Stuttgart, Germany

J. Peter Guthrie, Department of Chemistry, University of Western Ontario, London, Ontario, Canada

Assa Lifshitz, Institute of Chemistry, The Hebrew University of Jerusalem, Jerusalem, Israel

Lixia Ling, Research Institute of Special Chemicals, Taiyuan University of Technology, Taiyuan, China

Stephen F. Nelsen, Department of Chemistry, University of Wisconsin, Madison, WI, USA

Itai Panas, Department of Chemistry and Biotechnology, Energy and Materials, Chalmers University of Technology, Gothenburg, Sweden

Jack R. Pladziewicz, Department of Chemistry, University of Wisconsin, Eau Claire, WI, USA

Jiří Vaníček, Laboratory of Theoretical Physical Chemistry, Institut des Sciences et Ingénierie Chimiques, École Polytechnique Fédérale de Lausanne (EPFL), Lausanne, Switzerland

Baojun Wang, Key Laboratory of Coal Science and Technology, Taiyuan University of Technology, Ministry of Education and Shanxi Province, Taiyuan, China

Riguang Zhang, Key Laboratory of Coal Science and Technology, Taiyuan University of Technology, Ministry of Education and Shanxi Province, Taiyuan, China

PART I

METHODS

CHAPTER 1

Overview of Thermochemistry and Its Application to Reaction Kinetics

ELKE GOOS

Institute of Combustion Technology, German Aerospace Center (DLR), Stuttgart, Germany

ALEXANDER BURCAT

Faculty of Aerospace Engineering, Technion - Israel Institute of Technology, Haifa, Israel

1.1 HISTORY OF THERMOCHEMISTRY

Thermochemistry deals with energy and enthalpy changes accompanying chemical reactions and phase transformations and gives a first estimate of whether a given reaction can occur. To our knowledge, the field of thermochemistry started with the experiments done by Malhard and Le Chatelier [1] with gunpowder and explosives. The first of their two papers of 1883 starts with the sentence: "All combustion is accompanied by the release of heat that increases the temperature of the burned bodies." In 1897, Berthelot [2], who also experimented with explosives, published his two-volume monograph *Thermochimie* in which he summed up 40 years of calorimetric studies.

The first textbook, to our knowledge, that clearly explained the principles of thermochemical properties was authored by Lewis and Randall [3] in 1923.

Thermochemical data, actually heats of formation, were gathered, evaluated, and published for the first time in the seven-volume book *International Critical Tables of Numerical Data, Physics, Chemistry and Technology* [4] during 1926–1930 (and the additional index in 1933).

In 1932, the American Chemical Society (ACS) monograph No. 60 *The Free Energy of Some Organic Compounds* [5] appeared.

Rate Constant Calculation for Thermal Reactions: Methods and Applications, Edited by Herbert DaCosta and Maohong Fan.

In 1936 was published *The Thermochemistry of the Chemical Substances* [6] where the authors Bichowsky and Rossini attempted to standardize the available data and published them at a common temperature of 18°C (291K) and pressure of 1 atm.

In 1940, Josef Mayer and Nobel Prize winner Maria Mayer published their monograph *Statistical Mechanics* [7], in which the method of calculating thermochemical properties from spectroscopic data was explained in detail.

In 1947, Rossini et al. published their *Selected Values of Properties of Hydrocarbons* [8], which was followed by the famous NBS Circular 500 (1952) [9] that focuses on the thermochemistry of inorganic and organic species and lists not only the enthalpies of formation but also heat capacities (C_p), enthalpies ($H_T - H_0$), entropies (S), and equilibrium constants (K_c) as a function of temperature. Within the data, thermodynamic relations (e.g., through Hess's Law) between the same property of different substances or between different properties of the same substance were satisfied. During the 1950s, the loose leaf compendium of the Thermodynamic Research Center (TRC) [10] at A&M University in Texas appeared as a continuation of API Project 44. In this compendium, thermochemistry as a function of temperature is only a small part of their data that also include melting and boiling points, vapor pressures, IR spectra, and so on. Although their values are technically reliable, a very serious drawback is the lack of documentation of the data sources and the calculation methods.

In 1960, the first loose leaf edition of the Joint Army–Navy–Air Force (JANAF) thermodynamic tables appeared, but was restricted solely to U.S. government agencies. It is devoted to chemical species involving many elements; however, it contains only a very limited number of organic species. The publication, which became very famous when published as bound second edition in 1971 [11], set the standard temperature reference at 298.15K and published the enthalpy increments (also known as integrated heat capacities) as ($H_T - H_{298}$) instead of ($H_T - H_0$). This edition of the JANAF tables, with Stull as the main editor, for the first time described in detail methods of calculating thermochemical properties mainly based on the monograph of Mayer and Mayer [7]. It also set the upper temperature range limit of the tables up to 6000K in order to assist the needs and requests of the space research institutions and industry. Further editions published afterward [11] kept the many errors and wrong calculation results instead of correcting or improving them to include better available values.

Published in 1960, the report "Thermodynamic Data for Combustion Products" [12] by Gordon focused on high-performance solid rocket propellants.

In 1961, Duff and Bauer wrote a Los Alamos report [13], which was summarized in 1962 in the *Journal of Chemical Physics* [14], in which for the first time thermochemical properties of organic molecules, that are, enthalpies and free energies, were given as polynomials.

In 1963, McBride et al. published the "Thermodynamic Properties to 6000K for 210 Substances Involving the First 18 Elements," NASA Report SP-3001 [15]. This publication revealed for the first time to the public world the methods of calculating thermochemical data for monoatomic, diatomic, and polyatomic species. At that time, JANAF tables were accessible to only a very restricted number of people. The NASA publication lists, also for the first time, the thermochemical properties not only in table format but also as seven-coefficient polynomials. The NASA program to calculate thermochemical properties and these seven-term polynomials was published by McBride and Gordon in 1967 [16].

In 1965, the U.S. National Bureau of Standards (NBS) started publishing the Technical Note 270 [17] in a series of booklets where they presented heats of formation at 0, 273.15, and 298.15K.

In 1969, *The Chemical Thermodynamics of Organic Compounds* by Stull, Westrum and Sinke [18] was released, where the thermochemical properties of 741 stable organic molecules available until the end of year 1965 were published in the temperature range from 298 to 1000K.

In 1962, the first edition of *Thermodynamic Properties of Individual Substances* (TSIV) [19] appeared in Moscow. This monumental compendium became known worldwide as "Gurvich's Thermochemical Tables" from the further publications in 1978, 1979, 1982, and specifically the fourth edition of 1989 translated to English, which was also followed by further English editions in 1991, 1994, and 1997.

Other thermochemical properties mainly for solid species were published by Barin et al. [20] in 1973 and by Barin in 1995 [21].

Evaluations of heats of formation for organic molecules and radicals were published by Cox and Pilcher [22], Pedley and Rylance [23], Domalski and Hearing [24], and Pedley et al. [25].

1.2 THERMOCHEMICAL PROPERTIES

Malhard and Le Chatelier [1] observed that the interaction of substances (called reactants) results in new products, which was connected with release of heat Q ($Q < 0$ if heat is released and $Q > 0$ if heat is added). Thus, reactions that release heat will proceed more or less spontaneously (such as combustion), while those that absorb heat will not.

The heat released from producing 1 mol of a substance from its *reference elements* at a specified temperature T and at constant pressure P is defined as the *enthalpy of formation* $\Delta_f H_T$ of the product formed at this temperature.

The enthalpy of formation assigns a certain value, positive or negative, to each compound. By definition, all reference species (e.g., molecular gaseous hydrogen H_2, nitrogen N_2, oxygen O_2, chlorine Cl_2, fluorine F_2, crystal and liquid bromine Br_2, solid graphite $C_{graphite}$, white phosphorus P_{white}) in their standard states have each been assigned the value 0 to their enthalpy of formation $\Delta_f H_T$. Table 1.1 shows the standard enthalpies of formation for small gas-phase species relevant to combustion studies.

For a given material or substance, the standard state is the reference state for the substance's thermodynamic state properties such as enthalpy, entropy, Gibbs free energy, and so on.

According to the International Union of Pure and Applied Chemistry (IUPAC) [26], the standard state of a gaseous substance is the (hypothetical) state of the pure

TABLE 1.1 Standard Enthalpies of Formation in kJ/mol at 0 and 298.15K for Small Gas-Phase Species of Interest in Combustion

Species (g)	$\Delta_f H_{0K}$ (kJ/mol)	$\Delta_f H_{298.15K}$ (kJ/mol)	Uncertainty (kJ/mol)
C	711.38	716.87	±0.06
H	216.034	217.998	±0.0001
O	246.844	249.229	±0.002
N	470.57	472.44	±0.03
NO	90.59	91.09	±0.06
NO_2	36.83	34.02	±0.07
OH	37.26	37.50	±0.03
HO_2	15.18	12.27	±0.16
H_2O	−238.918	−241.822	±0.027
CH_4	−66.56	−74.53	±0.06
C_6H_5	350.6	337.3	±0.6
C_6H_6	100.7	83.2	±0.3
CO_2	−393.108	393.474	±0.014

Source: All species from Active Thermochemical Tables, version 1.110 (http://atct.anl.gov/); the listed uncertainties correspond to 95% confidence limits.

gaseous substance at standard pressure (1 bar), assuming ideal gas behavior. For a pure phase, a mixture, or a solvent in the liquid or solid state, the standard state is the state of the pure substance in the according phase at standard pressure.

It is not mandatory for the standard state of a substance to exist in nature. For instance, it is possible to calculate values for steam at 20°C and 1 bar, even though steam does not exist as a gas under these conditions. However, this definition results in the advantage of self-consistent tables of thermodynamic properties.

The *enthalpy of a reaction* $\Delta_r H_T$ is the sum of enthalpies of formation $\Delta_f H_T$ of all products minus the sum of enthalpies of formation of all reactants:

$$\Delta_r H_T = \sum_{\text{products}} \Delta_f H_T - \sum_{\text{react}} \Delta_f H_T \tag{1.1}$$

The enthalpy of reaction is negative if the reaction releases heat. This type of reaction is defined as an *exothermic reaction* and normally occurs instantaneously.

On the other hand, an *endothermic reaction* has a positive enthalpy of reaction. It can only take place if there is a particular amount of energy available to absorb, which is equal to or larger than the value of the enthalpy of reaction needed. The enthalpy itself is temperature dependent and called *sensible enthalpy* or *sensible heat*, and is defined as the amount of heat required for raising the temperature of a substance by 1K without changing its molecular structure.

The derivative of the enthalpy with respect to the temperature at constant pressure defines the *specific heat capacity* C_P of a substance:

$$C_p = \left(\frac{\partial H}{\partial T}\right)_p \tag{1.2}$$

It is usually easier to measure experimentally C_P rather than the *sensible enthalpy* H and therefore it is customary to calculate the *enthalpy* by integration of C_P; thus:

$$H_T(T) = H_{298\text{K}} + \int_{298\text{K}}^{T} C_p d\tilde{T} \tag{1.3}$$

and therefore

$$H_T - H_{298\text{K}} = \int_{298\text{K}}^{T} C_p d\tilde{T} \tag{1.4}$$

The "chemist's enthalpy" $H_T - H_{298\text{K}}$ is usually found in thermochemical tables [11,15,18,19,69].

In engineering practice, the *absolute enthalpy* is defined as

$$H_T(T) = \Delta_f H^{\circ}_{298\text{K}} + \int_{298\text{K}}^{T} C_p d\tilde{T} \tag{1.5}$$

which is equal to

$$H_T = \Delta_f H^{\circ}_{298} + (H_T - H_{298}) \tag{1.6}$$

This value is usually found in engineering thermodynamics books, in the NASA tables, and the NASA thermochemical polynomials [15].

Enthalpy is a *state function*; therefore, the heat change associated with a reaction does not depend on the reaction pathway. If the reaction proceeds from reactants to products in a single step or in a series of steps, the same enthalpy will be obtained. This is the basis of *Hess's Law*. A handy combination of reactions enables the calculation of enthalpies of formation of substances, which cannot be measured directly.

The term *combustion enthalpy* is used for the enthalpy of reaction for complete combustion of 1 mol of a substance into the products carbon dioxide and water.

Heat δQ added to a system in an infinitesimal process is used to increase the internal energy by dE and to perform an amount of work δW:

$$\delta Q = dE + \delta W \tag{1.7}$$

where E is a system property and δQ and δW are path-dependent properties. This is the state of the *first law of thermodynamics*.

The *second law of thermodynamics* says that a quantity called *entropy* S exists, and that for an infinitesimal process in a closed system the equation

$$TdS \geq \delta Q \tag{1.8}$$

is always fulfilled. For reversible processes, only the equality holds; for all natural processes, the inequality exists.

The entropy S is the hardest thermodynamic property to understand and to explain. It is a consequence of the second law of thermodynamics that states that we cannot produce energy from nothing, in other words, "it is impossible to build a Perpetuum Mobile." As a consequence, there is some energy content that we continually "waste" and entropy is a measurement of this "waste." In all natural processes, entropy increases and therefore the "world" entropy increases with time. It introduces the concept of irreversibility and defines a unique direction of time.

This can be explained on mixing phenomena. Two pure and unmixed substances have small entropy values. But with time, the substances tend to mix and the entropy of the system reaches its highest value at complete mixing of all contributing substances. Thus, entropy is a measurement of the disorder of a system or the measurement of the amount of energy in a system that cannot do work.

For pure substances, the entropy has a fixed value that is a function of temperature as all other thermochemical properties. The standard definition of entropy is

$$dS = \left(\frac{\partial Q}{T}\right)_{\text{reversible}} \tag{1.9}$$

and it can be calculated from partition functions using Eq. (1.20).

1.3 CONSEQUENCES OF THERMODYNAMIC LAWS TO CHEMICAL KINETICS

The second law of thermodynamics states that every closed isolated system will approach after infinite time an "equilibrium" state, where the properties of the system are independent of time.

Thermodynamics, however, is unable to predict the time required for reaching equilibrium or the system composition and its changes during the time needed to reach equilibrium.

On the other hand, the thermochemical properties are strong quantitative constraints on the kinetic parameters driving a time-varying system. The reason is that an equilibrium state is in reality a dynamic state in which, at the molecular level, chemical changes are still occurring, while at the macroscopic level these changes in composition are not noticeable because the rate of production of a given substance is equal to its rate of destruction.

It has been empirically found that the rate W by which a reaction $A + B \rightarrow C + D$ occurs is equal to

$$W = k_f \prod_i [C_i]^{m_i} \tag{1.10}$$

where k_f is the temperature-dependent reaction rate coefficient for the forward reaction and $[C_i]^{m_i}$ are the concentrations of the reactants i to the power of m.

The reaction rate coefficient can be described in Arrhenius form as

$$k_f = AT^n \exp\left(\frac{-E_a}{RT}\right) \tag{1.11}$$

with the pre-exponential factor A, the temperature exponent n, and the so-called activation energy E_a.

Thermochemistry can help us in finding good estimates for different values of the Arrhenius reaction coefficient, which are given by

$$k_f = \frac{k_B T}{h} \exp\left(-\frac{\Delta G^{\neq}}{RT} \right) \tag{1.12}$$

where k_B is the Boltzmann constant, h is Planck's constant, and $\Delta G^{\neq}$ is defined as the change in Gibbs energy G from reactants to transition state of the reaction under investigation:

$$\Delta G^{\neq} = \Delta H^{\neq} - T\Delta S^{\neq} \tag{1.13}$$

The transition state theory [27] of chemical kinetics assumes that the reaction rate is limited by the formation of a transient *transition state*, which is the point of maximum energy along the reaction pathway from reactants to products. The transition state is considered to be in quasi-equilibrium with the reactants. Differences between reactants and the transition state are denoted with a $\neq$ symbol. Therefore, $\Delta H^{\neq}$ is defined as the enthalpy difference between the transition state and the reactants:

$$\Delta H^{\neq} = H_{TS} - H_{react} \tag{1.14}$$

and the entropy and free Gibbs energy are defined accordingly.

Thus, reaction rate coefficients can be estimated from the "thermochemistry" of the transition states, whose molecular properties can be calculated with quantum chemical programs. In calculating reaction rate coefficients, the only negative second derivative of energy with respect to atomic coordinates (called "imaginary vibrational frequency") from the transition state is ignored, so that there are only $3N-7$ molecular vibrations in the transition structure ($3N-6$ if linear) and all internal and external symmetry numbers have to be included in the rotational partition functions (then any reaction path degeneracy is usually included automatically).

Detailed knowledge of thermodynamic data is needed to obtain both the endothermicity/exothermicity ΔH_r and endergonicity/exergonicity ΔG_r of a reaction, which determine the equilibrium composition of a reacting mixture. Accurate thermochemistry values or good estimates are needed, particularly at lower temperatures, in order to properly predict reaction rate coefficients and their temperature dependency.

For more complicated reaction systems with competing reaction pathways, an additional master equation modeling is necessary to calculate and predict reaction rate coefficients. This treatment [28] includes the collisional energy transfer between rotational and vibrational energy levels of the reactants through activation or collisional deactivation and the different energy amount needed to overcome the transition states.

Besides the calculation of reaction rate coefficients of unimolecular decomposition reactions such as the thermal decomposition of toluene [29] or methyl radicals [30], and of bimolecular reactions such as the reaction of CO with HO_2 to CO_2 and OH, which transforms a relatively stable radical HO_2 to a more reactive one OH [63] also the reaction rate coefficients and branching ratios of multiwell reactions [31] can be calculated with a lot of different product channels. These calculated reaction rate coefficients for elementary reactions can be used to build and evaluate chemical mechanisms for combustion models [32].

1.4 HOW TO GET THERMOCHEMICAL VALUES?

1.4.1 Measurement of Thermochemical Values

Using calorimetry, time-dependent heat changes of substances or chemical reaction systems can be measured in a closed chamber through monitoring temperature changes.

Since no work is performed in these constant volume chambers, the heat measured equals the change in internal energy U of the system. With known temperature change, the heat capacity C_V at constant volume V can be derived under the assumption that C_V is constant for the small temperature variation measured:

$$q = C_V \Delta T = \Delta U \tag{1.15}$$

Since the pressure is not kept constant, the heat measured does not represent the enthalpy change.

Improvement of measurement techniques allows the use of smaller amounts of stable species and substances with fewer impurities, which should yield more accurate experimental data.

1.4.2 Calculation of Thermochemical Values

1.4.2.1 Quantum Chemical Calculations of Molecular Properties For the calculation of atomic and molecular properties of chemical compounds, computational methods such as molecular mechanics, molecular dynamics, and semiempirical and *ab initio* molecular orbital methods are available.

Due to the developments of computer hardware in combination with developments in the quantum chemical calculation methods, thermochemistry calculations for small molecules are now possible with accuracy in sub-kilojoule per mole.

In the last few decades, semiempirical methods [33], implemented in programs such as MOPAC [34], were superseded by density functional theory (DFT) and more accurate *ab initio* methods, which are available in program packages such as Columbus [35], DGauss [36], GAMESS (US) [37], GAMESS-UK [38], Gaussian [39], MOLPRO [40], NWChem [41], Q-Chem [42], and other electronic structure computational programs.

Among the methods that calculate the species electronic structure, DFT has gained an important position. Specifically, the Becke exchange functional [43] coupled with the Lee–Yang–Parr functional [44], which is widely known as B3LYP, is often used because it was one of the first to allow calculations for large molecules.

The composite G3 method [45] and its variant G3B3 [46] are able to achieve good accuracy (with a 95% confidence limit that is generally around ±2 kcal/mol or better) for calculation of thermochemical values, without requiring an exorbitant computational effort.

The composite G3B3 [46] method optimizes the geometry and calculates the vibrational frequencies and rotational constants using DFT method with B3LYP functional. The results compare very well with experimental UV-VIS, IR, and Raman spectra.

The molecular energy is then calculated using a composite approach that performs a sequence of calculations at various levels of theory and with various basis sets, effectively estimating the energies at QCISD(T) level using a large basis set (G3Large).

The molecular properties, such as geometry, vibrational frequencies, and rotational constants, are needed to compute thermodynamic properties such as enthalpy, entropy, and Gibbs free energy through calculation of the partition functions of the substances using statistical mechanics methods.

Nowadays, it is well known that a density functional (DF) performing well for a certain property is not necessarily adequate for computing completely different types of molecular systems or molecular properties. Actual research continues to develop DFs that are equally well applicable to a variety of different properties.

Pople and coworkers [47] have first realized the benefit of evaluating quantum chemical methods by benchmarking them against accurate experimental measurements. Their work mainly focused on atomization energies, which were used to calculate the heats of formation for around 150 molecules having well-established enthalpies of formation at 298K and were summarized in the so-called G2/97 benchmark test set [48] and later enhanced to the benchmark versions G3/99 [49] and G3/05 [50], where electron and proton affinities and ionization potentials of small molecules played an additional minor role.

The idea of benchmarking quantum chemical methods by introducing databases covering a wide variety of different properties, for example, atomization energies, spectroscopic properties, barrier heights and reaction energies of diverse reactions, proton affinities, interaction energies of noncovalent bond systems, transition metal systems, and catalytic processes, was extended by Truhlar and coworkers [51]. They were the first to carry out overall statistical analyses of combinations of different test sets to obtain an overall mean absolute deviation (MAD) number for each tested quantum chemical method, which made a comparison with other approaches more feasible.

Later on, Goerigk and Grimme further improved density functionals and enhanced the range of benchmarked parameters and the size of the calculated molecules [52].

On the other hand, computational thermochemistry values in the sub-kilojoule per mole accuracy range are now possible only for small molecules. They can be calculated through the highly accurate extrapolated *ab initio* thermochemistry (HEAT) [53] approach developed by an international group of researchers and by the Weizmann-4 (W4) method [54] from Martin's group, which was benchmarked on atomization energies of 99 small molecules [55]. They further developed an economical post-CCSD(T) computational thermochemistry protocol [56] that decreased the demanding amount of computer resources needed and therefore were able to apply these methods to small aromatic systems with less than 10 heavy atoms.

In addition, simple and efficient CCSD(T)-F12x approximations ($x = a$, b) [57] were proposed and benchmarked [58] by Werner's group. They obtained improvements in basis set convergence for calculations of equilibrium geometries, harmonic vibrational frequencies, atomization energies, electron affinities, ionization potentials, and reaction energies of open- and closed-shell reaction systems, where chemical accuracy of total reaction energies was obtained for the first time using valence double-zeta basis sets.

High-level benchmarked quantum chemical calculation results have been reached or are now more accurate than experimental accuracy, and spectroscopic and thermodynamic properties of molecules, such as radicals, which are otherwise very hard to measure experimentally, can be predicted.

As of now, quantum chemical methods with high accuracy are very demanding on computer resources and have been applied only to smaller molecules. But with improvements in computer resources, faster writing/reading speeds of data storage units, and further development of quantum chemical methods, it will be possible in the future to predict chemical properties of molecules with larger size with high accuracy.

1.4.2.2 Calculation of Thermodynamic Functions from Molecular Properties The calculation methods for thermodynamic functions (entropy S, heat capacities C_p and C_V, enthalpy H, and therefore Gibbs free energy G) for polyatomic systems from molecular and spectroscopic data with statistical methods through calculation of partition functions and its derivative toward temperature are well established and described in reference books such as Herzberg's *Molecular Spectra and Molecular Structure* [59] or in the earlier work from Mayer and Mayer [7], who showed, probably for the first time in a comprehensive way, that all basic thermochemical properties can be calculated from the partition function Q and the Avagadro's number N. The calculation details are well described by Irikura [60] and are summarized here. Emphasis will be placed on calculations of internal rotations.

The *partition function* Q can be computed from all the molecule's specific energy levels ε_i and the Boltzmann constant k_B:

$$Q(T) = \sum_i \exp\left(\frac{-\varepsilon_i}{k_B T}\right) \tag{1.16}$$

Ideal gas values for the heat capacity, enthalpy increment, and entropy can be computed from the partition function Q.

The equation for calculation of heat capacity at constant volume is

$$C_v = RT \frac{\partial^2}{\partial T^2}(T \ln Q) \tag{1.17}$$

The enthalpy difference relative to absolute temperature of 0K can be calculated from the heat capacity C_p at constant pressure

$$C_p = C_v + R \tag{1.18}$$

through

$$H_T - H_0 = \int_0^T C_p dT = \frac{RT^2}{Q}\frac{\partial Q}{\partial T} + RT \tag{1.19}$$

The entropy S is computable as

$$S = R\left[\frac{\partial}{\partial T}(T \ln Q) - \ln N + 1\right] \tag{1.20}$$

However, a complete set of molecular energy levels needed for calculation of the partition function (Eq. (1.16)) is not available in most cases. The arising problem can be simplified through the approximation that the different types of motion such as vibration, rotation, and electronic excitations are on a different timescale and therefore are unaffected by each other and can be treated as decoupled motions. This leads to a separation of Q into factors that correspond to separate partition functions for electronic excitations, translation, vibration, external molecular rotation, and hindered and free internal rotation:

$$Q = Q_{\text{elect}} Q_{\text{trans}} Q_{\text{vib}} Q_{\text{rot\_external}} Q_{\text{rot\_hindered\_internal}} Q_{\text{rot\_free\_internal}} \tag{1.21}$$

The partition function for electronic excitation contributions to the thermochemical properties will be

$$Q_{\text{elec}} = \sum g_i \exp\left(\frac{-\varepsilon_i}{k_{\text{B}}T}\right) \tag{1.22}$$

where g_i is the degeneracy of the electronic state with the energy ε_i.

The partition function for all translational modes is

$$Q_{\text{trans}} = (2\pi m k_{\text{B}} T)^{3/2} h^{-3} V \tag{1.23}$$

and for all vibrational modes it is

$$Q_{\text{vib}} = \prod_i \left[1 - \exp\left(\frac{-h\nu_i}{k_{\text{B}}T}\right)\right]^{-1} \tag{1.24}$$

For external rotation of a nonlinear molecule, the partition function results in

$$Q_{\text{rot}}^{\text{nonlinear}} = \frac{8\pi^2}{\sigma \cdot h^3}(2\pi k_{\text{B}} T)^{3/2}\sqrt{I_{\text{A}} I_{\text{B}} I_{\text{C}}} = \left(\frac{k_{\text{B}}T}{h}\right)^{3/2}\sqrt{\frac{\pi}{ABC}}\frac{1}{\sigma} \tag{1.25}$$

with the symmetry number σ, the moments of inertia I_{A}, I_{B}, and I_{C}, and the rotational constants A, B, and C.

In most quantum chemical program packages, these equations are used only to calculate the temperature dependence of thermodynamic properties. Internal free and hindered rotation contributions to the partition functions are normally neglected or implicitly use the pseudo-vibration approach for the internal rotor.

In molecules or radicals, such as ethyl, internal rotations around bonds such as CH_3–Ʊ–CH_2 occur. Accordingly, the partition function for a free rotor is defined as

$$Q_{\text{free rotor}} = \frac{\sqrt{8\pi^3 I_{\text{int}} k_B T}}{\sigma_{\text{int}} h} \tag{1.26}$$

with

$$I_{\text{int}} = I_{\text{top}} - I_{\text{top}}^2 \left(\frac{\alpha^2}{I_A} + \frac{\beta^2}{I_B} + \frac{\gamma^2}{I_C} \right) \tag{1.27}$$

where I_{top} is the moment of inertia of the rotating fragment about the axis of internal rotation:

$$I_{\text{top}} = \sum m_i r_i^2$$

The internal symmetry number σ_{int} equals the number of minima (or maxima) in the torsional potential energy curve, which can be calculated with quantum chemical programs by scans along the internal rotor coordinate.

The rotational barrier V for the aforementioned rotation around the C–C bond in ethyl is below 1 kJ/mol. Since it is much less than kT, the rotor can be considered as freely rotating.

In ethane, the rotational barrier is around 12 kJ/mol and, therefore, it is necessary to treat it as a hindered rotation.

If the torsional potential has the simple form

$$U(\phi) = \frac{V(1 - \cos \sigma_{\text{int}} \phi)}{2} \tag{1.28}$$

with the barrier V and the internal symmetry number σ, then the tables of Pitzer and Gwinn [61] can be used to compute the contribution of the hindered rotor to the thermodynamic functions.

A popular method is to represent the hindered rotor potential by an expansion introduced by Laane and coworkers [62], who used, for example, a six-term summation such as

$$V = \frac{1}{2} \sum_{n=1}^{6} V_n (1 - \cos (n\phi)) \tag{1.29}$$

But especially in cases where the hindered rotational potential is asymmetric (see Figure 1.1), the calculation of the partition function needs to take into account the different barrier heights and the according rotation angle as delimiter of the integral.

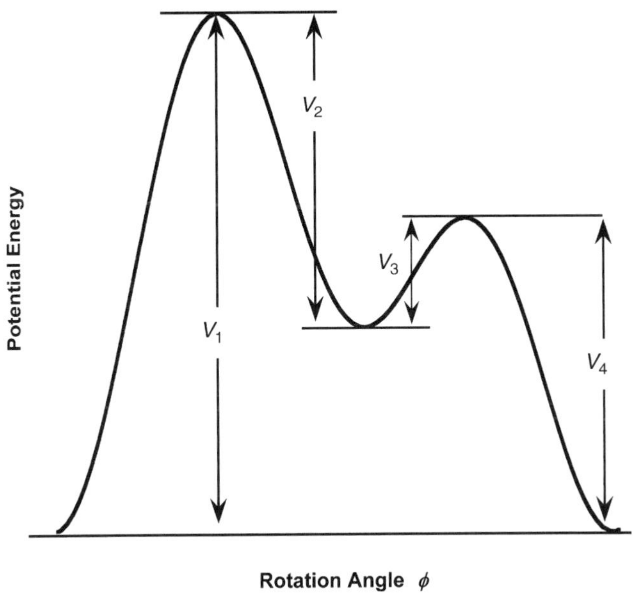

FIGURE 1.1 An example of an asymmetric, hindered rotational potential.

Applied to Figure 1.1, the partition function is

$$Q_{\text{rot\_hindered}}(T) = \frac{1}{2\pi}\left(\frac{\pi k_B T}{B}\right)^{1/2} \sum_{l=1}^{4} \int_{(i-1)\pi/2}^{i\pi/2} d\phi e^{-V/k_B T} \tag{1.30}$$

The further treatment for an asymmetric, hindered internal rotation with different barrier heights is shown, for example, in Ref. [63], where the calculation was needed for the rotation about the HOO–C*O bond and the HO–OC*O bond in the transition states of the reaction $CO + HO_2 \rightarrow CO_2 + OH$. This reaction is very important in syngas (H_2, CO) combustion at high pressures due to the fact that a relatively stable radical HO_2 is converted to a more reactive radical OH.

The effect of using different internal rotor treatments (harmonic oscillator or free rotator approximations) instead of hindered rotor treatment on the calculated reaction rate coefficient is also shown there [63].

Many scientists in the fields of thermodynamics and computational software use the rigid rotor–harmonic oscillator approximation or other shortcuts due to the relatively small contribution of the internal rotations to the whole enthalpy and entropy values.

This is however a potential point of error (having a tendency to affect the computed entropy somewhat more visibly than the corresponding enthalpy increment or heat capacity), and the user is warned about this simplification, which is often used, for example, to convert 0K enthalpy of formation to 298K value.

1.5 ACCURACY OF THERMOCHEMICAL VALUES

1.5.1 Standard Enthalpies of Formation

Standard enthalpies ("heats") of formation of all species can be divided into three categories:

(a) Those that were experimentally measured either by combustion calorimetry or by determining the enthalpy of a reaction involving the target (and other) species
(b) Those estimated on the basis of experimental values of other (similar or related) compounds
(c) Those estimated on the basis of other estimated compounds or structural groups

Here, we would like to make a few cautionary comments on the state of affair with respect to traditional sources.

Overall, the number of species important in combustion for which experimental values of standard enthalpies of formation can be assigned is comparably small. All are based on chemical reactions to which enthalpy changes of reaction can be assigned with high accuracy either calorimetrically or from the temperature dependence of equilibrium constants. As far as stable molecules of the elements carbon, hydrogen, oxygen, and nitrogen are concerned, it is fortunate that combustion reactions themselves serve for this purpose as the standard enthalpies of formation of the combustion products. Carbon dioxide and water have been painstakingly evaluated and reactions can usually be arranged to occur with accurately measured stoichiometry [22].

Even for the most favorable cases, however, the error bars that have to be accepted are larger than one would wish. This is illustrated in Table 1.2, adapted from Cohen and Benson [64] who give references to the archival literature. Here one sees that the "best available" standard enthalpy of formation values for the small hydrocarbons come with error ranges that imply significant uncertainty in equilibrium constants (a $\pm$ 1 kJ/mol uncertainty in the enthalpy or Gibbs free energy change of a reaction at 1000K implies an uncertainty of $\pm 12\%$ in its equilibrium constant).

The uncertainty ranges asserted by the evaluators are larger than one would wish. But more difficult is the fact that the differences between the experimental values obtained with the two most trustworthy calorimetric techniques differ from one

TABLE 1.2 Standard Enthalpies of Formation in kJ/mol at 298.15K for Small Hydrocarbons

Species	Bomb Calorimeter	Flame Calorimeter	ATcT [65]
CH_4	-74.85 ± 0.29	-74.48 ± 0.42	-74.53 ± 0.06
C_2H_6	-84.68 ± 0.50	-83.85 ± 0.29	-83.79 ± 0.2
C_3H_8	-103.89 ± 0.59	-104.68 ± 0.50	-104.68 ± 0.6
n-C_4H_{10}	-127.03 ± 0.67	-125.65 ± 0.67	-125.86 ± 0.38
i-C_4H_{10}	-135.60 ± 0.54	-134.18 ± 0.63	-134.35 ± 0.4

Source: Calorimetry values from Ref. [64].

another by more than the sum of the stated uncertainty ranges for two of the five cases. Apart from these discrepancies, which can now be successfully treated and resolved via the Thermochemical Network (TN) analysis of Active Thermochemical Tables (ATcTs) [66], the asserted experimental uncertainty ranges are mostly larger than those for the values obtained with TN analysis for stable molecules.

The thermochemistry values are less well known for most of the other stable species of interest in combustion, and still less well known for unstable ones. Among the unstable species, the thermochemistry of free radicals has attracted particular interest in combustion modeling because of their roles as chain centers.

An overview of current knowledge of the standard enthalpies of formation of some of the common radicals is given in Table 1.3. The values were evaluated by an IUPAC Task Force about Critical Evaluation of Thermochemical Properties of Selected Radicals [65].

In contrast to the stable hydrocarbons, where the standard enthalpy of formation is based on one or another of the direct calorimetrical methods, values for radicals come from all sorts of very difficult measurements ranging from photoionization mass spectroscopy to measurements of reaction rates. It is no surprise that the results are more contentious and less accurate. In Table 1.3, the uncertainty ranges can be seen to be typically an order of magnitude greater than those for stable hydrocarbon values except where the Active Thermochemical Tables can help.

TABLE 1.3 Standard Enthalpies of Formation in kJ/mol at 298.15K for Common Radicals

Species (g)	$\Delta_f H_{298}$ IUPAC	$\Delta_f H_{298}$ ATcT
OH	37.3 ± 0.3	37.5 ± 0.03
CH	595.8 ± 0.6	596.30 ± 0.25
CN	438.68 ± 2	438.81 ± 0.52
NH		358.78 ± 0.37
SH	141.87 ± 0.52	
CH_2OH	-17.0 ± 0.7	-17.18 ± 0.37
CH_3O	21.0 ± 2.1	20.257 ± 0.42
HO_2	12.296 ± 0.25	12.27 ± 0.16
CHO	42.3 ± 2	42.296 ± 0.3
CH_2	391.2 ± 1.6	391.465 ± 0.27
CH_3	146.7 ± 0.3	146.582 ± 0.1
C_2O	291.04 ± 63	385.68 ± 1.9
C_2H	567.4 ± 1.5	568.06 ± 0.31
C_2H_3		297.27 ± 0.45
C_2H_5	119.7 ± 0.7	
C_3H_3(*CH_2-CCH)	339 ± 4	351.5 ± 0.5
C_3H_5	171 ± 3	
n-C_3H_7	101.32 ± 1	
i-C_3H_7	90.19 ± 2	
C_6H_5	339.7 ± 2.5	337.3 ± 0.6

Values for hydrocarbons accepted by IUPAC Task Force for Thermochemistry of Radicals of Relevance in Combustion and Atmospheric Chemistry [65] and ATcT values [66,84].

For hydrocarbons and their various derivatives containing oxygen and nitrogen atoms, a long history of thermochemical investigation has left a legacy of experimental standard enthalpy of formation values (approximately 3000 have been compiled by Pedley et al. [25]).

The uncertainty level of this legacy varies considerably because of the fluctuating care given to the (mostly) calorimetric measurements and problems of reagent purity and reaction stoichiometry. Early on there have been successful efforts to systematize the database in terms of molecular structure (reviewed in detail by Cox and Pilcher [22]). As a result, one can compute a standard enthalpy of formation value for "ordinary" compounds (without strained rings, partially delocalized structures) that have not been studied experimentally with almost the same confidence that one can place in the experimental values themselves.

Unfortunately, many of the most interesting molecules and radicals used in combustion modeling have highly strained rings or electronic structures that are not well represented in the experimental database used for setting group additivity parameters. For such molecules and radicals, we recommend to abstain from use of group additivity values estimated from experimental data. We prefer instead to do *ab initio* calculations or, if that appears infeasible, as in case of large species, semiempirical or semitheoretical molecular electronic structure calculations or use group additivity values that were calculated with according quantum chemical methods.

Standard enthalpies of formation are quoted by different authors, making it sometimes challenging to find out to which of the three categories the quoted $\Delta_f H^\circ_{298}$ values belong. When the measured values of individual compounds change with time due to better experimental systems or errors found in previous measurements, it causes a need to change $\Delta_f H^\circ_{298}$ values of compounds whose determination or estimation was based on those values. However, there were no convenient means to perform these corrections other than tedious and continuous manual examination of each individual $\Delta_f H^\circ_{298}$ value. The differences in the auxiliary values used to extract the enthalpy of formation of the species from the measured quantity are frequently at the core of disputes between groups of researchers claiming a different heat of formation for an important species.

These types of problems, together with other disadvantages connected to the traditional sequential approach to evolving enthalpies of formation, are being currently successfully addressed by ATcT approach [66].

1.5.2 Active Thermochemical Tables

ATcT is a new paradigm that catapults thermochemistry into the twenty-first century. As opposed to traditional sequential thermochemistry, ATcT provides reliable, accurate, and internally consistent thermochemistry by utilizing TN approach [66].

The traditional approach is geared up to determine the enthalpies of formation of the target species using a sequential procedure. In this procedure, only one species is examined during each step. The available measurements (and computations) that link the target species only to those determined in previous steps are examined. From these, the "best" determination (or, occasionally, the average of a few determinations that appear to be of similar quality) is selected and used to obtain the enthalpy of formation of the target species at one temperature. Spectroscopic data (vibrational frequencies and rotational constants) are then used to compute the temperature dependence of the enthalpy and the remaining complement of thermochemical functions pertinent to the target species. At that point, the thermochemical properties of the target species are "frozen" and the procedure moves on to a new target species.

The primary disadvantage is that the resulting tabulation of enthalpies of formation stores for any species only the final value for the enthalpy, which is in reality connected to other enthalpies across the table via a maze of hidden progenitor–progeny relationships, making it next to impossible to update the resulting data with new information. Namely, even if, for example, a newly measured bond dissociation energy is used to revise the enthalpy of formation of some species, there are generally other species in the table that are pegged to the old value and would also need to be revised. Which are those species is not clear without investing a very laborious manual effort that examines each and every species in the tabulation.

In addition, the uncertainties obtained in the traditional approach typically do not properly reflect the complete knowledge that was available at the time the tabulation was created. For example, some of the existing knowledge is simply ignored (or taken only as a secondary check) because it did not make it into the subset of "best" determinations. Since there is no feedback to values obtained in the previous steps, the relevant dependencies that are used in later steps in the procedure (and involve directly or indirectly the species that were determined in previous steps) do not contribute to the quantification of the uncertainties in earlier steps nor do they help improve the reliability of values that are already frozen. In short, available knowledge is used only partially.

As opposed to the sequential approach, ATcT tables use TN approach. TN does not store enthalpies of formation of various species as such; rather, it stores the various relationships between the enthalpies as given by the actual measurements and computations, creating a network of thermochemical interdependencies. In order to obtain the desired enthalpies of formation, TN is solved simultaneously for all the species it describes, producing a complete set of thermochemical values that are entirely mutually consistent. Furthermore, the dependencies stored in TN are not based on the selected "best" subset of determinations. Rather, *all* available determinations from the literature are stored in the network. Since these are not necessarily self-consistent (because some of the quoted uncertainties are "optimistic," i.e., some determinations are not as correct as the uncertainty might imply, or are even "wrong"), TN solution is preceded by a statistical analysis and evaluation of the determinations that span and define TN. The statistical evaluation of the determinations in TN is made possible by redundancies in TN, such as competing measurements of the same enthalpy of reaction and alternate network pathways that interrelate participating chemical species. The statistical analysis produces a self-consistent TN, from which the optimal thermochemical values are obtained by simultaneous solution in error-weighted space, thus allowing the best possible use of all knowledge present in TN. This results in significantly better values than the traditional sequential approach since it uses efficiently *all* the available knowledge and also relies on a statistical analysis.

The significantly increased reliability and accuracy of the values obtained from TN approach manifests itself through uncertainties (which are given as 95% confidence limits, as customary in thermochemistry) that are typically several times smaller than the equivalent sequential values that could be obtained by the traditional sequential approach.

Besides the dramatically improved reliability, accuracy, and consistency of the resulting thermochemical values, ATcT tables offer a number of features that are neither present nor possible in the traditional sequential approach. With ATcT, new knowledge can be painlessly propagated through all affected thermochemical values. Namely, a new measurement can be simply added to TN, followed by the automatic analysis and solution of TN, producing a new (revised) complement of thermochemical values for all the species present in the network, thus fully propagating the consequences of the new measurement through all the affected values.

ATcT approach also allows hypothesis testing and evaluation, as well as discovery of weak links in TN. The latter provides pointers to new experimental or theoretical determinations that will most efficiently improve the underlying thermochemical knowledge.

The knowledge base of ATcT is organized in a series of "Libraries." The Main Library contains the Core (Argonne) Thermochemical Network that is currently being developed. TN contains fully networked data on about 1000 species, containing H, O, C, N, and halogens, connected through more than 10,000 thermochemically relevant determinations, and it is growing on a daily basis.

Most of the initial species included in this TN are relatively small and play the role of "hubs" in the network (significantly overlapping with the notion of "key" CODATA species [67]), but as the network grows, larger species are being introduced. Besides TN, the Main Library also contains the relevant spectroscopic data for gas-phase species and tabulated data for condensed-phase species that are needed to compute the heat capacity, enthalpy, enthalpy increment, entropy, the temperature dependence of the enthalpy, Gibbs energy of formation, and so on. As the new data are introduced in TN in the Main Library, a new set of solutions of TN is periodically computed, producing a new version and storing the prior version into the archives (following an elaborate archival system).

Auxiliary libraries (e.g., CODATA Library [67], Gurvich Library [19], JANAF Library [11], etc.) are more static in nature and contain non networked data needed to reproduce the values in various historical tabulations for ready-reference purposes.

Although at its beginning, ATcT has already produced for a number of "key" species significantly more accurate thermochemical values, thus considerably increasing the number of species known to very high accuracy. Nevertheless, in general, only a small minority of species of interest in combustion can be assigned standard enthalpies of formation with uncertainty limits so narrow that for combustion modeling purposes they may be taken to be exact [22,67,68]. The most accurately known of all (apart from the elements in their reference states, for which the value 0.0 is defined to be exact) are those based on carefully recorded molecular electronic spectra supplemented by quantum mechanical analysis. Among these, the hydrogen atom stands out, and a few diatomic and triatomic species whose electronic spectra have been successfully analyzed to establish the dissociation limit also belong to the exact category. An overview of the uncertainties of the standard enthalpies of formation of key combustion-relevant atomic or small species that have been exhaustively studied by calorimetric and spectroscopic methods is given in Tables 1.1–1.3. This benchmark group of species is setting a standard of what can be achieved in accuracy for standard enthalpies of formation by using available information about measured and computed thermochemical values and its connections such as the Hess's Law. The new ATcT values are expected to bring about significant overall improvements in the accuracy and reliability of the available thermochemistry values.

1.6 REPRESENTATION OF THERMOCHEMICAL DATA FOR USE IN ENGINEERING APPLICATIONS

The thermodynamic data of pure substances can be provided to the users in different ways.

1.6.1 Representation in Tables

Traditionally, printed versions of tables were supplied where the heat content, the chemical enthalpy, the entropy, the Gibbs energy, the enthalpy of formation, and the equilibrium constant are listed as a function of temperature such as in the following compendia:

1. The Gurvich Russian thermochemical compendium [19]
2. The TRC loose leaf thermochemical data collection [10], later replaced by the books of Frenkel et al., *Thermodynamics of Organic Compounds in the Gas State*, Vols I and II [69]
3. Stull et al., *The Chemical Thermodynamics of Organic Compounds* [18]
4. The JANAF thermochemical tables [11]
5. Barin et al.'s thermochemical tables books [20,21]

1.6.2 Representation with Group Additivity Values

A second way of presenting thermochemical data was ascribed by Benson and Buss [70], but is originating to a few groups, whose work is described in an earlier review from Janz, "The estimation of thermodynamic properties for organic compounds and chemical reactions" [71].

Instead of describing thermochemistry properties of pure substances, thermochemistry properties of groups of atoms are defined, where additively a few groups can form together the properties of ideal gas compounds. The values of the groups are derived from experiments or in the last years by quantum chemical calculations. In the group additivity method, the values of C_p are supplied as a function of temperature and the heat of formation and entropy values are supplied at 300K and sometimes at 298K.

The reason for this is that there was no reason to differentiate between these temperatures since the difference between the properties at 298 and 300K was less than the experimental error or uncertainty of the estimates.

Since no specific compounds were included in this kind of representation, the group additivity (GA) thermoproperties [70] are used as inputs for computer programs that calculate at request the thermochemical properties for many pure gaseous compounds. Some programs are as follows:

1. Stein's NIST Structures and Properties, Version 2.0, Computerized Database 25 [72]—this program is also implemented on NIST Chemistry WebBook database [73]
2. Ritter and Bozzelli's THERM program [74]
3. Muller et al.'s THERGAS program [75]
4. Green et al.'s Thermodata generator within RMG - Reaction Mechanism Generator [88]

1.6.3 Representation as Polynomials

The third kind of representation is in the form of polynomials of the properties of pure substances.

Polynomials were mentioned for the first time by Lewis and Randall in 1923 [3] as a means to present thermochemical properties such as heat capacity (C_p), enthalpy, and so on, as a function of temperature. The publication of elaborate tables of properties was very problematic in a world where computers were not even imagined. Polynomials seemed a compact way to publish a lot of numbers and also a good way to smooth out scatter of the data. Despite the advantages, polynomials were not used abundantly before the proliferation of computers starting about 1965.

U.S. government agencies such as NASA and National Laboratories had computers by the end of the 1950s, and therefore started using polynomials in order to get thermochemical properties as a function of temperature. The functions were needed in order to calculate equilibrium compositions of reacting mixtures, which were extensively used before kinetic simulation programs were available. This was the reason for the publication of Duff and Bauer's paper [14], which included extensive equilibrium calculations.

They used two different sets of polynomials to compute heat capacity (C_p) and the free energy function (F).

In addition, they found out that the full temperature range of 298.15–6000K cannot be represented by a single polynomial. Therefore, they were the first who published for each of the two functions two polynomials (two branches) for the temperature range of 298.15–1000K and for 1000–6000K. But the two polynomials were not coinciding at any temperature and their use in 1000K region included a discontinuity.

The thermodynamic group at NASA Lewis Center in Cleveland, led by Gordon, undertook a long study in order to investigate the problem of chemical equilibrium [76,77]. As a result, a close scrutiny of the polynomialization of the thermodynamic data was also undertaken, and they proposed a solution with two important features:

(a) Single set of coefficients could be used for as many properties as possible for a single compound.
(b) The same polynomial form fits all thermodynamic data for gases, liquids, and solids for all possible chemical compounds.

Zeleznik and Gordon [77] invented the method of simultaneous regression of the thermochemical properties so that more than one property can be approximated by a single polynomial. This work ended up with the famous NASA seven-term polynomials first published by Zeleznik and Gordon [77] and McBride et al. [15], which cover heat capacity C_p, enthalpy, and entropy.

In their first version, the polynomials were fit for two temperature ranges. The first polynomial was fit for the temperature region important for combustion, that is, 1000–6000K. The second polynomial was fit for the lower temperature region, that is, 300–1000K. The two polynomials were "pinned" at 1000K. They were constrained to reproduce exactly 1000K value, thus ensuring that both branches will match at 1000K without discontinuity. The consequence of this method was that the values at the standard reference temperature of 298.15K, which were not used to create any constraints, were always reproduced with some small error, depending on the polynomial fit. Later, in 1982, following users' requests, the fitting of the polynomials was slightly changed: the lower branch was extended to 200K, and the pinning of the polynomials was transferred to 298.15K value, while the two branches were still constrained to have the same value at 1000K.

Because of the need of NASA to calculate properties beyond 6000K limit for shuttle reentry problems, the research into the polynomials was extended, and in 1987 a new set of NASA nine-term polynomials was adopted. The foremost quality of these polynomials is that new branches can be added above and below the original temperature range; in addition, their error of reproducing the fitted data was improved between 1 and 2 orders of magnitude. The maximum error at peak temperature of the seven-term polynomials is typically in the range of one-tenth of 1% to 1%, while the typical fitting error of the new nine-term polynomials is in the range of one-thousandth of 1% to one-hundredth of 1%.

The program to calculate thermochemical properties (called PAC for properties and coefficients) and the corresponding seven-term polynomials were published by McBride and Gordon in 1967 [16], and a new version that calculates the nine-term polynomials was published in 1992 [78].

Other types of polynomials were also proposed. For example, the Wilhoit [79] polynomials were intended to allow the extrapolation of TRC thermochemical properties beyond 1000 or 1500K temperature range. The Wilhoit polynomials are used internally for extrapolation by PAC [16] and THERM programs [74].

The NIST WebBook site [73] prefers the Shomate polynomials [80], which are defined in the following way with the coefficients A–H:

$$C_p^\circ = A + B\left(\frac{T}{1000}\right) + C\left(\frac{T}{1000}\right)^2 + D\left(\frac{T}{1000}\right)^3 + \frac{E}{(T/1000)^2} \qquad (1.31)$$

$$H^{\circ} - H^{\circ}_{298.15\mathrm{K}} = A\left(\frac{T}{1000}\right) + B\frac{(T/1000)^2}{2} + C\frac{(T/1000)^3}{3} + D\frac{(T/1000)^4}{4} - \frac{E}{T/1000} + F - H \tag{1.32}$$

$$S^{\circ} = A\ln\left(\frac{T}{1000}\right) + B\left(\frac{T}{1000}\right) + C\frac{(T/1000)^2}{2} + D\frac{(T/1000)^3}{3} - \frac{E}{2(T/1000)^2} + G \tag{1.33}$$

The Gurvich polynomial [19] for the partition function is seldom used in the western part of the world. Various series of negative powers of the temperature were also proposed in the past. However, none of them got the wide acceptance and extensive use of the seven-term NASA polynomials, mainly due to the existence of big free databases of these polynomials.

The thermochemical properties can be calculated in general with confidence in the fourth and fifth digits in the units of kcal/mol in the range of 150–3000K. But since many engineering problems require the knowledge of data above and below this range, they exist in the form of seven-term polynomials to 6000K and as nine-term polynomials from 50 to 6000K, and are sometimes extended to 20,000K for reentry problems of satellites and shuttles [85].

The seven-coefficient NASA polynomials can be used to calculate the following functions:

$$\frac{C_p^{\circ}}{R} = a_1 + a_2T + a_3T^2 + a_4T^3 + a_5T^4 \tag{1.34}$$

$$\frac{H_T^{\circ}}{RT} = a_1 + \frac{a_2T}{2} + \frac{a_3T^2}{3} + \frac{a_4T^3}{4} + \frac{a_5T^4}{5} + \frac{a_6}{T} \tag{1.35}$$

$$\frac{S_T^{\circ}}{R} = a_1\ln T + a_2T + \frac{a_3T^2}{2} + \frac{a_4T^3}{3} + \frac{a_5T^4}{4} + a_7 \tag{1.36}$$

$$\frac{G_T^{\circ}}{RT} = \frac{H_T^{\circ}}{RT} - \frac{S_T^{\circ}}{R} = a_1(1 - \ln T) - \frac{a_2T}{2} - \frac{a_3T^2}{6} - \frac{a_4T^3}{12} - \frac{a_5T^4}{20} + \frac{a_6}{T} - a_7 \tag{1.37}$$

It should be noted that the value H_T° obtained from the polynomials is the "engineering enthalpy" defined as

$$H_T^{\circ} = \Delta_f H_{298}^{\circ} + \int_{298}^{T} C_p^{\circ} dT \tag{1.38}$$

Similarly, the G°/RT functions of the molecules in a reaction can be used directly to compute the reaction's equilibrium "constant" in terms of concentrations through

$$K_c = (RT)^{-\Delta\nu} \exp\left(\Delta a_1 (\ln T - 1) + \frac{\Delta a_2 T}{2} + \frac{\Delta a_3 T^2}{6} + \frac{\Delta a_4 T^3}{12} + \frac{\Delta a_5 T^4}{20} - \frac{\Delta a_6}{T} + \Delta a_7 \right) \tag{1.39}$$

where the change in the number of moles during reaction is $\Delta\nu = \Sigma\nu_j$ and the coefficient changes are $\Delta a_I = \Sigma\nu_j a_{ij}$. The summations are over all the reactant and product species j with the stoichiometric coefficients ν_j being positive for products and negative for reactants.

The seven-term polynomials actually include 15 constants. The first set of 7 constants belongs to 1000–6000K polynomial, the second set of 7 constants belongs to 200–1000K polynomial, and 15th constant is $H_{298}/R \equiv \Delta_f H_{298}/R$. The latter value (and the corresponding position within the polynomial format) is not used by most chemical kinetic programs, such as Cantera [81], CHEMKIN [82], and COSILAB [83], and therefore does not interfere with their calculations.

The nine-term polynomials can be used to calculate the functions

$$\frac{C_p^\circ}{R} = a_1 T^{-2} + a_2 T^{-1} + a_3 + a_4 T + a_5 T^2 + a_6 T^3 + a_7 T^4 \tag{1.40}$$

$$\frac{H_T^\circ}{RT} = -a_1 T^{-2} + \frac{a_2 \ln T}{T} + a_3 + \frac{a_4 T}{2} + \frac{a_5 T^2}{3} + \frac{a_6 T^3}{4} + \frac{a_7 T^4}{5} + \frac{a_8}{T} \tag{1.41}$$

$$\frac{S_T^\circ}{R} = -\frac{a_1 T^{-2}}{2} - a_2 T^{-1} + a_3 \ln T + a_4 T + \frac{a_5 T^2}{2} + \frac{a_6 T^3}{3} + \frac{a_7 T^4}{4} + a_9 \tag{1.42}$$

$$\begin{aligned} \frac{G_T^\circ}{RT} = \frac{H_T^\circ}{RT} - \frac{S_T^\circ}{R} = & -\frac{a_1 T^{-2}}{2} + \frac{2a_2(1 - \ln T)}{T} + a_3(1 - \ln T) \\ & - \frac{a_4 T}{2} - \frac{a_5 T^2}{6} - \frac{a_6 T^3}{12} - \frac{a_7 T^4}{20} + \frac{a_8}{T} - a_9 \end{aligned} \tag{1.43}$$

and also K_c, following a similar philosophy as given before for the seven-term polynomials.

1.6.3.1 *How to Change Δ_f H$_{298K}$* Without Recalculating NASA Polynomials Sometimes better enthalpies of formation values are available for a substance and the polynomials need to be adapted. This can be done in the polynomial form without changing the remaining of the thermodynamic data of the species, if the other molecular properties of the substance need no update. In this case, we shall refer to Eq. (1.35) for NASA seven-term polynomials. The term that includes the information on the enthalpy of formation is a_6. To change, write

$$a_{6\text{new}} = a_{6\text{old}} + \frac{\Delta_f H_{298\text{new}} - \Delta_f H_{298\text{old}}}{R} \tag{1.44}$$

Both temperature intervals have to be adjusted; therefore, coefficients a_6 and a_{13} have to be recalculated in the same way. Do not forget to change coefficient a_{15} that is $\Delta_f H_{298\text{K}}/R$.

It is possible to apply the same procedure to NASA nine-term polynomials using a_8 in Eq. (1.41) and accordingly each eigth coefficient in the polynomials for all temperature ranges.

1.7 THERMOCHEMICAL DATABASES

There are basically a few databases that include thermochemical data in polynomial form:

(a) The Extended Third Millennium Thermochemical Database [84] authored by Goos, Ruscic and Burcat that includes NASA-type polynomials of seven and nine terms for easy use in kinetic modeling and computational fluid dynamics software.

It contains data mostly of gaseous compounds, liquids, and solids, ranging from pure elements, metals, and ions to inorganic substances, organic stable compounds such as hydrocarbons, and reactive species such as radicals. It also contains all inert gases and a limited number of other elements such as Al, B, Bi, Br, Cl, F, I, H, D, T, Cr, Cu, Fe, Ge, Hg, K, Ir, Mg, Mn, Mo, N, Na, Ni, O, Os, P, Pb, Pd, Pt, S, Sb, Si, Sn, W, Zn, Zr, and their compounds.

The thermochemical and spectroscopic data of more than 3000 substances are mainly used to model and optimize combustion processes and to understand or model the impact on the atmosphere, in addition to optimizing chemical processes.

The database is reviewed and available reference literature values (calculated and measured ones) and own calculation results are provided. Today this database is the biggest collection of ATcT and G3B3 calculated values that were provided for about half of the included species. In addition, the accuracy of the data and the used values are shown in detail to make the calculation results traceable and or correctable, if better data are available (e.g., quantum chemical results such as spectroscopic properties like vibrations and rotational constants, additional data used to calculate the partition functions, and finally the deviations of the fits to obtain NASA polynomial data from the temperature-dependent thermochemical properties).

The database is also updated and enhanced on a regular basis and on user requests.

(b) The NASA Glenn thermodynamic database, last updated in 2002 [85], which is dedicated to general use and contains only nine-term NASA polynomials. Tables of thermodynamical functions such as heat capacity C_p, entropy S, enthalpy, and the log K can be calculated online from the provided coefficients.

(c) The gas-phase thermochemical database of Sandia National Laboratories maintained by Allendorf containing seven-term NASA polynomials [86] and the according spectroscopic information of nearly 1000 compounds of mostly inorganic and organometallic systems containing the elements Al, B, Be, C, Ca, Cl, Cr, F, Fe, H, In, K, Li, Mg, Mn, N, Na, O, Sb, Si, and Sn. The thermochemical data are mainly predicted from calculated molecular energies using MP4 method. Sometimes they were corrected using empirically derived bond additivity corrections (BACs) [87].

1.8 CONCLUSION

This chapter presented a short noncomprehensive overview of some of the aspects of thermochemistry that are relevant to scientific and engineering applications. It explained the calculation methods used to obtain thermodynamic data and showed how to evaluate the soundness of different methods and some of their pitfalls. In addition, it provided the sources of accurate thermodynamic data and explanation of the formats used. This makes the use of these data in engineering applications such as computational fluid dynamic (CFD) simulation relatively easy.

REFERENCES

1. (a) M. M. Malhard and H. Le Chatelier, *Ann. Mines* **4**, 274–378 (1883). (b) M. M. Malhard and H. Le Chatelier, *Ann. Mines* **4**, 379 (1883).
2. M. Berthelot, *Thermochimie*, Vols 1–2, Gauthier-Villars, Paris (1897).
3. G. N. Lewis and M. Randall, *Thermodynamics and the Free Energy of Substances*, McGraw Hill, New York (1923).
4. E. W. Washburn (ed.), *International Critical Tables of Numerical Data, Physics, Chemistry and Technology*, Vols I–VII, McGraw Hill, New York (1926–1930) Index (1933).
5. G. S. Parks and H. M. Huffman, *The Free Energies of Some Organic Compounds*, Monograph No. 60, Chemical Catalog Co., New York (1932).
6. F. R. Bichowsky and F. D. Rossini, *The Thermochemistry of the Chemical Substances*, Reinhold Publishing Co., New York (1936).
7. J. E. Mayer and M. Göppert Mayer, *Statistical Mechanics*, John Wiley & Sons, Inc., New York (1940) (10th Reprint 1963).
8. F. D. Rossini, K. S. Pitzer, W. J. Taylor, J. P. Ebert, J. E. Kilpatrick, C. W. Beckett, M. G. Williams, and H. G. Werner, *Selected Values of Properties of Hydrocarbons* (American Petroleum Institute Research Project 44), Circular of the National Bureau of Standards (NBS) 461, U.S. Government Printing Office, Washington, DC (1947).
9. F. D. Rossini, D. D. Wagman, W. H. Evans, S. Levine, and I. Jaffe, *Selected Values of Chemical Thermodynamic Properties*, Circular of National Bureau of Standards (NBS) 500, U.S. Government Printing Office, Washington, DC (1952).
10. Thermodynamic Research Center (TRC), *Thermodynamic Tables of Hydrocarbons,* formerly American Petroleum Institute Project 44 (see Ref. [8]) (published as loose leaf sheets by Texas A&M University, College Station, TX; presently NIST-TRC Boulder Co.).
11. (a) D. R. Stull and H. Prophet, *JANAF Thermochemical Tables*, 2nd edition, NSRDS-NBS 37, National Standard Reference Data System (NSRDS), National Bureau of Standards (NBS), Washington, DC (1971). (b) M. W. Chase, Jr., C. A. Davies, J. R. Downey, Jr., D. J. Frurip, R. A. McDonald, and A. N. Syverud, *JANAF Thermochemical Tables*, 3rd edition (Parts I and II), *J. Phys. Chem. Ref. Data* **14** (Suppl. 1), 1–1856 (1985). (c) M. W. Chase,

Jr. (ed.), *NIST-JANAF Thermochemical Tables*, 4th edition (Part I (Al-Co) and Part II (Cr-Zr)), *J. Phys. Chem. Ref. Data Monograph* **9**, 1–1951 (1998).

12. J. S. Gordon, *Thermodynamic Data for Combustion Products,* REP RMD 210-E3, Thiokol Chemical Corporation (1960).
13. R. E. Duff and S. H. Bauer, The Equilibrium Composition of the C/H/System at Elevated Temperature, Los Alamos Report 2556 (1961).
14. R. E. Duff and S. H. Bauer, *J. Chem. Phys.* **36**, 1754 (1962).
15. B. J. McBride, S. Heimel, J. G. Ehlers, and S. Gordon, *Thermodynamic Properties to 6000K for 210 Substances Involving the First 18 Elements,* NASA-SP-3001 (1963).
16. B. J. McBride and S. Gordon, *FORTRAN IV Program for Calculation of Thermodynamic Data,* NASA TN-D 4097 (1967).
17. D. D. Wagman, W. H. Evans, I. Halow, V. S. Parker, S. M. Bailey, and R. H. Schumm, *Selected Values of Chemical Thermodynamic Properties*, U.S. National Bureau of Standards, Technical Note 270-1 to Technical Note 270-8 (1965–1981), superseded by D. D. Wagman, W. H. Evans, V. B. Parker, R. H. Schumm, I. Halow, S. M. Bailey, K. L. Churney, and R. L. Nuttall, *The NBS Tables of Chemical Thermodynamic Properties: Selected Values for Inorganic and C_1 and C_2 Organic Substances in SI Units, J. Phys. Chem. Ref. Data* 11 (Suppl. 2) (1982).
18. D. R. Stull, E. F. Westrum, Jr., and G. C. Sinke, *The Chemical Thermodynamics of Organic Compounds*, John Wiley & Sons, Inc., New York (1969).
19. (a) L. V. Gurvič,TSIV, Termodinamičeskie svojstva individual'nych veščestv (Thermodynamic Properties of Individual Substances), Vols 1–4, Nauka, Moskva (in Russian) (1962, 1978, 1979, 1982). (b) L. V. Gurvich, I. V. Veyts, and C. B. Alcock, *Thermodynamic Properties of Individual Substances*, 4th edition, Vol. 1, Hemisphere Publishing Co. (1989) (Vol. 2 (1991), Vol. 3 (1994), Vols 4–6 (1997)).
20. I. Barin, O. Knacke, and O. Kubaschewski, *Thermochemical Properties of Inorganic Substances*, Vol. 1, Springer, Berlin (1973).
21. (a)O. Knacke, O. Kubaschewski, and K. Hesselmann, *Thermochemical Properties of Inorganic Substances*, Vol. 2, Springer, Berlin (1991). (b) I. Barin, *Thermochemical Data of Pure Substances, Part 1: Ag–Kr, Part 2: La–Zr*, 3rd edition, Wiley-VCH, Weinheim (1995).
22. J. D. Cox and G. Pilcher, *Thermochemistry of Organic and Organometallic Compounds*, Academic Press, London (1970).
23. J. B. Pedley and J. Rylance, *Computer Analysed Thermochemical Data: Organic and Organometallic Compounds*, University of Sussex, Brighton, UK (1977).
24. E. S. Domalski and E. D. Hearing, *J. Phys. Chem. Ref. Data* **17**, 1637 (1988).
25. J. B. Pedley, R. D. Naylor, and S. P. Kirby, *Thermochemical Data of Organic Compounds*, 2nd edition, Chapman and Hall, New York (1986).
26. International Union of Pure and Applied Chemistry (IUPAC), *Pure Appl. Chem.* **54** (6), 1239–1250 (1982). Available at http://media.iupac.org/publications/pac/1982/pdf/5406x1239.pdf.
27. (a) H. Eyring, *J. Chem. Phys.* **3**, 107 (1935). (b) M. G. Evans and M. Polyani, *Trans. Faraday Soc.* **31**, 875 (1935). (c) D. G. Truhlar and B. C. Garett, *Annu. Rev. Phys. Chem.* **35**, 159 (1984).
28. (a) J. A. Miller and S. J. Klippenstein, *J. Phys. Chem. A* **110**, 10528 (2006). (b) Y. Georgievskii and S. J. Klippenstein, *J. Chem. Phys.* **122**, 194103 (2005). (c) S. J.

Klippenstein and J. A. Miller, *J. Phys. Chem. A* **106**, 9267 (2002). (d) A. Fernandez-Ramos, J. A. Miller, S. J. Klippenstein, and D. G. Truhlar, *Chem. Rev.* **106**, 4518 (2006).

29. R. A. Eng, A. Gebert, E. Goos, H. Hippler, and C. Kachiani, *Phys. Chem. Chem. Phys.* **4**, 3989 (2002).
30. R. A. Eng, A. Gebert, E. Goos, H. Hippler, and C. Kachiani, *Phys. Chem. Chem. Phys.* **3**, 2258 (2001).
31. J. A. Miller, J. P. Senosiain, S. J. Klippenstein, and Y. Georgievskii, *J. Phys. Chem. A* **112**, 9429 (2008).
32. M. J. Pilling, *Proc. Combust. Inst.* **32**, 27 (2009).
33. (a) J. J. P. Stewart, *J. Comput. Chem.* **10**, 209 (1989). (b) J. J. P. Stewart, *J. Comput. Chem.* **10**, 221 (1989). (c) J. J. P. Stewart, in K. B. Lipkowitz, and D. B. Boyd (eds), *Reviews of Computational Chemistry*, VCH, New York (1990), p. 45.
34. MOPAC2009, James J. P. Stewart, Stewart Computational Chemistry, Colorado Springs, CO, USA, http://openmopac.net (2008).
35. http://www.univie.ac.at/columbus/.
36. DGauss 4.1, *Oxford Molecular* (1998–1999). Available at http://www.cache.fujitsu.com/cache/dgauss/index.shtml.
37. M.W. Schmidt, K. K. Baldridge, J. A. Boatz, S. T. Elbert, M. S. Gordon, J. H. Jensen, S. Koseki, N. Matsunaga, K. A. Nguyen, S. J. Su, T. L. Windus, M. Dupuis, and J. A. Montgomery, *J. Comput. Chem.* **14**, 1347–1363 (1993). Available at http://www.msg.ameslab.gov/GAMESS/GAMESS.html.
38. (a) GAMESS-UK is a package of ab initio programs written by M. F. Guest, J. H. van Lenthe, J. Kendrick, K. Schöffel, and P. Sherwood,with contributions from R. D. Amos, R. J. Buenker, H. J. J. van Dam, M. Dupuis, N. C. Handy, I. H. Hillier, P. J. Knowles, V. Bonacic-Koutecky, W. von Niessen, R. J. Harrison, A. P. Rendell, V. R. Saunders, A. J. Stone, and A. H. de Vries. The package is derived from the original GAMESS code due to M. Dupuis, D. Spangler, and J. Wendoloski, *NRCC Software Catalog,* Vol. 1, Program No. QG01 (GAMESS) (1980). (b) M. F. Guest, J. Kendrick, J. H. van Lenthe, and P. Sherwood, *GAMESS 6.2*, Computing for Science Ltd., UK (1999). (c) http://wserv1.dl.ac.uk/CFS/.
39. M. J. Frisch, G. W. Trucks, H. B. Schlegel, G. E. Scuseria, M. A. Robb, J. R. Cheeseman, J. A. Montgomery, Jr., T. Vreven, K. N. Kudin, J. C. Burant, J. M. Millam, S. S. Iyengar, J. Tomasi, V. Barone, B. Mennucci, M. Cossi, G. Scalmani, N. Rega, G. A. Petersson, H. Nakatsuji, M. Hada, M. Ehara, K. Toyota, R. Fukuda, J. Hasegawa, M. Ishida, T. Nakajima, Y. Honda, O. Kitao, H. Nakai, M. Klene, X. Li, J. E. Knox, H. P. Hratchian, J. B. Cross, C. Adamo, J. Jaramillo, R. Gomperts, R. E. Stratmann, O. Yazyev, A. J. Austin, R. Cammi, C. Pomelli, J. W. Ochterski, P. Y. Ayala, K. Morokuma, G. A. Voth, P. Salvador, J. J. Dannenberg, V. G. Zakrzewski, S. Dapprich, A. D. Daniels, M. C. Strain, O. Farkas, D. K. Malick, A. D. Rabuck, K. Raghavachari, J. B. Foresman, J. V. Ortiz, Q. Cui, A. G. Baboul, S. Clifford, J. Cioslowski, B. B. Stefanov, G. Liu, A. Liashenko, P. Piskorz, I. Komaromi, R. L. Martin, D. J. Fox, T. Keith, M. A. Al-Laham, C. Y. Peng, A. Nanayakkara, M. Challacombe, P. M. W. Gill, B. Johnson, W. Chen, M. W. Wong, C. Gonzalez, and J. A. Pople, *Gaussian 03, Revision B.02*, Gaussian, Inc., Pittsburgh, PA (2003).
40. MOLPRO is a package of *ab initio* programs designed by H.-J. Werner and P. J. Knowles. The authors are H.-J. Werner, P. J. Knowles, M. Schütz, R. Lindh, P. Celani, T. Korona, G. Rauhut, R. D. Amos, A. Bernhardsson, A. Berning, D. L. Cooper, M. J. O. Deegan, A. J. Dobbyn, F. Eckert, C. Hampel, G. Hetzer, A. W. Lloyd, S. J. McNicholas, F. R. Manby, W.

Meyer, M. E. Mura, A. Nicklass, P. Palmieri, R. Pitzer, U. Schumann, H. Stoll, A. J. Stone R. Tarroni, and T. Thorsteinsson.

41. M. Valiev, E. J. Bylaska, N. Govind, K. Kowalski, T. P. Straatsma, H. J. J. van Dam, D. Wang, J. Nieplocha, E. Apra, T. L. Windus, and W. A. de Jong, *Comput. Phys. Commun.* **181**, 1477 (2010). Available at http://www.nwchem-sw.org/index.php/Main_Page.
42. http://www.q-chem.com/index.htm.
43. A. D. Becke, *J. Chem. Phys.* **98**, 5648 (1993).
44. C. Lee, W. Yang, and R. G. Parr, *Phys. Rev. B* **37**, 785 (1988).
45. (a) L. A. Curtiss, K. Raghavachari, P. C. Redfern, V. Rassolov, and J. A. Pople, *J. Chem. Phys.* **109**, 7764 (1998). (b) L. A. Curtiss, P. C. Redfern, K. Raghavachari, V. Rassolov, and J. A. Pople, *J. Chem. Phys.* **110**, 4703 (1999).
46. A. G. Baboul, L. A. Curtiss, P. C. Redfern, and K. Raghavachari, *J. Chem. Phys.* **110**, 7650 (1999).
47. L. A. Curtiss, K. Raghavachari, G. W. Trucks, and J. A. Pople, *J. Chem. Phys.* **94**, 7221 (1991).
48. L. A. Curtiss, K. Raghavachari, P. C. Redfern, and J. A. Pople, *J. Chem. Phys.* **106**, 1063 (1997).
49. L. A. Curtiss, K. Raghavachari, P. C. Redfern, and J. A. Pople, *J. Chem. Phys.* **112**, 7374 (2000).
50. L. A. Curtiss, P. C. Redfern, and K. Raghavachari, *J. Chem. Phys.* **123**, 124107 (2005).
51. (a) B. J. Lynch and D. G. Truhlar, *J. Phys. Chem. A* **107**, 8996 (2003). (b) Y. Zhao and D. G. Truhlar, *J. Phys. Chem. A* **110**, 10478 (2006). (c) Y. Zhao and D. G. Truhlar, *J. Chem. Theory Comput.* **1**, 415 (2005). (d) Y. Zhao and D. G. Truhlar, *J. Phys. Chem. C* **112**, 6860 (2008). (e) Y. Zhao, B. J. Lynch, and D. G. Truhlar, *J. Phys. Chem. A* **108**, 2715 (2004). (f) Y. Zhao, N. González-García, and D. G. Truhlar, *J. Phys. Chem. A* **109**, 2012 (2005). (g) Y. Zhao, O. Tishchenko, J. R. Gour, W. Li, J. J. Lutz, P. Piecuch, and D. G. Truhlar, *J. Phys. Chem. A* **113**, 5786 (2009). (h) Y. Zhao and D. G. Truhlar, *Theor. Chem. Acc.* **120**, 215 (2008). (i) Y. Zhao and D. G. Truhlar, *J. Chem. Theory Comput.* **5**, 324 (2009).
52. L. Goerigk and S. Grimme, *J. Chem. Theory Comput.* **6**, 107 (2010).
53. (a) A. Tajti, P. G. Szalay, A. G. Császár, M. Kállay, J. Gauss, E. F. Valeev, B. A. Flowers, J. Vázquez, and J. F. Stanton, *J. Chem. Phys.* **121**, 11599 (2004). (b) Y. J. Bomble, J. Vázquez, M. Kállay, C. Michauk, P. G. Szalay, A. G. Cászár, J. Gauss, and J. F. Stanton, *J. Chem. Phys.* **125**, 064108 (2006). (c) M. E. Harding, J. Vázquez, B. Ruscic, A. K. Wilson, J. Gauss, and J. F. Stanton, *J. Chem. Phys.* **128**, 114111 (2008).
54. A. Karton, E. Rabinovich, and J. M. L. Martin, *J. Chem. Phys.* **125**, 144108 (2006).
55. A. Karton, A. Tarnopolsky, J. F. Lamere, G. C. Schatz, and J. M. L. Martin, *J. Phys. Chem. A* **112**, 12868 (2008).
56. A. Karton, I. Kaminker, and J. M. L. Martin, *J. Phys. Chem. A* **113**, 7610 (2009).
57. T. B. Adler, G. Knizia, and H.-J. Werner, *J. Chem. Phys.* **127**, 221106 (2007).
58. G. Knizia, T. B. Adler, and H.-J. Werner, *J. Chem. Phys.* **130**, 054104 (2009).
59. (a) G. Herzberg, *Molecular Spectra and Molecular Structure, I. Spectra of Diatomic Molecules*, 2nd edition (sixth printing), D. Van Nostrand Company, Princeton, NJ (1959) (1st print 1950) (partition functions for diatomic molecules, p. 466 ff). (b) G. Herzberg, *Molecular Spectra and Molecular Structure, II. Infrared and Raman Spectra of Polyatomic Molecules*, Krieger, Malabar, FL (1989). (c) G. Herzberg, *Molecular Spectra and*

Molecular Structure, III. Electronic Spectra and Electronic Structure of Polyatomic Molecules, Reprint edition (1991) (Krieger Publishing Company) with corrections of 2nd edition (1966) (D. Van Nostrand Company, Princeton, NJ, originally 1950, partition functions for diatomic molecules, p. 466 ff).

60. K. K. Irikura, in K. K. Irikura and D. J. Frurip (eds), *Computational Thermochemistry: Prediction and Estimation of Molecular Thermodynamics* (ACS Symposium Series 677), American Chemical Society, Washington, DC (1998).
61. (a) K. S. Pitzer and W. D. Gwinn, *J. Chem. Phys.* **10**, 428 (1942). (b) K. S. Pitzer, *J. Chem. Phys.* **14**, 239 (1946).
62. J. D. Lewis, T. B. Malloy, Jr., T. H. Chao, and J. Laane, *J. Mol. Struct.* **12**, 427 (1972).
63. X. You, H. Wang, E. Goos, C.-J. Sung, and S. J. Klippenstein, *J. Phys. Chem. A* **111**, 4031 (2007).
64. N. Cohen and S. W. Benson, in S. Patai and Z. Rappaport (eds), *The Chemistry of Alkanes and Cycloalkanes*, Chapter 6, John Wiley & Sons, Inc., New York (1992).
65. B. Ruscic, J. E. Boggs, A. Burcat, A. G. Császár, J. Demaison, R. Janoschek, J. M. L. Martin, M. L. Morton, M. J. Rossi, J. F. Stanton, P. G. Szalay, P. R. Westmoreland, F. Zabel, and T. Bérces, *J. Phys. Chem. Ref. Data* **34**, 573 (2005).
66. (a) Active Thermochemical Tables. Available at http://atct.anl.gov/. (b) B. Ruscic, R. E. Pinzon, M. L. Morton, G. von Laszewski, S. Bittner, S. G. Nijsure, K. A. Amin, M. Minkoff, and A. F. Wagner, *J. Phys. Chem. A* **108**, 9979 (2004). (c) B. Ruscic, R. E. Pinzon, G. Von Laszewski, D. Kodeboyina, A. Burcat, D. Leahy, D. Montoy, and A. F. Wagner, *J. Phys. Conf. Ser.* **16**, 561 (2005). (d) B. Ruscic, R. E. Pinzon, M. L. Morton, N. K. Srinivasan, M.-C. Su, J. W. Sutherland, and J. V. Michael, *J. Phys. Chem. A* **110**, 6592 (2006). (e) B. Ruscic, *McGraw-Hill 2005 Yearbook of Science and Technology*, pp. 3–7, McGraw Hill, New York (2004).
67. J. D. Cox, D. D. Wagman, and V.A. Medvedev, *CODATA Key Values for Thermodynamics*, Hemisphere Publishing Corporation, New York (1989). Available at http://www.codata.org/resources/databases/key1.html.
68. N. Cohen, *J. Phys. Chem. Ref. Data* **25**, 1411 (1996).
69. M. Frenkel, G. J. Kabo, K. N. Marsh, G. N. Roganov, and R. C. Wilhoit, *Thermodynamics of Organic Compounds in the Gas State*, TRC Data Series, CRC Press Inc., College Station, TX (1994).
70. (a) S. W. Benson and J. H. Buss, *J. Chem. Phys.* **29**, 546–572 (1958). (b) S. W. Benson, *Thermochemical Kinetics*, John Wiley & Sons, Inc., New York (1976).
71. G. J. Janz, *Q. Rev. Chem. Soc.* **9**, 229–254 (1955).
72. (a) S. E. Stein, *NIST Structures and Properties, Version 2.0, Computerized Database 25*, NIST, Gaithersburg, MD (1994). (b) S. E. Stein, J. M. Rukkers, and R. L. Brown, *NIST Structures & Properties Database and Estimation Program*, NIST, Gaithersburg, MD (1991). (c) H. Y. Afeefy, J. F. Liebmann, and S. E. Stein, in W. G. Mallard and P. J. Linstrom (eds), *Neutral Thermochemical Data in NIST Standard Reference Database Number 69* NIST, Gaithersburgh, MD (2005).
73. *NIST Chemistry WebBook*. Available at http://webbook.nist.gov/chemistry/.
74. (a) E. R. Ritter and J. W. Bozzelli, THERM: Thermodynamic Property Estimation for Gas Phase Radicals and Molecules, 12th International CODATA Conference, July 1990. (b) E. R. Ritter, THERM User's Manual, Department of Chemical Engineering, New Jersey Institute of Technology, Newark, NJ (1990).

75. C. Muller, V. Michel, G. Scacchi, and G. M. Chatome, *J. Chimie Phys.* **92**, 1154 (1995).
76. V. N. Huff, S. Gordon, and V. Morrell, General Method and Thermodynamic Tables for Computation of Equilibrium Composition and Temperature of Chemical Reactions, NASA Report 1037 (1951).
77. (a) F. J. Zeleznik and S. Gordon, *An Analytical Investigation of Three General Methods of Calculating Chemical Equilibrium Compositions*, NASA TN-D 473 (1960). (b) F. J. Zeleznik and S. Gordon, *Simultaneous Least Square Approximation of a Function and* its *First Integrals, with Application to Thermodynamic Data*, NASA-TN-D-767 (1961). (c) F. J. Zeleznik and S. Gordon, *A General IBM704 or 7090 Computer Program for Computation of Chemical Equilibrium Compositions, Rocket Performance and Chapman–Jouguet Detonations*, NASA TN-D 1454 (1962).
78. B. J. McBride and S. Gordon, *Computer Program for Calculating and Fitting Thermodynamic Functions*, NASA Report 1271 (1992).
79. R. C. Wilhoit, Ideal gas thermodynamic functions, *TRC Current Data News 3*, No. 2 (1975).
80. C. H. Shomate, *J. Phys. Chem.* **58**, 368 (1954).
81. http://code.google.com/p/cantera/.
82. http://www.sandia.gov/chemkin/index.html.
83. http://www.rotexo.com/cms/index.php.
84. A. Burcat and B. Ruscic,Third Millennium Ideal Gas and Condensed Phase Thermochemical Database for Combustion with Updates from Active Thermochemical Tables, ANL-Report-05/20 and TAE Report 960 (2005) (now authored by E. Goos, B. Ruscic and A. Burcat, data available on request from Elke.Goos@dlr.de).
85. B. J. McBride, M. J. Zehe, and S. Gordon, NASA Glen Coefficients for Calculating Thermodynamic Properties of Individual Species NASA/TP-2002-211556 (2002). Available at http://cea.grc.nasa.gov as "NASA Thermo Build".
86. Gas Phase Thermochemical Database provided by Mark D. Allendorf from Sandia National Laboratories. Available at http://www.ca.sandia.gov/HiTempThermo//.
87. C. F. Melius and M. D. Allendorf, *J. Phys. Chem. A* **104**, 2168 (2000).
88. W. H. Green, J. W. Allen, R. W. Ashcraft, G. J. Beran, C. A. Class, C. Gao, C. F. Goldsmith, M. R. Harper, A. Jalan, G. R. Magoon, D. M. Matheu, S. S. Merchant, J. D. Mo, S. Petway, S. Raman, S. Sharma, J. Song, K. M. Van Geem, J. Wen, R. H. West, A. Wong, H.-W. Wong, P. E. Yelvington, J. Yu; RMG - Reaction Mechanism Generator v3.3, 2011, http://rmg.sourceforge.net/.

CHAPTER 2

Calculation of Kinetic Data Using Computational Methods

FERNANDO P. COSSÍO

Departamento de Química Orgánica I, Universidad del País Vasco-Euskal Herriko Unibertsitatea, San Sebastián-Donostia, Spain

2.1 INTRODUCTION

Prediction of kinetic data is one of the most important contributions of computational chemistry to the understanding of chemical transformations [1–5]. There are many accurate studies involving calculation of kinetic constants of reactions in which unstable, highly reactive species or intermediates play a significant role [6]. Many of these reactions are carried out in the gas phase [7–11]. However, most of these detailed calculations deal with small ensembles of reactive species involving up to 10–15 atoms. In this chapter, relatively less accurate computational methods will be presented, in which transformations involving medium-size molecules in solution are involved. In these cases, observables related with reactivity and selectivity (especially stereoselectivity) are analyzed rather than the absolute magnitudes of the corresponding kinetic constants. These computational analyses are useful to unveil the reasons underlying the kinetic control of many complex reactions that are closer to the kind of systems usually studied by synthetic chemists.

Most of the computational methods for the calculation of kinetic constants rely on the empirical Arrhenius rate law, discovered in 1899:

$$k(T) = A\, e^{-E_a/RT} \tag{2.1}$$

where $k(T)$ is the temperature-dependent rate constant, A is the pre-exponential factor, and E_a is the activation energy:

$$E_a = -R\frac{d \ln k(T)}{d(1/T)} = RT^2 \frac{d \ln k(T)}{dT} \tag{2.2}$$

Rate Constant Calculation for Thermal Reactions: Methods and Applications, Edited by Herbert DaCosta and Maohong Fan.

In Eq. (2.1), the experimental values of A and E_a can be obtained by measuring kinetic constants at different temperatures (Arrhenius plots).

The computation of kinetic constants without relying on experimental data can be achieved by means of different methods. Perhaps the most important one is the transition state theory, which permits the calculation of $k(T)$ and the interpretation of the Arrhenius law in terms of an activated complex. Since its formulation in 1935 by Eyring and Polanyi [12,13], this theory has been refined [14] leading to very accurate estimates of $k(T)$ values. Before discussing the most relevant aspects of this theory, a brief survey will be presented about the computational methods available to obtain relative energies of stationary points along any given reaction coordinate.

2.2 STATIONARY POINTS AND POTENTIAL ENERGY HYPERSURFACES

The energy of a molecule (or supermolecule) can be calculated by means of the Schrödinger equation:

$$\hat{H}\Psi(r,R,t) = ih\frac{\partial}{\partial t}\Psi(r,R,t) \tag{2.3}$$

where $\hat{H}$ is the Hamiltonian operator ($\hat{H} = \hat{T} + \hat{V}$, both right-hand terms being the kinetic and potential energy operators, respectively) and $\Psi(r,R,t)$ is the total wave function of the system, which depends on the electronic and nuclear coordinates (r—*including spin*—and R, respectively) and on time (t).

For stationary states, the time-independent Schrödinger equation holds:

$$\hat{H}\Psi(r,R) = E\Psi(r,R) \tag{2.4}$$

In Eq. (2.4), $\Psi(r,R)$ is the time-independent wave function and E is the total energy of the system.

For a molecule containing N nuclei and n electrons, the Hamiltonian operator can be separated in its nuclear ($\hat{H}_N$) and electronic ($\hat{H}_e$) contributions:

$$\hat{H} = \hat{H}_N + \hat{H}_e \tag{2.5}$$

The nuclear part is the kinetic energy associated with the nuclei and is expressed in atomic units as

$$\hat{H}_N = -\sum_{i=1}^{N}\frac{1}{2M_i}\nabla_i^2 \tag{2.6}$$

where M_i is the mass of nucleus i. The electronic part includes the kinetic energy of the electrons, the electron–electron potential energies, the nucleus–electron Coulombic potential energies, and the nucleus–nucleus Coulombic potential energies, respectively, and can be expressed in atomic units as

$$\hat{H}_e = -\sum_{\lambda=1}^{n}\frac{1}{2}\nabla_\lambda^2 + \sum_{\lambda>\mu}\frac{1}{r_{\lambda\mu}} - \sum_{\lambda=1}^{n}\sum_{i=1}^{N}\frac{Z_i}{R_{\lambda i}} + \sum_{i>j}\frac{Z_iZ_j}{R_{ij}} \tag{2.7}$$

where $r_{\lambda\mu}$ is the distance between electrons λ and μ, Z_i is the atomic number of nucleus i, $R_{\lambda i}$ is the distance between electron λ and nucleus i, and R_{ij} is the internuclear distance corresponding to nuclei i and j. The last term of Eq. (2.7), namely, the Coulombic repulsion between the nuclei, is actually associated with nuclear interactions, but is included in $\hat{H}_e$ for convenience (see below).

Equation (2.4) can be solved more easily by describing the total wave function as a product of the electronic (Ψ_e) and nuclear (Ψ_N) wave functions:

$$\Psi(r,R) = \Psi_e(r,R)\Psi_N(R) \tag{2.8}$$

Note that in Eq. (2.8), the nuclear wave function depends only on the nuclear coordinates, whereas the electronic wave function depends *explicitly* on the electronic coordinates and *implicitly* on the nuclear coordinates. This simplification is known as the Born–Oppenheimer approximation [15] and assumes that electrons, much lighter than nuclei, can adapt their position instantaneously to any change of the nuclear coordinates. Under these conditions, the *electronic* time-independent Schrödinger equation can be written as

$$\hat{H}_e\Psi_e(r,R) = E_e(R)\Psi_e(r,R) \tag{2.9}$$

Consequently, Eq. (2.3) can be solved as

$$[\hat{H}_N + E_e(R)]\Psi_N(R) = E_{total}\Psi_N(R) \tag{2.10}$$

In Eq. (2.9), $E_e(R)$ is a function (scalar field) that associates an energy value with a given ensemble of nuclear coordinates. The resulting hypersurface is usually named potential energy surface and it constitutes an essential tool for the understanding of chemical reactivity. The familiar energy versus reaction coordinate plots are bidimensional projections of the Born–Oppenheimer approximation onto the reaction coordinate associated with the chemical transformation under study (Figure 2.1). In particular, they provide a physical interpretation of the Arrhenius equation and the activation energy.

The positions on the potential energy surface of reactants or products (or reaction intermediates in a stepwise process) and the transition structure are intrinsically different. In both cases, the gradient of energy is zero, that is, the derivatives of $E_e(R)$ with respect to all the reaction coordinates are zero:

$$\vec{\nabla}E_e(R) = \sum_{i=1}^{3N-6} \frac{\partial E_e(R)}{\partial R_i}\vec{i} = \vec{0} \tag{2.11}$$

where the summation is extended to $3N - 6$ degrees of freedom of a nonlinear molecule containing N nuclei. Both local minima (reactants, products, or reaction intermediates) and transition structures fulfill Eq. (2.11) and are called *stationary points*. It can be demonstrated that the structure of these stationary points does not depend on the choice of the coordinate system.

In case of local minima, the second derivatives of the electronic energy along the nuclear coordinates are always positive or close to zero in case of supermolecules

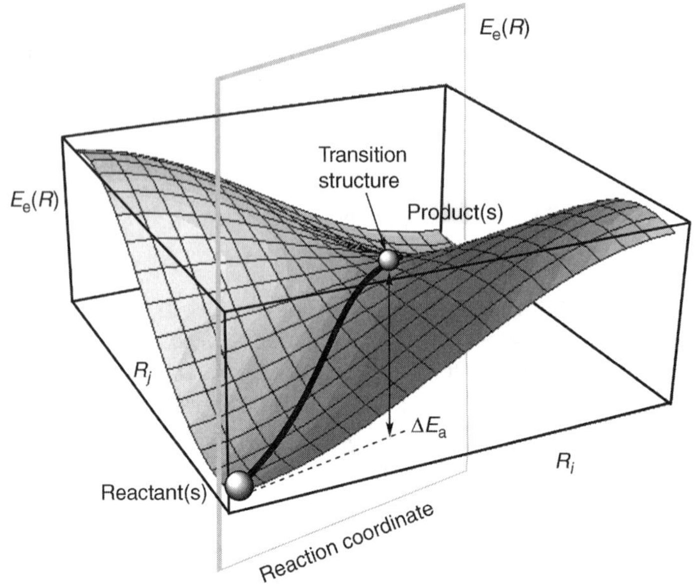

FIGURE 2.1 Potential energy surface $E_e(R)$ projected onto nuclear coordinates R_i and R_j, showing the difference between the transition structure (saddle point) and the reactant(s), product(s), or intermediate(s) (local minima). The transition structure is represented as the maximum of the energy along the reaction coordinate that connects reactant(s) and product(s). The activation energy (ΔE_a) is shown as the difference of energy between reactant(s) and transition structure along the reaction path within the plane determined by $E_e(R)$ and the reaction coordinate.

including all the reactive species and solvent molecules:

$$\frac{\partial^2 E_e(R)}{\partial R_i^2} \geq 0 \tag{2.12}$$

for any $i = 1, 2, \ldots, 3N - 6$.

In case of a transition structure, Eq. (2.12) applies to $i = 1, 2, \ldots, q - 1, q + 1, \ldots, 3N - 6$ nuclear coordinates. However, there is one and only one coordinate R_q for which

$$\frac{\partial^2 E_e(R)}{\partial R_q^2} < 0 \tag{2.13}$$

For a harmonic oscillator of reduced mass μ and force constant k along coordinate x, it is known that the force constant and the potential energy $V(x)$ are connected by the following equation:

$$k = \frac{d^2 V(x)}{dx^2} \tag{2.14}$$

The frequency associated with this vibration is

$$\upsilon = \frac{1}{2\pi}\sqrt{\frac{k}{\mu}} \tag{2.15}$$

Inspection of Eqs (2.13) and (2.14) shows that within the harmonic oscillator approximation, the second derivatives of the Born–Oppenheimer energy of a local minimum determine the force constants of the chemical bonds of the molecule under study. In general, the matrix formed by the second derivatives with respect to the nuclear coordinates is named *force constant matrix* or *Hessian*. The diagonalized Hessian conveys the information of the force constants and the frequencies of the corresponding normal modes of vibration. Since all the terms of the Hessian of a local minimum fulfill Eq. (2.12), it is said that its Hessian is *positive definite*. In case of a transition structure, Eq. (2.13) holds and, according to Eq. (2.15), there is one and only one imaginary frequency that must be associated with the reaction coordinate (Figure 2.1). It is said that a transition structure has an imaginary frequency or, in a concise form, that for a transition structure, number of imaginary frequencies (NIMAG) = 1. Of course, for a local minimum on the Born–Oppenheimer surface, it is found that NIMAG = 0.

Optimization—that is, minimization of the energy gradient—of a reactant, reaction intermediate (if existing), or product is usually straightforward and the only caveat is to make sure that its corresponding Hessian is positive definite. However, characterization and optimization of a transition structure constitute more difficult tasks and different algorithms [16,17] have been developed to this end. The located transition structure must fulfill the following four requirements, also known as *McIver–Komornicki conditions* [18]:

1. It must be a stationary point and therefore invariant with respect to the coordinate system used.
2. Its force constant matrix must have one and only one negative eigenvalue.
3. It must be the highest energy point on a continuous line connecting reactants and products. The imaginary frequency associated with this maximum must be associated with nuclear motion along the reaction coordinate under study.
4. It must be the lowest energy point that satisfies the above three conditions.

Although the first two conditions can be checked easily, the third and fourth ones must be analyzed carefully by studying the possible conformations energetically accessible and analyzing the vibrational mode or by performing intrinsic reaction coordinate (IRC) calculations [19,20].

It must be emphasized that the transition *structure* as defined above is closely related, but not identical, to the transition *state* as formulated in Eyring's theory [21]. We will discuss this point subsequently.

2.3 CALCULATION OF REACTION AND ACTIVATION ENERGIES: LEVELS OF THEORY AND SOLVENT EFFECTS

2.3.1 Hartree–Fock and Post-Hartree–Fock Methods

Evaluation of the energy in the time-independent Schrödinger equation requires the solution of an eigenvalue–eigenvector problem [22]. For an electronic wave function satisfying Eq. (2.9), an eigenvector—the total electronic energy—will be found. A possible polyelectronic wave function for n electrons could have the form of a *Hartree product*:

$$\Psi_e(r,R) = \prod_{i=1}^{n} \phi_i(i) = \phi_1(1)\phi_2(2)\ldots\phi_n(n) \tag{2.16}$$

where $\phi_i(i)$ is a monoelectronic function operating on electron i. This simple product is not valid for an ensemble of electrons since the electronic wave function must fulfill the antisymmetry principle (the subscript "e" of the electronic wave function will be omitted hereafter):

$$\Psi(1,2,\ldots,i,\ldots,p,\ldots,n) = -\Psi(1,2,\ldots,p,\ldots,i,\ldots,n) \tag{2.17}$$

According to Eq. (2.17), exchange of two electrons must yield the same function with the opposite sign. The simplest way to fulfill this condition is to antisymmetrize the possible Hartree products by means of a *Slater determinant*:

$$\Psi(1,2,\ldots,n) = \frac{1}{\sqrt{n!}}\begin{vmatrix} \phi_1(1) & \phi_2(1) & \ldots & \phi_n(1) \\ \phi_1(2) & \phi_2(2) & \ldots & \phi_n(2) \\ \vdots & \vdots & \vdots & \vdots \\ \phi_1(n) & \phi_2(n) & \ldots & \phi_n(n) \end{vmatrix} \tag{2.18}$$

Within the restricted Hartree–Fock (HF or RHF) approach, electrons occupy the product elements by couples differing only in the spin coordinate (otherwise the value of the determinant would be zero):

$$\Psi(1,2,\ldots,n) = \frac{1}{\sqrt{n!}}\begin{vmatrix} \phi_1(1) & \bar{\phi}_1(1) & \ldots & \phi_{n/2}(1) & \bar{\phi}_{n/2}(1) \\ \phi_1(2) & \bar{\phi}_1(2) & \ldots & \phi_{n/2}(2) & \bar{\phi}_{n/2}(2) \\ \vdots & \vdots & \vdots & \vdots & \vdots \\ \phi_1(n) & \bar{\phi}_1(n) & \ldots & \phi_{n/2}(n) & \bar{\phi}_{n/2}(n) \end{vmatrix} \tag{2.19}$$

where the superscript "bar" indicates the change of the spin coordinate from $\alpha(\uparrow)$ to $\beta(\downarrow)$ or vice versa. In Eqs (2.18) and (2.19), the $1/\sqrt{n!}$ term is a factor required to fulfill the normalization condition:

$$\int \Psi^*(1,2,\ldots,n)\Psi(1,2,\ldots,n)d\tau_1 d\tau_2 \ldots d\tau_n = \langle\Psi|\Psi\rangle = 1 \tag{2.20}$$

where the sign * means the conjugate function of Ψ.

For an electronic wave function as given in Eq. (2.19) fulfilling condition (2.20), the value of the energy is given by

$$E = \int \Psi^*(1, 2, \ldots, n)\hat{H}\Psi(1, 2, \ldots, n)d\tau_1 d\tau_2 \ldots d\tau_n = \langle\Psi|\hat{H}|\Psi\rangle \qquad (2.21)$$

which can be expressed in compact form as

$$E = \sum_{i=1}^{n} h_i + \frac{1}{2}\sum_{i=1}^{n}\sum_{i>j}(J_{ij} - K_{ij}) + V_{NN} \qquad (2.22)$$

In Eq. (2.22), V_{NN} is the internuclear repulsion defined in Eq. (2.7). The monoelectronic terms h_i are given by

$$h_i = \int \phi_i^*(1)\left[-\frac{1}{2}\nabla^2 - \sum_{\alpha=1}^{N}\frac{Z_\alpha}{R_{i\alpha}}\right]\phi_i(1)d\tau_1 \qquad (2.23)$$

The bielectronic terms J_{ij} are the *Coulomb integrals* and reflect the electrostatic repulsion between electrons:

$$J_{ij} = \iint \phi_i^*(1)\phi_j^*(2)r_{12}^{-1}\phi_i(1)\phi_j(2)d\tau_1 d\tau_2 \qquad (2.24)$$

Bielectronic terms K_{ij} are the *exchange integrals* and have no classical counterpart:

$$K_{ij} = \iint \phi_i^*(1)\phi_j^*(2)r_{12}^{-1}\phi_i(2)\phi_j(1)d\tau_1 d\tau_2 \qquad (2.25)$$

These integrals constitute about 25% stabilizing correction to the total electronic repulsion.

A convenient way to solve Eq. (2.22) is to describe the molecular spin orbitals ϕ_i as a linear combination of atomic orbitals (LCAO) χ_α:

$$\phi_i = \sum_{\alpha} c_{\alpha i}\chi_\alpha \qquad (2.26)$$

For example, the Coulomb integrals are transformed under the LCAO approach into

$$\begin{aligned} J_{ij} &= \iint \left(\sum_{\alpha} c_{\alpha i}\chi_i^*(1)\right)\left(\sum_{\gamma} c_{\gamma j}\chi_j^*(2)\right)r_{12}^{-1}\left(\sum_{\beta} c_{\beta i}\chi_i(1)\right)\left(\sum_{\delta} c_{\delta j}\chi_j(2)\right)d\tau_1 d\tau_2 \\ &= \sum_{\alpha}\sum_{\beta}\sum_{\gamma}\sum_{\delta} c_{\alpha i}c_{\beta i}c_{\gamma j}c_{\delta j}\iint \chi_i^*(1)\chi_j^*(2)r_{12}^{-1}\chi_i(1)\chi_j(2)d\tau_1 d\tau_2 \end{aligned} \qquad (2.27)$$

The expansion coefficients $c_{\alpha i}$ are optimized to fulfill the condition

$$\frac{\partial E}{\partial c_{\alpha i}} = 0, \quad \text{for any value of } \alpha \text{ and } i \qquad (2.28)$$

Under these constraints, mono- and bielectronic integrals required to solve Eq. (2.22) are reduced to integrals involving atomic orbitals, which are in principle Slater-type orbitals (STOs). Integration of these STOs is not well suited for computer algorithms and therefore the STOs are usually described [23] as linear combinations of Gaussian-type orbitals (GTOs). One usual implementation of this idea consists of the description of electrons using *split-valence Gaussian basis sets*. One of the most frequently used basis sets of this type is 6-31G*. In this basis set, the core electrons are described as linear combinations of six GTOs, whereas the valence electrons are in turn described as a contribution of an inner part described by three GTOs and an outer part described by means of a single GTO. In addition, for all elements except hydrogen, an additional set of GTOs of higher angular momentum is added. This means that for each carbon, oxygen, and nitrogen atom, a set of d-type GTOs is added. This feature is reflected by the asterisk (*) after the description of the split-valence basis set.

Using the above-mentioned approaches and techniques, it is possible to obtain values for the energies of different stationary points on chemically relevant potential energy surfaces. However, under these conditions within the HF approach and even using extremely accurate basis sets, a value higher than the exact energy will always be obtained. The difference between the exact energy and the HF limit energy is called *correlation energy* [24]. This energy is significant in absolute value and, most importantly, is usually different in magnitude for transition structures and local minima. Therefore, correlation energies must be evaluated at least in an approximate and consistent form in order to estimate accurately enough activation energies and related thermodynamic magnitudes (see below).

One possibility consists of expanding the reference HF wave function in terms of monoexcitations, diexcitations, and polyexcitations in a monoreference *configuration interaction* (CI) wave function:

$$\Psi_{\mathrm{CI}} = a_0 \Psi_{\mathrm{HF}} + \sum_{i}^{\mathrm{occ}} \sum_{a}^{\mathrm{virt}} a_i^a \Psi_i^a + \sum_{i<j}^{\mathrm{occ}} \sum_{a<b}^{\mathrm{virt}} a_{ij}^{ab} \Psi_{ij}^{ab} + \cdots \tag{2.29}$$

Although a_0 is the most important term along the CI expansion, the other terms are by no means negligible and account for a significant part of the correlation energy. These CI wave functions and other related CI techniques [25] can result in accurate estimates of the correlation energies, but are quite expensive for "real" reactants and products usually studied in synthetically useful chemical transformations.

The Møller–Plesset (MP) method [26] constitutes a relatively less expensive alternative and is conceptually related to Rayleigh–Schrödinger perturbation theory. The lowest term in the MPn series is MP2. The MP2 energy can be efficiently calculated as

$$E_{\mathrm{MP2}} = E_{\mathrm{HF}} + \sum_{i}^{\mathrm{occ}} \sum_{i<j}^{\mathrm{occ}} \sum_{a}^{\mathrm{virt}} \sum_{a<b}^{\mathrm{virt}} \frac{[(ij|ab) - (ia|jb)]^2}{\varepsilon_i + \varepsilon_j - (\varepsilon_a + \varepsilon_b)} \tag{2.30}$$

where ε_i, ε_j and ε_a, ε_b are the orbital energies of occupied and virtual MOs, respectively. The bielectronic integrals in Eq. (2.30) are given by

$$(ij|ab) = \iint \phi_i^*(1)\phi_j^*(2)r_{12}^{-1}\phi_a(1)\phi_b(2)d\tau_1 d\tau_2 \quad (2.31)$$

and

$$(ia|jb) = \iint \phi_i^*(1)\phi_a^*(2)r_{12}^{-1}\phi_j(1)\phi_b(2)d\tau_1 d\tau_2 \quad (2.32)$$

In order to obtain very accurate energies for different stationary points, several computational schemes such as G2 [27], G3 [28], and G4 [29] and different variants have been proposed. However, once again, the cost of these calculations for realistic structures and complicated kinetic schemes is often too high.

2.3.2 Methods Based on Density Functional Theory

Perhaps density functional theory (DFT) [30] is nowadays the best compromise between computational cost and reliability to obtain both accurate geometries and a wide range of stationary points. In this approach, energy is described as a *functional* (i.e., a function whose argument is another function) of the electron density $E = E[\rho(r)]$. Within this scheme, electron density includes the total number of electrons of a molecule, namely,

$$\int \rho(r)dr = n \quad (2.33)$$

and in turn electron density can be built up from the contributions of the different atoms by means of the square of an ensemble of Kohn–Sham (KS) orbitals $\phi_i(r)$:

$$\rho(r) = \sum_{i=1}^{N} |\phi_i(r)|^2 \quad (2.34)$$

These KS orbitals provide an interpretative framework similar to or even better than that provided by the canonical HF MOs since they include electron correlation [31] and are consistent with the exact electron density.

The DFT energy is given by the Kohn–Sham equation:

$$E_{\mathrm{DFT}}[\rho] = T_{\mathrm{S}}[\rho] + E_{\mathrm{ne}}[\rho] + J[\rho] + E_{\mathrm{XC}}[\rho] + V_{\mathrm{NN}} \quad (2.35)$$

In Eq. (2.35), V_{NN} (i.e., the internuclear repulsion Coulomb energy) has the same meaning as that in Eqs. (2.7) and (2.22). $T_{\mathrm{S}}[\rho]$ is the electronic kinetic energy, described as

$$T_{\mathrm{S}}[\rho] = \sum_{i=1}^{n} \left\langle \phi_i \middle| -\frac{1}{2}\nabla^2 \middle| \phi_i \right\rangle \quad (2.36)$$

$E_{\mathrm{ne}}[\rho]$ is the electron–nuclei attractive term and it can be described in terms of electron density as

$$E_{\mathrm{ne}}[\rho] = \sum_{\alpha=1}^{N} \iint \frac{Z_\alpha \rho(r)}{|R_\alpha - r|} dr \quad (2.37)$$

Both terms $T_{\mathrm{S}}[\rho]$ and $E_{\mathrm{ne}}[\rho]$ are very similar to their HF analogues included in the h_i terms described by Eq. (2.22) and demand a similar computational effort. However,

$T_S[\rho]$ refers to a *fictitious* ensemble of noninteracting electrons whose density is adequately described by KS orbitals. $J[\rho]$ is DFT equivalent of the ensemble of Coulomb integrals described in Eq. (2.24):

$$J[\rho] = \frac{1}{2}\iint \frac{\rho(r)\rho(r')}{|r-r'|} dr dr' \tag{2.38}$$

Finally, $E_{XC}[\rho]$ in Eq. (2.35) is exchange–correlation functional. This term is the most difficult to model and evaluate and different approaches have been proposed to compute it [32]. It must be emphasized that the accuracy of a DFT calculation is determined by the accuracy of the evaluation of the exchange–correlation energy since the other terms of Eq. (2.35) are in principle exact [33]. The two main approaches to evaluate $E_{XC}[\rho]$ are the local density approximation (LDA) and the generalized gradient approximation (GGA). The last one is better suited for the treatment of bond breaking and bond forming processes associated with chemical transformations. According to this methodology, the exchange–correlation energy depends on both the electron density and the norm of its gradient:

$$E_{XC}^{GGA} = \int f(\rho(r), |\nabla\rho(r)|) dr \tag{2.39}$$

One of the most extended implementations within this approach is a hybrid methodology usually denoted as B3LYP after the three-parameter exchange–correlation functional developed by Becke [34]:

$$E_{XC}^{B3LYP} = (1-a_0)E_X^{LSDA} + a_0 E_X^{HF} + a_X E_X^{B88} + a_C E_C^{LYP} + (1-a_C)E_C^{VWN} \tag{2.40}$$

The values of the three parameters in Eq. (2.40) are $a_0 = 0.20$, $a_X = 0.72$, and $a_C = 0.81$. The exchange functional E_X^{LSDA} depends only on electron density of electron of spins α and β and is given by

$$E_X^{LSDA} = -2^{1/3}\frac{3}{4}\left(\frac{3}{\pi}\right)^{1/3}\int \left(\rho_\alpha^{4/3} + \rho_\beta^{4/3}\right) dr \tag{2.41}$$

E_X^{HF} is the exact Hartree–Fock exchange and the E_X^{B88} exchange functional was defined by Becke in 1988 [35] and depends on both electron density and the norm of its gradient:

$$E_X^{B88} = E_X^{LDA} - \beta \sum_\sigma \int \rho_\sigma^{4/3} \frac{\left(|\nabla\rho_\sigma|/\rho_\sigma^{4/3}\right)^2}{1 + 6\beta\left(|\nabla\rho_\sigma|/\rho_\sigma^{4/3}\right)\sinh^{-1}\left(|\nabla\rho_\sigma|/\rho_\sigma^{4/3}\right)} dr \tag{2.42}$$

In Eq. (2.42), the summation on σ runs over $\uparrow$ and $\downarrow$ spin electrons, β is a parameter obtained by least-squares fit to yield $\beta = 0.0042$ a.u. and E_X^{LDA} is the exchange-only functional within the local density approximation:

$$E_X^{LDA} = -\frac{3}{2}\left(\frac{3}{4\pi}\right)^{1/3}\sum_\sigma \int \rho_\sigma^{4/3} dr \tag{2.43}$$

In Eq. (2.40), E_C^{LYP} is the correlation functional defined by Lee et al. [36] and is given by the following expression based on a three-parametric fitting procedure developed by Colle and Salvetti [37] using HF orbital of helium atom:

$$E_C^{LYP} = -0.049 \int \frac{1}{1 + 0.349\rho^{-1/3}} [\rho + 0.132\rho^{-2/3}\Omega \exp(-0.2533\rho^{-1/3})]dr \tag{2.44}$$

where Ω term is an estimate of the kinetic correlation energy density given by

$$\Omega = C_F\rho^{5/3} - 2T_W + \left(\frac{1}{9}T_W + \frac{1}{18}\nabla^2\rho\right) \tag{2.45}$$

where $C_F\rho^{5/3}$ is the Thomas–Fermi kinetic energy density, with $C_F = (3/10)(3\pi^2)^{2/3}$, and T_W is a local kinetic energy density described as

$$T_W = \frac{1}{8}\left(\frac{|\nabla\rho(r)|^2}{\rho(r)} - \nabla^2\rho(r)\right) \tag{2.46}$$

The last contribution to correlation energy in Eq. (2.40) is E_C^{VWN}, defined by Vosko et al. for both closed-shell and open-shell systems [38] using different expressions [39]. The default option of this term in the *Gaussian* suite of programs [40] for closed-shell systems is

$$E_C^{VWN} = \int A\left[\ln\frac{\rho}{\rho + b\rho^{1/2} + c} + \frac{2b}{(4c - b^2)^{1/2}}\tan^{-1}\frac{(4c - b^2)^{1/2}}{2\rho^{1/2} + b} - \varepsilon(x_0, \rho)\right]\rho dr \tag{2.47}$$

where the term $\varepsilon(x_0, \rho)$ is given by

$$\varepsilon(x_0, \rho) = \frac{bx_0}{x_0^2 + bx_0 + c}\left[\ln\frac{(\rho^{1/2} - x_0)^2}{\rho + b\rho^{1/2} + c} + \frac{2(b + 2x_0)}{(4c - \rho)^{1/2}}\tan^{-1}\frac{(4c - \rho)^{1/2}}{2\rho^{1/2} + b}\right] \tag{2.48}$$

The best set of parameters found by Vosko et al. for singlet closed-shell species is $A = 0.0621814$, $x_0 = -0.409286$, $b = 13.0720$, and $c = 42.7198$.

B3LYP and other functionals usually yield reliable data at a reasonable computational cost. Actually, the three parameters of this popular hybrid functional were optimized to fit the G2 data set. However, recent research has reported that in several cases, B3LYP treatment of noncovalent interactions is not so accurate and the performance of this DFT method can be problematic [41], for instance, for the treatment of highly branched or substituted systems. Recently, M06 [42] and M06-2X functionals have been proposed by Zhao and Truhlar to solve this problem [43]. These methods can be useful to obtain reliable relative energies when highly substituted reactants are studied.

2.3.3 Computational Treatment of Solvent Effects

Since most of the chemical transformations are carried out in solution, an adequate treatment of solvent effects is required to obtain realistic relative energies and even realistic reaction profiles [44]. Many different methods for the computational treatment of solvent effects have been developed [45]. One possibility consists of including solvent molecules around the reactants as discrete molecules (Figure 2.2a). These solvent molecules can be treated at the molecular mechanics (MM) level,

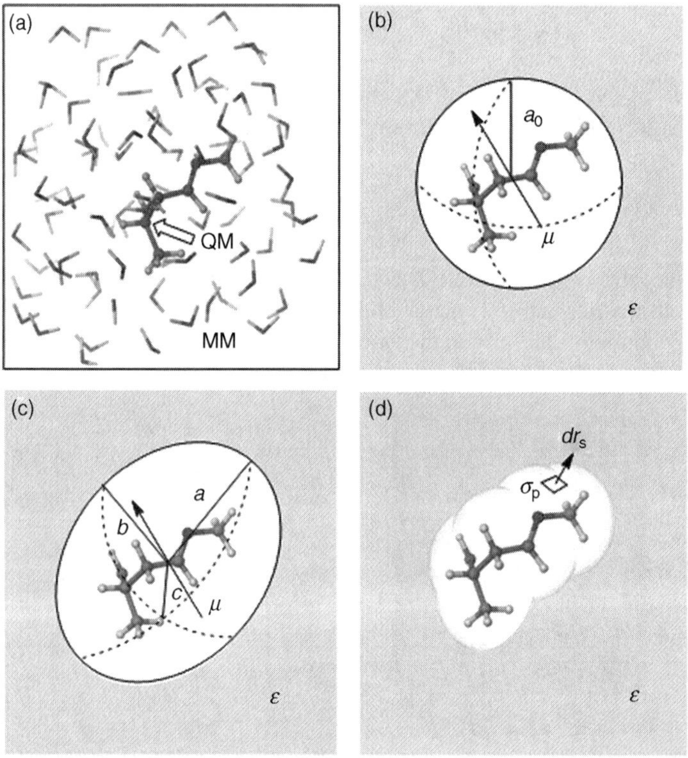

FIGURE 2.2 Different computational treatments of solute–solvent interactions. (a) Solvent treated as an ensemble of discrete molecules. In this case, solvent molecules (in *tube* representation) can be treated at MM level, whereas the solute (in *ball-and-stick* representation) is treated at QM level. (b) Treatment of the solvent as a dielectric continuum. The solute is modeled as a sphere of radius a_0 and dipole moment μ according to the Onsager model. (c) Same as (b) but the solute is in an ellipsoidal cavity defined by axes a, b, and c. (d) Treatment of the solute–solvent interactions according to the polarization continuum model (PCM).

whereas active solvent molecules (i.e., those coordinated to active centers and involved in bond forming or bond breaking) can be treated using quantum mechanics (QM). There are many examples in the literature on the study of reaction mechanisms using this QM/MM treatment [46–48].

The most used QM treatments are those that describe implicitly the solvent as a continuum dielectric (Figure 2.2b–d) [49–51]. In these methods, the usual Hamiltonian is modified to include the solvation energy as an additional term in which the solute–solvent system is described by means of the interaction between the dipole μ of the solute and a dielectric continuum of electric permittivity ε, thus generating a reactive field g:

$$\hat{H} = \hat{H}_0 - \mu g \langle \psi | \mu | \psi \rangle \tag{2.49}$$

In Eq. (2.49), $\hat{H}_0$ is the gas-phase Hamiltonian defined in Eq. (2.7). Note that now the electronic wave function must be self-consistent with respect to the gas-phase Hamiltonian and simultaneously with respect to the dipole of the solute and the dielectric continuum. Therefore, this approach is termed the self-consistent reaction field (SCRF) method. The electronic Schrödinger equation is

$$[\hat{H}_0 - \mu g \langle \psi | \mu | \psi \rangle]\psi = E(R)\psi \tag{2.50}$$

and therefore the potential energy surface can be scanned taking into account solvent effects.

The simplest implementation of SCRF theory is the Onsager model [52,53], in which the reactive field is

$$g = \frac{2}{a_0^3}\left(\frac{\varepsilon - 1}{2\varepsilon + 1}\right) \tag{2.51}$$

Within this model, the solute is described as a sphere of radius a_0 with a dipole μ at the center of the cavity. The solvent is a dielectric isotropic medium described by its permittivity ε (Figure 2.2b). Under this scheme, the Gibbs solvation energy is approximated as

$$\Delta G_{\mathrm{s}} = -\frac{1}{2}\left[\frac{2(\varepsilon - 1)}{2\varepsilon + 1}\right]\frac{\mu^2}{a_0^3} \tag{2.52}$$

This expression can be generalized [54] by expanding the dipole in a multipole series (usually up to the sixth term) and by describing the cavity generated by the solute as an ellipsoid of axes a, b, and c (Figure 2.2c):

$$\Delta G_{\mathrm{S}} = -\frac{1}{2}\sum_{l,l'=1}^{6}\sum_{m,m'=-l,l'}^{+l,l'} M_l^m f_{l,l'}^{m,m'} M_{l'}^{m'} \tag{2.53}$$

where M_l^m represents the (l, m) component of the multipole expansion computed at the center of the cavity and $f_{l,l'}^{m,m'}$ are the reaction field coefficients, which depend only on the axes a, b, and c and on ε.

If the multipole expansion is truncated at the charge level (i.e., the lowest one) and the center of the expansion is shifted to the different nuclei, a generalized Born approach is achieved. Thus, the Born formula for the free energy of an ion of charge q and radius a in a dielectric of permittivity ε, namely,

$$\Delta G = -\left(\frac{\varepsilon - 1}{\varepsilon}\right)\frac{q^2}{2a} \tag{2.54}$$

is transformed for a molecule of N atoms into

$$\Delta G_s = -\left(\frac{\varepsilon - 1}{\varepsilon}\right)\sum_{i=1}^{N}\sum_{j>i}^{N}\frac{q_i q_j}{2f_{GB}} = \left(\frac{\varepsilon - 1}{\varepsilon}\right)\sum_{i=1}^{N}\sum_{j>i}^{N} q_i q_j \gamma_{i,j} \tag{2.55}$$

where q_i and q_j are the partial charges of atoms i and j, respectively, and $1/f_{GB}$ is an empirical Coulombic operator [55] acting on all pairs (i, j). According to Cramer and Truhlar [56], the $(2f_{GB})^{-1}$ term can be expressed as an effective Coulomb integral $\gamma_{i,j}$ in the form

$$\gamma_{i,j} = \frac{1}{\sqrt{r_{i,j}^2 + b_i b_j \exp[-r_{i,j}^2 (d_{i,j} b_i b_j)^{-1}]}} \tag{2.56}$$

In Eq. (2.56), $r_{i,j}$ is the internuclear distance between centers i and j, b_i and b_j are the respective effective atomic Born radii, and $d_{i,j}$ is a parameter for the pair (i, j). The most recent version of this approach is SM8 [56].

Another important SCRF approximation is based on a simplified Poisson equation in which the electrostatic potential $V(r)$ inside the space occupied by a cavity representing the solute is given by

$$-\nabla^2 V(r) = 4\pi\rho_M(r) \tag{2.57}$$

whereas outside the cavity and within the dielectric medium of permittivity ε, Eq. (2.58) holds:

$$-\varepsilon\nabla^2 V(r) = 0 \tag{2.58}$$

Within this approach, the popular polarization continuum model (PCM) method defined by Tomasi and coworkers [57,58] computes the solvation energy of a molecule containing n electrons and N nuclei as

$$\Delta G_S = \frac{1}{2}\left[-\sum_{i=1}^{n}\oint_{S_c}\frac{\sigma(s_c)ds_c}{|r_i - r_{S_c}|} + \sum_{\alpha=1}^{N} Z_\alpha \oint_{S_c}\frac{\sigma(s_c)ds_c}{|r_\alpha - r_{S_c}|}\right] \tag{2.59}$$

where S_c is the surface of the cavity generated by the solute, defined as an ensemble of intersecting spheres (Figure 2.2d), Z_α is the atomic number of nucleus α, and $\sigma(s_c)$ is the superficial charge density defined [59] as

$$\sigma(s_c) = \frac{\varepsilon - 1}{4\pi\varepsilon}\frac{\partial}{\partial n_+}(V_M + V_\sigma)_{in} \tag{2.60}$$

In this expression, n_+ is a unit vector parallel to ds_c, namely, pointing toward the dielectric medium (Figure 2.2d), V_M is the electrostatic potential generated by the charge distribution of the solute inside the cavity, and V_σ is the reaction potential generated by the polarization of the dielectric medium.

Besides the electrostatic part of the solvation energy, other contributions can be taken into account, such as cavitation (cav) and dispersion (disp) energies, and possible changes in the local structure of the solvent (solv). These additional terms are often evaluated by means of a combined expression [60] based on the solvent-accessible surface (SAS):

$$G_{cav} + G_{disp} + G_{solv} = \sum_{k=1}^{N} (\sigma_k + \sigma_M)\mathrm{SAS}_k \tag{2.61}$$

where σ_k and σ_M are atomic and molecular surface tensions.

2.4 ESTIMATE OF RELATIVE FREE ENERGIES: STANDARD STATES

Although the solvation free energies can be obtained from the methods commented above, the electronic and total energies obtained from Eqs (2.9) and (2.10) are internal energies. Therefore, since Gibbs free energies are required to calculate kinetic constants, the energies obtained according to the computational methods and levels of theory outlined above must be transformed into the corresponding Gibbs free energies at a given temperature:

$$\Delta G(T) = \Delta H(T) - T\,\Delta S(T) \tag{2.62}$$

Assuming ideal gas-phase behavior of 1 mol of substance, the enthalpy at a certain temperature T from 0K results from the following contributions:

$$\Delta H(T) = E_{total} + H_{trans}(T) + H_{rot}(T) + H_{vib}(T) + RT \tag{2.63}$$

For a nonlinear stationary point, the translational and rotational terms are (for a linear species, the rotational contribution is RT)

$$H_{trans}(T) + H_{rot}(T) = \frac{3}{2}RT + \frac{3}{2}RT = 3RT \tag{2.64}$$

The vibrational contribution is given by

$$\Delta H_{vib}(T) = H_{vib}(T) - \mathrm{ZPVE} = N_A h \sum_{i=1}^{3n-m} \frac{\nu_i}{\exp(h\nu_i/k_B T) - 1} - \mathrm{ZPVE} \tag{2.65}$$

where N_A is the Avogadro number, h is the Planck constant, k_B is the Boltzmann constant, and ν_i is the corresponding vibrational frequency. $3n - m$ is the number of degrees of freedom of the n-atom containing species under consideration. For a local minimum (reactant, intermediate, or product), $m = 6$, and for a transition structure

$m = 7$, since the imaginary frequency is not included in the summation (see above). ZPVE is the zero-point vibrational energy and is given by

$$\mathrm{ZPVE} = \frac{1}{2} h \sum_{i}^{3n-m} \nu_i \tag{2.66}$$

The entropy contribution to the Gibbs energy of 1 mol of molecules is given by

$$S = S_{\mathrm{trans}} + S_{\mathrm{vib}} + S_{\mathrm{rot}} + S_{\mathrm{el}} - R[\ln(N_{\mathrm{A}}) - 1] \tag{2.67}$$

The translational, rotational, vibrational, and electronic terms are given by the following expressions, respectively:

$$S_{\mathrm{trans}} = R\left\{\frac{3}{2} + \ln\left[\left(\frac{2\pi M_{\mathrm{w}} k_{\mathrm{B}} T}{h^2}\right)^{3/2} \left(\frac{RT}{P}\right)\right]\right\} \tag{2.68}$$

$$S_{\mathrm{rot}} = R\left\{\frac{3}{2} + \ln\left[\frac{\sqrt{\pi I_{\mathrm{A}} I_{\mathrm{B}} I_{\mathrm{C}}}}{\sigma} \left(\frac{8\pi^2 k_{\mathrm{B}} T}{h^2}\right)^{3/2}\right]\right\} \tag{2.69}$$

$$S_{\mathrm{vib}} = R \sum_{i=1}^{3n-m} \left[\frac{h\nu_i}{k_{\mathrm{B}} T} (e^{h\nu_i/k_{\mathrm{B}}T} - 1)^{-1} - \ln(1 - e^{-(h\nu_i/k_{\mathrm{B}}T)})\right] \tag{2.70}$$

$$S_{\mathrm{el}} = R \ln \omega_{\mathrm{el}} = R \ln(2S_{\mathrm{total}} + 1) \tag{2.71}$$

where M_{w} is the molecular weight, I_{A}, I_{B}, and I_{C} are the three principal moments of inertia, σ is the symmetry number (for C_1 and C_s symmetric species $\sigma = 1$), and ω_{el} is the electronic ground state multiplicity (for closed-shell species, $S_{\mathrm{el}} = 0$ since the total electronic spin S_{total} is zero). Once again, for a local minimum, $m = 6$ and for a transition structure, $m = 7$.

Inclusion of solvent effects in the calculations involves, besides the computation of the solvation free energies, conversion from the gas-phase standard state ($P = 1$ atm) to the solution standard state (1 M). The relation between both states is given by [61–63]

$$\Delta G_{\mathrm{s}}^{*} = \Delta G_{\mathrm{s}}^{\circ} - RT \ln\left(\frac{V^{\circ}}{V^{*}}\right) \tag{2.72}$$

Superscripts * and ° refer to 1 M and 1 atm standard states, respectively. In Eq. (2.72), V° is the volume occupied by an ideal gas at a pressure of 1 atm and V^{*} is the volume occupied by an ideal gas at a concentration of 1 mol/L. It is found that at

298.15K, the correction term is

$$RT \ln\left(\frac{V^{\circ}}{V^{*}}\right) = RT \ln(24.45) = 1.9\ (\text{kcal/mol}) \tag{2.73}$$

From Eqs (2.1) and (2.2), we can calculate the free energies of activation in solution at 298.15K and under both standard states:

$$\Delta G^{*}_{a_s} = \Delta Ga^{\circ}_{s} - 1.9\Delta n\ (\text{kcal/mol}) \tag{2.74}$$

where $\Delta G^{*}_{a_s}$ and $\Delta G^{\circ}_{a_s}$ are the activation Gibbs energies in solution (in kcal/mol) at 1 M and 1 atm standard states, respectively, and Δn is a stoichiometric term given by

$$\Delta n = 1 - \sum_{i}^{\text{reactants}} n_i \tag{2.75}$$

Therefore, for an elementary process of type $A^{\rightarrow} \rightarrow TS$, the activation energy in solution at 1 M standard state will be identical to that computed at 1 atm standard state at the same theoretical level since in this case $\Delta n = 0$. However, for an elementary process of type $A + B^{\rightarrow} \rightarrow TS$, the stoichiometric correction is $\Delta n = -1$ and therefore

$$\Delta G^{*}_{a_s} = \Delta G^{\circ}_{a_s} + 1.9\ (\text{kcal/mol}) \tag{2.76}$$

Similarly, the Gibbs energies of reaction in solution at 298.15K under both standard states can be converted as follows:

$$\Delta G^{*}_{\text{rxn}_s} = \Delta G^{\circ}_{\text{rxn}_s} - 1.9\Delta n\ (\text{kcal/mol}) \tag{2.77}$$

In the latter case, the stoichiometric term is given by

$$\Delta n = \sum_{i}^{\text{products}} n_i - \sum_{j}^{\text{reactants}} n_j \tag{2.78}$$

Therefore, for reactions of type $A^{\rightarrow} \rightarrow B$ or $A + B^{\rightarrow} \rightarrow C + D$, no stoichiometric corrections are required. Otherwise, the reaction energies at 1 M standard state will require corrections with respect to $P = 1$ atm standard state.

This conversion of total electronic energies into Gibbs free energies assumes implicitly that the geometries of the actual reactants, reaction intermediates, products, and transition states do not differ significantly from those of the corresponding stationary points. We must remember that these stationary points were obtained by minimizing the electronic energies with respect to the nuclear coordinates. Therefore, a stationary (zero-gradient) point on the Born–Oppenheimer potential energy surface is not necessarily a stationary point in the Gibbs free energy surface (Figure 2.3). In other words, by computing the Gibbs energies and the corresponding kinetic constants, we expect that both energy surfaces are close enough to each other.

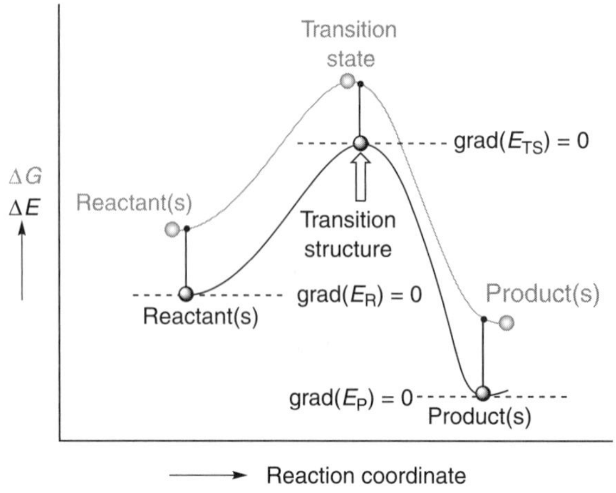

FIGURE 2.3 Stationary points on the Born–Oppenheimer potential energy surface and on the Gibbs free energy surface (gray). Note the difference between transition structure and transition state.

2.5 THEORETICAL APPROXIMATE KINETIC CONSTANTS AND TREATMENT OF DATA

Although activation energies can be calculated by computational methods (see above), a theoretical estimate of the pre-exponential factor$A$ is required to obtain kinetic constants. The transition state theory provides a link between the Arrhenius equation and this factor. According to this theory, assuming that the transition state (TS) is in equilibrium with the reactant (R), namely,

$$K^{\dagger} = \frac{[\mathrm{TS}]}{[\mathrm{R}]} = e^{-(G_{\mathrm{TS}}-G_{\mathrm{R}})/k_{\mathrm{B}}T} \approx \frac{Q_{\mathrm{TS}}}{Q_{\mathrm{R}}} e^{-(U_{\mathrm{TS}_0}-U_{\mathrm{R}_0})/k_{\mathrm{B}}T} \tag{2.79}$$

where U_{TS_0} and U_{R_0} are the internal energies (i.e., total electronic energies plus ZPVEs) of the TS and the reactant, respectively, and Q_{TS} and Q_{R} are the respective partition functions.

The reaction rate associated with the conversion from the reactants to the product through TS according to Eq. (2.80)

$$\mathrm{R} \underset{k_{\leftarrow}}{\overset{k_{\rightarrow}}{\rightleftarrows}} \mathrm{TS} \overset{k^{\dagger}}{\longrightarrow} \mathrm{P} \tag{2.80}$$

fulfills the condition

$$k_1[\mathrm{R}] = k^{\dagger}[\mathrm{TS}] \tag{2.81}$$

and according to Eq. (2.79),

$$k_1 \approx k^{\dagger} \frac{Q_{\mathrm{TS}}}{Q_{\mathrm{R}}} e^{-(U_{\mathrm{TS}_0}-U_{\mathrm{R}_0})/k_{\mathrm{B}}T} \tag{2.82}$$

After eliminating $k^{\dagger}$, the following expression is obtained:

$$k_1 = \frac{k_B T}{h} \frac{Q_{TS}}{Q_R} e^{-(U_{TS_0} - U_{R_0})/k_B T} \tag{2.83}$$

Similarly, for a reaction of type

$$A + B \underset{k_{\leftarrow}}{\overset{k_{\rightarrow}}{\rightleftarrows}} TS \xrightarrow{k^{\dagger}} P \tag{2.84}$$

we can write

$$k_1 = \frac{k_B T}{h} \frac{Q_{TS}}{Q_A Q_B} e^{-[U_{TS,0} - (U_{A,0} + U_{B,0})]/k_B T} \tag{2.85}$$

After algebraic manipulation of the partition functions and the standard states, the Eyring equation is obtained for an $R \xrightarrow{k_1} P$ transformation:

$$k_1 = \frac{k_B T}{h} e^{-\Delta G^{\neq}/k_B T} \tag{2.86}$$

Combination of Eqs (2.1) and (2.86) leads to the following expressions for the activation energy and the pre-exponential factor:

$$E_a = \Delta H^{\neq^0} + RT \tag{2.87}$$

and

$$A = \frac{k_B T}{h} e^{(1 + \Delta S^{\neq^0})/R} \tag{2.88}$$

Since all the quantities in Eqs (2.86)–(2.88) can be obtained from electronic structure calculations and the exploration of the potential energy surface for a given transformation is also feasible, it is possible to obtain *in silico* kinetic data both in the gas phase and in solution. In addition, it is possible to test mechanistic hypotheses and compare the computational results with the experimental outcome in order to understand the observed behavior and, at least in favorable cases, to *predict* the behavior of untested reactants or catalysts.

2.6 SELECTED EXAMPLES

In the following three sections, we present recent results obtained in our group as examples of the utility of the above-described computational techniques to solve practical issues such as relative reactivities of different substrates and the origins of the stereocontrol in complex stepwise cycloadditions. In all these examples, approximate evaluations of the relative energies were made since the reactants and products were quite close, if not identical, to the reactants used in the experimental studies.

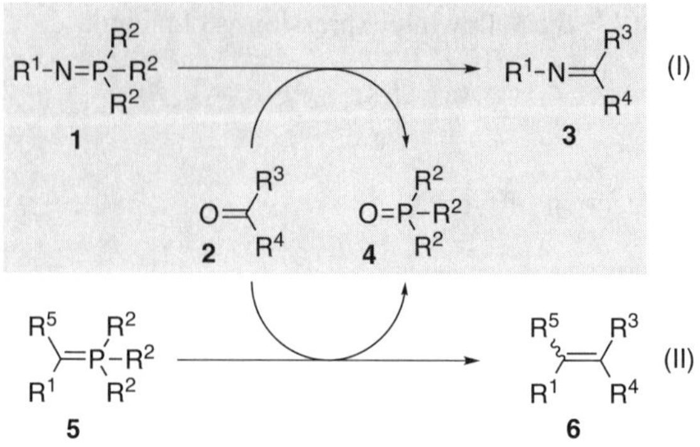

SCHEME 2.1 General aza-Wittig (I) and (II) reactions.

2.6.1 Relative Reactivities of Phosphines in Aza-Wittig Reactions

The aza-Wittig reaction consists of the reaction between a phosphazene and a carbonyl compound (frequently an aldehyde) to yield an imine (Scheme 2.1). This reaction is closely related to the well-known Wittig reaction for the synthesis of alkenes and uses phosphazenes **1** instead of phosphorus ylides **4**. In both reactions, the by-product is the oxide of the starting phosphine.

The formation of phosphazenes was discovered by Staudinger and Meyer in 1919 [64,65]. These authors found that when an azide was treated with a phosphine, an iminophosphorane (phosphazene) was formed. Reaction of this reactive species with carbonyl compounds resulted in the formation of the corresponding imines. The aza-Wittig reaction has been used for the synthesis of different types of azadienes [66–69] and valuable compounds [70], including natural products such as (+)-hamacanthin B [71] and mycosporine [72].

According to DFT calculations, the mechanism of this reaction is stepwise [73] and consists of a tandem [2+2] cycloaddition–cycloreversion sequence (Scheme 2.2). Both processes take place via *supra–supra* mechanisms that are thermally allowed because of the interplay between $\pi^{-}\sigma^{-}$ and lone pair localized orbitals. The corresponding transition structures **TS1** and **TS2** are nonaromatic as indicated by the low negative values of the nucleus-independent chemical shifts (NICS), thus discarding true pericyclic processes [74,75]. In one special case, a highly hindered 1,3,2-λ^5-oxaphosphazetidine **INT** (Scheme 2.2) reaction intermediate was charac-

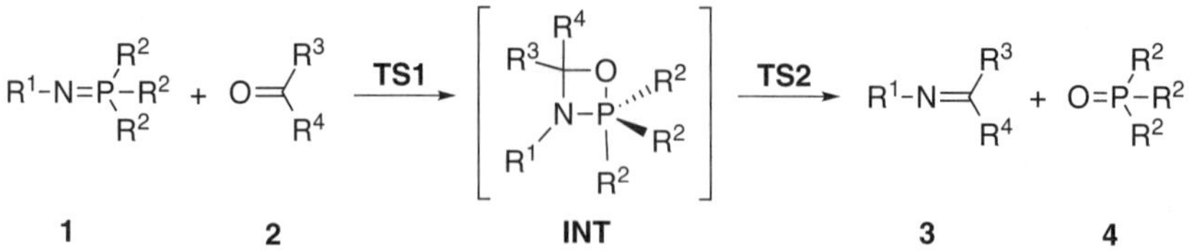

SCHEME 2.2 General mechanism for the aza-Wittig reaction.

terized by X-ray diffraction analysis [76] and the structural data were found to be in good agreement with the computed ones for more reactive intermediates.

One goal of the mechanistic studies was to elucidate the relative reactivities of trialkyl and triaryl phosphines. To this end, DFT calculations were carried out at B3LYP/6-31G* level of theory. The model structures analyzed were phosphazenes **1b** ($R^1 = H$, $R^2 = CH_3$) and **1c** ($R^1 = H$, $R^2 = C_6H_5$) and formaldehyde **2a** ($R^3 = R^4 = H$). Therefore, the aza-Wittig reactions were **1b** + **2a** → [**INTba**] → **3a** + **4b** and **1c** + **2a** → [**INTca**] → **3a** + **4c** (Figure 2.4).

The relative reaction profiles are gathered in Figure 2.1. According to these results, in both cases the second step, namely, the [2+2] cycloreversions of 1,3,2-λ^5-oxaphosphazetidines **INTba** and **INTca** are the limiting ones. The relative activation barriers are opposite for the two reactions. In case of **1b**, derived from trimethylphosphine, the first step is faster than in the case of phosphazene **1c**, obtained from triphenylphosphine. However, the second steps follow the reverse trend (Figure 2.1).

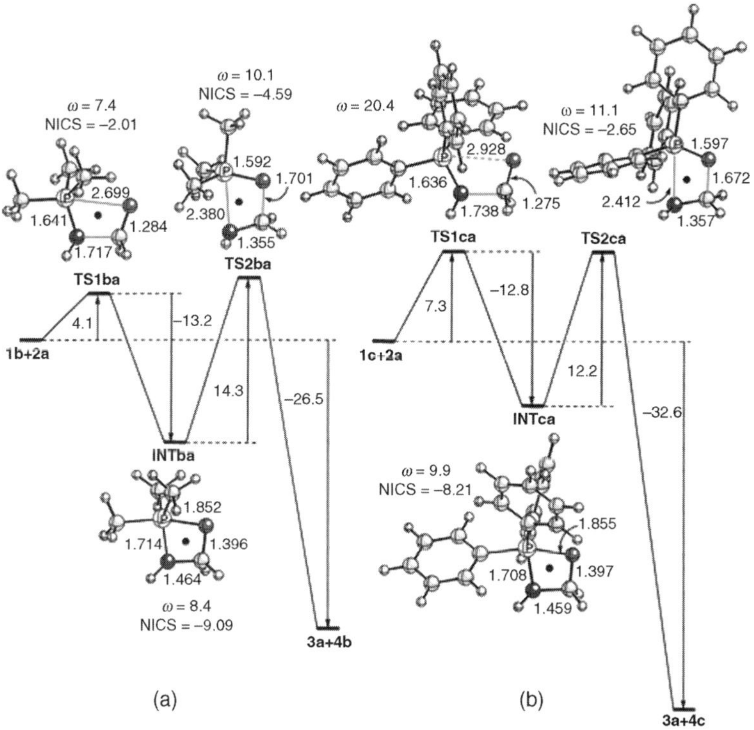

FIGURE 2.4 Stationary points (B3LYP/6-31G* level) and energy barriers (B3LYP/6-31G*+ΔZPVE level) for **1b** + **2a** → **3a** + **4b** (a) and **1c** + **2a** → **3a** + **4c** (b) reactions. Bond distances and dihedral ω(P–O–C–N) angles (in absolute value) are given in angstrom and degrees, respectively. Energy barriers are reported in kcal/mol. The nucleus-independent chemical shifts (NICS) calculated at the ring points (small black spheres) are given in ppm (see text). Reproduced from Ref. [70] with permission of the American Chemical Society.

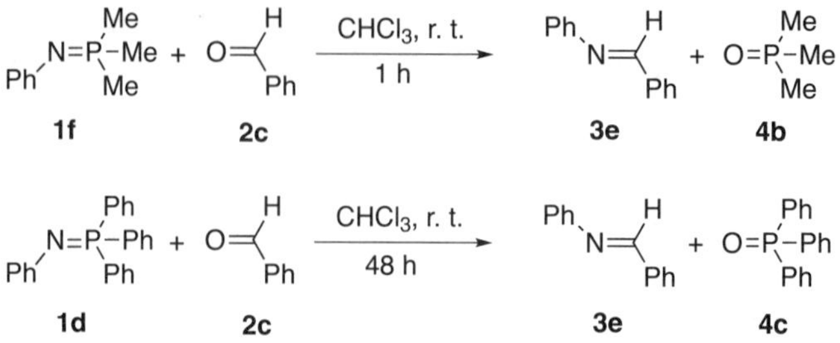

SCHEME 2.3 Aza-Wittig reactions involving phosphazenes derived from trimethylphosphine and triphenylphosphine.

The relative reaction rates can be calculated by applying the steady state to **INTba** and **INTca**:

$$\frac{d[\mathbf{INTba}]}{dt} = \frac{d[\mathbf{INTca}]}{dt} \approx 0 \tag{2.89}$$

For both the methyl and phenyl series, the kinetic scheme can be expressed as

$$1 + 2 \underset{k_{-1}}{\overset{k_1}{\rightleftarrows}} \mathbf{INT} \xrightarrow{k_2} 3 + 4 \tag{2.90}$$

Under these conditions and from the computed relative energies, the following expression can be obtained:

$$\frac{v^{\mathrm{Me}}}{v^{\mathrm{Ph}}} \approx \frac{k_1^{\mathrm{Me}} k_2^{\mathrm{Me}}}{k_1^{\mathrm{Ph}} k_2^{\mathrm{Ph}}} \left(\frac{k_{-1}^{\mathrm{Ph}} + k_2^{\mathrm{Ph}}}{k_{-1}^{\mathrm{Me}} + k_2^{\mathrm{Me}}} \right) \approx 40.9 \tag{2.91}$$

Therefore, the kinetic calculations predict that the phenyl series is about 41 times less reactive than the methyl series, despite the relatively lower second barrier for the former case. It was found experimentally that the relative reaction rates for the related reactions shown in Scheme 2.3 were found to be about 48:1, in qualitative agreement with the calculations performed on the simpler case.

2.6.2 Origins of the Stereocontrol in the Staudinger Reaction Between Ketenes and Imines to Form β-Lactams

Since its discovery in 1907, the Staudinger reaction is one of the most useful methods for the synthesis of β-lactams (2-oxazetidinones) [77,78]. The importance of these strained heterocycles relies not only on the chemical synthesis of β-lactam antibiotics [79] but also on other valuable compounds in medicinal chemistry [80] such as the acyclic chain of taxol and bestatin [81] or ecteinascidin-743 (Scheme 2.4) [82].

Despite its formal simplicity, the mechanism of the Staudinger reaction is actually quite complex [83,84]. There is now wide experimental and computational evidence to conclude that the reaction is stepwise. The first step consists of a nucleophilic attack

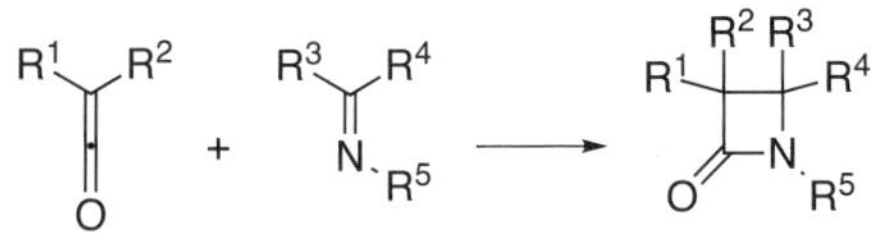

7 **3** **8**

SCHEME 2.4 General Staudinger reaction between ketenes and imines to form β-lactams.

of the imine lone pair to the sp-hybridized carbon atom of the ketene to yield a zwitterionic intermediate whose conrotatory electrocyclic reaction leads to the corresponding β-lactam cycloadduct **8**.

If the reaction between monosubstituted ketenes **7a** and (*E*)-aldimines **3b** is considered, formation of *cis*- and *trans*-β-lactams can be observed (Scheme 2.5). After careful analysis, Xu and coworkers [85,86] concluded that the stereoselectivity of the reaction depends on the rate of the second step leading to the *cis*-cycloadducts versus the rate associated with rotation about the N1–C4 bond of intermediates **9** to yield isomeric zwitterions **10**:

$$\frac{[cis\text{-}\mathbf{8a}]}{[trans\text{-}\mathbf{8a}]} = \frac{k_1 \int [\mathbf{9}]dt}{k_2 \int [\mathbf{9}]dt - [\mathbf{10}]} \approx \frac{k_1}{k_2} \tag{2.92}$$

Banik et al. [87,88] reported that when imines derived from condensed aromatic amines react with monosubstituted ketenes, only the corresponding *trans*-β-lactams are obtained. However, similar Staudinger reaction between *N*-phenylimines occurs under the same conditions—only the *cis*-cycloadducts are obtained with complete stereocontrol (Scheme 2.6). In addition, these *trans*-β-lactams exhibited interesting anticancer activities [89,90].

In order to understand the reasons for this unexpected outcome, we performed a complete computational analysis on the systems indicated in Scheme 2.7 at B3LYP

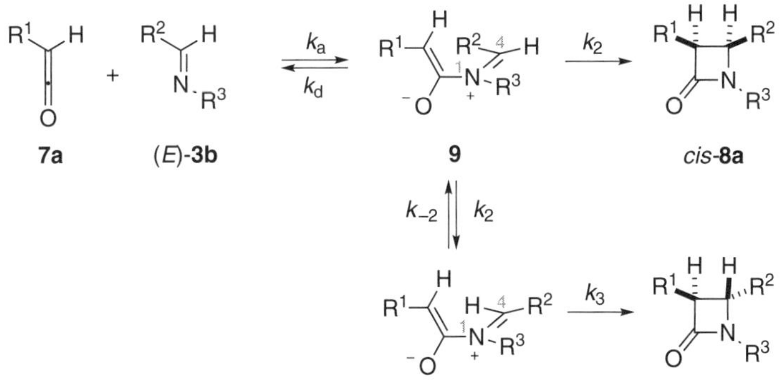

7a (*E*)-**3b** **9** *cis*-**8a**

10 *trans*-**8a**

SCHEME 2.5 Kinetic scheme for the formation of *cis*- and *trans*-β-lactams to form monosubstituted ketenes and (*E*)-imines, according to Xu and coworkers [85].

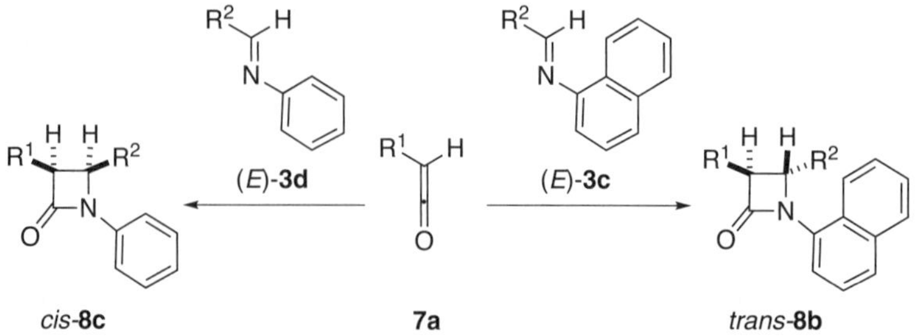

SCHEME 2.6 Stereodivergent outcome in the Staudinger reaction between monosubstituted ketenes and (*E*)-imines.

(PCM)/6-31G* level [91]. The four possible kinetic routes were analyzed and the corresponding Gibbs energy profiles were obtained. Besides the isomerization pathway described by Xu et al. and our group [92], we also considered the isomerization of the starting imines, thus leading to an even more complicated kinetic scheme. Since our calculations revealed that the second steps leading to the cycloadducts are irreversible, the formation of *cis*- and *trans*-β-lactams could be described as

$$\frac{d}{dt}[cis\text{-}\mathbf{14a,b}] = k_2^c[\mathbf{12a,b}] \tag{2.93}$$

and

$$\frac{d}{dt}[trans\text{-}\mathbf{14a,b}] = k_2^t[\mathbf{13a,b}] \tag{2.94}$$

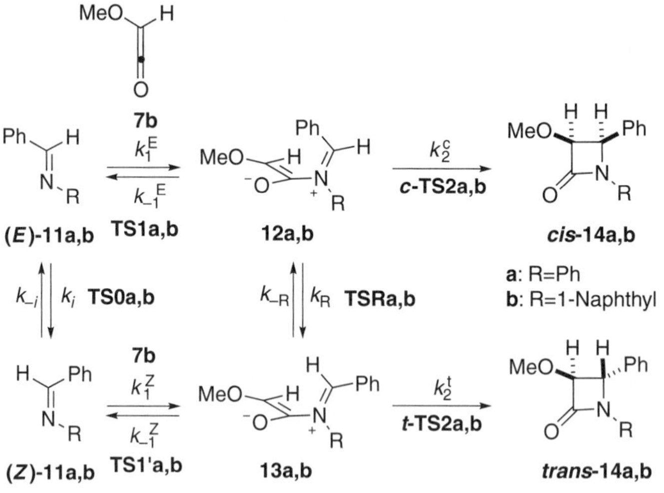

SCHEME 2.7 A more general kinetic scheme for the formation of *cis*- and *trans*-β-lactams from monosubstituted ketenes and (*E*)-imines.

The evolution of reactants and reaction intermediates is described by the following equations:

$$-\frac{d}{dt}A = KA \tag{2.95}$$

where A is the matrix of reactants and reaction intermediates:

$$A = \begin{pmatrix} [(E) - \mathbf{11a,b}] \\ [(Z) - \mathbf{11a,b}] \\ [\mathbf{12a,b}] \\ [\mathbf{13a,b}] \end{pmatrix} \tag{2.96}$$

and K is the interaction matrix given by

$$K = \begin{pmatrix} k_i + k_1^E[\mathbf{7b}] & -k_{-i} & -k_{-1}^E & 0 \\ -k_i & k_{-i} + k_1^Z[\mathbf{7b}] & 0 & k_{-1}^Z \\ -k_1^E[\mathbf{7b}] & 0 & k_{-1}^E + k_R + k_2^c & -k_{-R} \\ 0 & -k_1^Z[\mathbf{7b}] & k_R & k_{-1}^Z + k_{-R} + k_2^t \end{pmatrix} \tag{2.97}$$

Numerical solution of Eqs (2.93)–(2.97) forms the relative energies and Eq. (2.86) showed that the computational kinetic analysis reproduced perfectly the stereodivergent outcome observed by Banik et al. (Figure 2.5).

In addition, our computational study revealed that the origin of this surprising behavior relies on the relative rates associated with the $(E) \rightarrow (Z)$ isomerization of imines **11a,b** and with the nucleophilic attack on methoxyketene **7b**. The

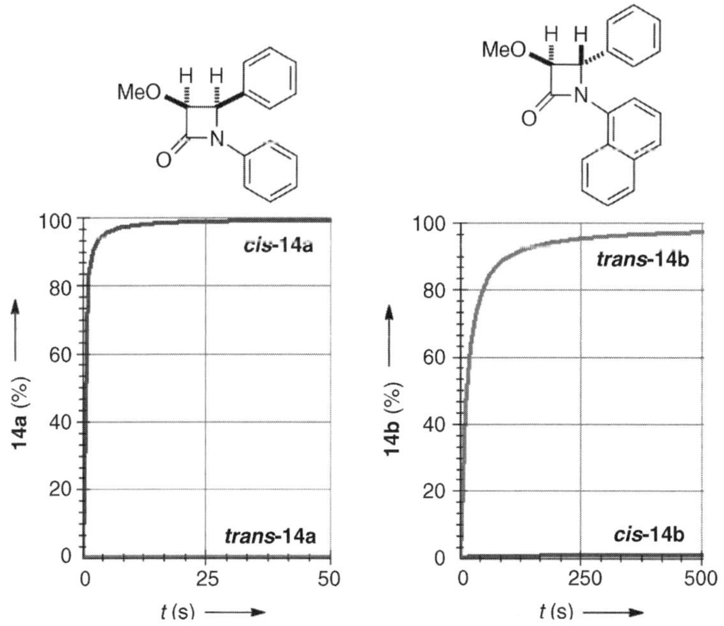

FIGURE 2.5 Simulated stereochemical outcome associated with the reaction between methoxyketene **7b** and imines (*E*)-**11a,b** to yield β-lactams *cis*- and *trans*-**14a,b**. This distribution of isomers was obtained from Eqs (2.93)–(2.95).

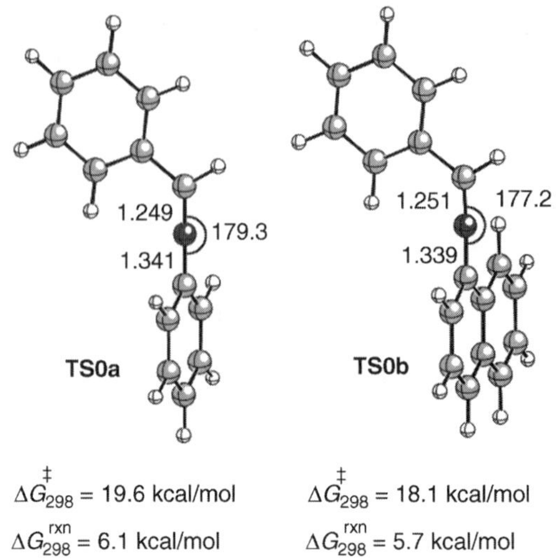

FIGURE 2.6 Fully optimized structures of the transition structures associated with the (*E*) → (*Z*) isomerization of imines (*E*)-**1a,b** at B3LYP(PCM)/6-31+G*//B3LYP(PCM)/6-31G*+ΔZPVE level of theory. Bond distances and angles are given in angstrom and degrees, respectively. Reproduced from Ref. [91] with permission.

isomerization of the starting imines takes place via aza-allenic transition structures **TS0a,b** [93] in which the negative charge is more efficiently delocalized in case of *N*-(1-naphthyl) moiety (Figure 2.6). The resulting (*Z*)-imine is less stable than its (*E*)-isomer and is more nucleophilic toward **7b**. The combination of these features results in the preferential formation of *trans*-**14b** since in this case $k_i > k_1^E < k_1^Z$. Therefore, the final stereoselectivity of the whole reaction is determined by an isomerization process prior to the Staudinger reaction itself. In case of intramolecular reactions involving ketenimines, the reaction profiles can be described within the steady-state approximation [94,95] and the source of the stereocontrol lies mainly in the first step, namely, the formation of the N1–C2 bond.

2.6.3 Origins of the Stereocontrol in the Reaction Between Imines and Homophthalic Anhydride

The reaction between cyclic anhydrides and imines [96–98] (Scheme 2.8) to yield carboxylic acids is an important tool for the synthesis of natural products [99] and

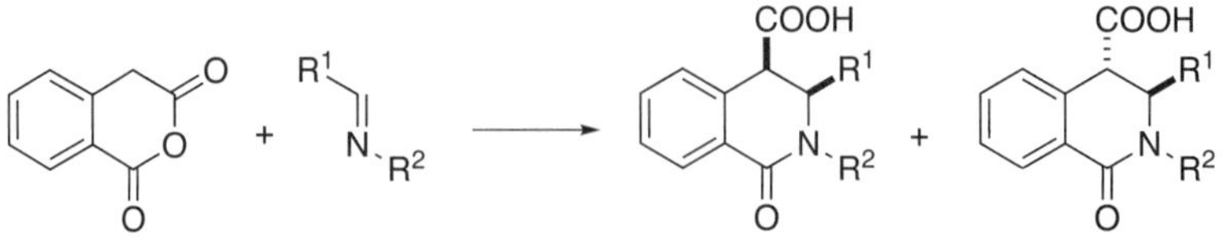

15 **3** ***cis*-16** ***trans*-16**

SCHEME 2.8 General reaction between homophthalic anhydride and (*E*)-imines.

valuable compounds such as topoisomerase I [100] or HOXA13 inhibitors [101]. We found that when (*E*)-imines and homophthalic anhydride are used, preferential formation of *trans*-1,2,3,4-tetrahydro-1-oxoquinoline-4-carboxylic acids **15** is observed in the presence of $TiCl_4$ and a tertiary amine [102].

A preliminary computational analysis of the reaction via titanium(IV) enolates showed a stepwise mechanism in which intermediate (*E*)-enolates yield preferentially *trans*-compounds (Figure 2.7). Thus, transition structure*trans*-**TS3** leading to *trans*-**16a** has chair–sofa geometry, more stable than the boat geometry of *cis*-**TS3**, in which destabilizing interactions between one chlorine ligand and the imine are also observed. According to this scheme, preferential formation of *cis*-compounds should be expected in the reaction between **15** and cyclic (*Z*)-imines.

However, it was observed [103] that reaction of titanium(IV) enolates derived from homophthalic anhydride and five-, six- and seven-membered cyclic imines also yield *trans*-adducts instead of the expected *cis*-isomers (Scheme 2.9).

In order to understand the origins of this unexpected behavior, we evaluated the different reaction rates at M06(SM8)//B3LYP(PCM)+ΔZPVE level of theory of the enolate derived from **15** and model cyclic imine **17d**, a simpler analogue of **17b** (Scheme 2.10). As expected, **TS4s** has chair–sofa geometry and is more stable than its diastereomer **TS4b**. However, the second step of the reaction, associated with formation of the tetrahedral intermediate corresponding to the formation of the

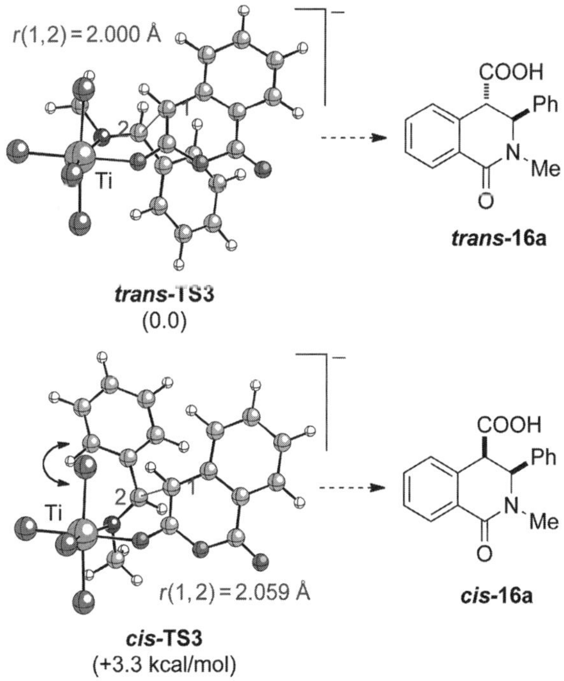

FIGURE 2.7 B3LYP/LANL2DZ&6-31G* optimized structures of transition structures associated with the formation of *cis*- and *trans*-**16a**. Reproduced form Ref. [102] with permission of the American Chemical Society.

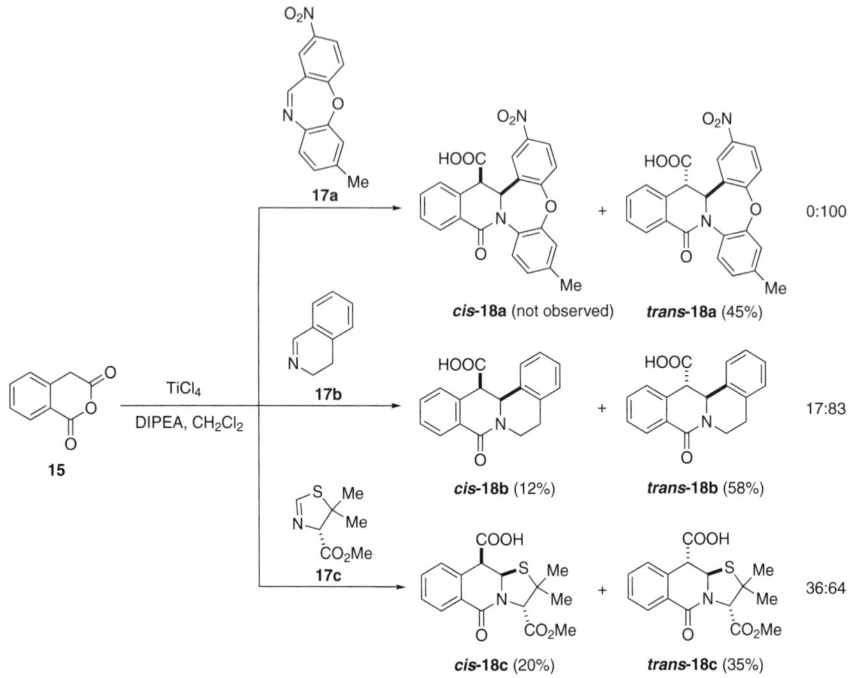

SCHEME 2.9 Reaction between homophthalic anhydride and cyclic imines in the presence of $TiCl_4$ and DIPEA.

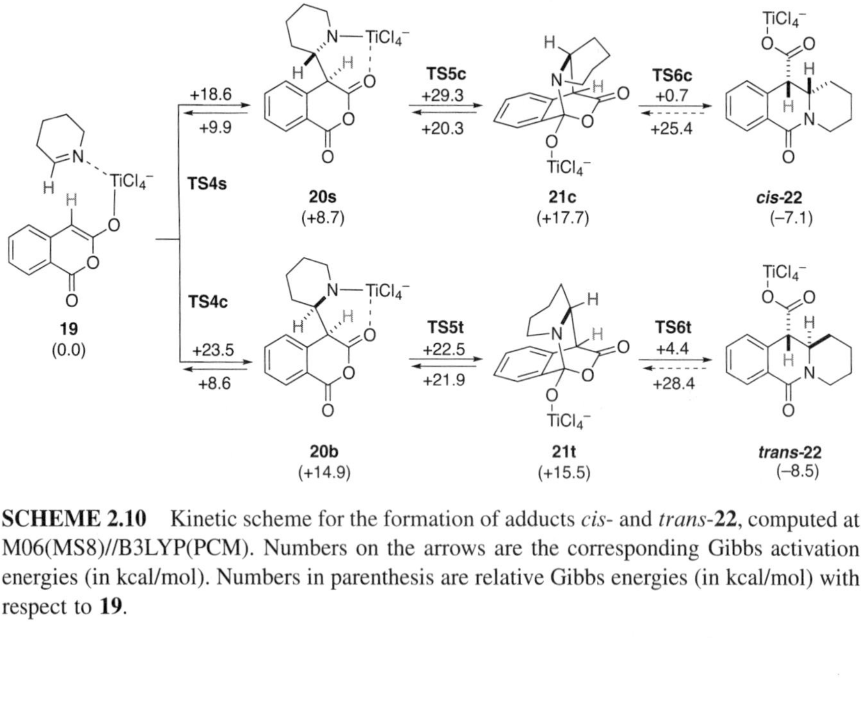

SCHEME 2.10 Kinetic scheme for the formation of adducts *cis*- and *trans*-**22**, computed at M06(MS8)//B3LYP(PCM). Numbers on the arrows are the corresponding Gibbs activation energies (in kcal/mol). Numbers in parenthesis are relative Gibbs energies (in kcal/mol) with respect to **19**.

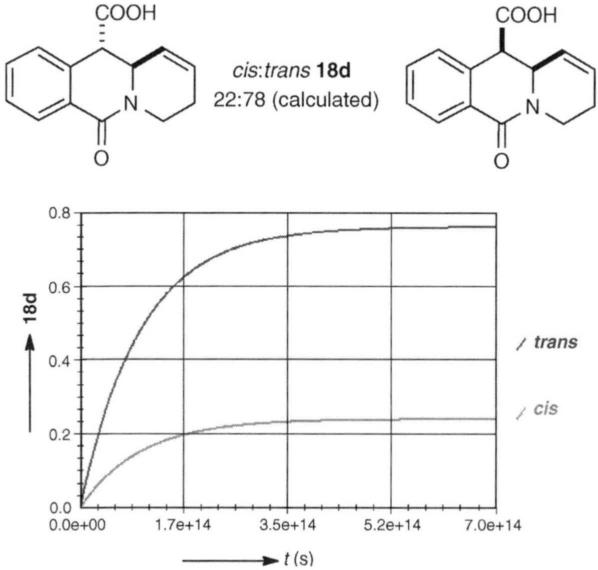

FIGURE 2.8 Simulated stereochemical outcome associated with the conversion of complex **19** into *cis-* and *trans-*cycloadducts **18d**.

amide bond, follows the opposite trend. In this case, **TS5c**, which leads to the *cis*-adduct, is calculated to be about 7 kcal/mol higher in energy than its isomeric saddle point **TS5t**, which in turn is associated with the formation of *trans*-**22**. After numerical integration of the corresponding kinetic equations, we observed that these opposite profiles result in the preferential formation of *trans*-**18d**, in a kinetic ratio close to the experimental value found for the analogue cycloadduct *trans*-**18b** (Figure 2.8).

2.7 CONCLUSIONS AND OUTLOOK

In the first part of this chapter, we have reported a brief survey of the computational tools available to evaluate the geometries and relative energies of the stationary points (reactants, transition structures, and reaction intermediates) associated with the elementary steps of any transformation under the Born–Oppenheimer approximation. These computational data permit the calculation of individual kinetic constants associated with each step. In the second part, we have commented on several results obtained in our laboratory. These kinetic calculations involved up to 40 atoms and 160 electrons and in many cases included solvent effects. Under these conditions, the individual rate constants associated with each elementary step were expected not to be very accurate. However, since the error magnitudes along the different reaction coordinates were similar enough, the final relative rate constants reproduced surprisingly well the experimental observations, even when very subtle effects were studied. In addition, the possibility of evaluating each step of the different reactions permitted us to identify the main variables that govern the stereochemical outcome of different complex reactions. In several cases, the computational findings challenge conventional wisdom about stereoselectivity under kinetic control. For example, we found that, at least in two cases, the stereocontrol of the whole reaction lies at steps not associated with the generation of chiral centers. We think that these unexpected and contraintuitive findings obtained from computational results are not exceptional. Therefore, extension of these methodologies to other reactions could permit a better understanding of the origins of the stereocontrol and hence the design of novel reaction conditions and new catalysts for the corresponding chemical transformations.

REFERENCES

1. C. J. Cramer, *Essentials of Computational Chemistry*, John Wiley & Sons, Inc., New York (2006).
2. J. I. Steinfeld, J. S. Francisco, and W. L. Hase, *Chemical Kinetics and Dynamics*, 2nd edition, Prentice Hall, Upper Saddle River, NJ (1999).
3. E. V. Anslyn and D. A. Dougherty, *Modern Physical Organic Chemistry*, University Science Books, Sausalito, CA (2006).
4. A. R. Leach, *Molecular Modelling: Principles and Applications*, 2nd edition, Pearson Education Ltd., Harlow, Essex (2001).
5. B. K. Carpenter, *Determination of Organic Reaction Mechanisms*, John Wiley & Sons, Inc., New York (1984).
6. N. P. Gritsan and M. S. Platz, *Chem. Rev.* **106**, 3844 (2006).
7. T. V. Albu, J. Espinosa-García, and D. G. Truhlar, *Chem. Rev.* **107**, 5101 (2007).
8. R. Becerra, I. W. Carpenter, M. S. Gordon, L. Rostop, and R. Walsh, *Phys. Chem. Chem. Phys.* **9**, 2121 (2007).
9. J. Espinosa-García, *Phys. Chem. Chem. Phys.* **10**, 1277 (2008).
10. C. Carballotti, R. Rota, and S. Carrà, *J. Phys. Chem. A* **106**, 7769 (2002).
11. X.-J. Jia, Y.-J. Liu, J.-Y. Sun, H. Sun, F. Wang, Z.-M. Su, X.-M. Pan, and R.-S. Wang, *J. Comput. Chem.* **31**, 2263 (2010).
12. D. G. Truhlar, W. L. Hase, and J. T. Hynes, *J. Phys. Chem.* **87**, 2657 (1983).
13. I. H. William, *Chem. Soc. Rev.* **22**, 277 (1993).
14. D. G. Truhlar, B. C. Garret, and S. J. Klippenstein, *J. Phys. Chem.* **100**, 12771 (1996).
15. P. W. Atkins and R. S. Friedman, *Molecular Quantum Mechanics*, 3rd edition, Oxford University Press, Oxford (1997).
16. H. B. Schlegel, *J. Comput. Chem.* **3**, 214 (1982).
17. (a) F. Bernardi and M. A. Robb, in K. P. Lawley (ed.), *Advances in Chemical Physics: Ab Initio Methods in Quantum Chemistry*, Part I, Vol. 67, John Wiley & Sons, Inc., New York (2007). (b) H. B. Schlegel, in K. P. Lawley (ed.), *Advances in Chemical Physics: Ab Initio Methods in Quantum Chemistry*, Part I, Vol. 67, John Wiley & Sons, Inc., New York (2007).
18. J. W. McIver, Jr. and A. Komornicki, *J. Am. Chem. Soc.* **94**, 2625 (1972).

19. K. Fukui, *Pure Appl. Chem.* **54**, 1825 (1982).
20. C. González and H. B. Schlegel, *J. Phys. Chem.* **94**, 5523 (1990).
21. K. N. Houk, Y. Li, and J. D. Evanseck, *Angew. Chem., Int. Ed. Engl.* **31**, 682 (1992).
22. J. M. Mercero, J. M. Matxain, X. Lopez, D. M. York, A. Largo, L. A. Eriksson, and J. M. Ugalde, *Int. J. Mass Spectrom.* **240**, 37 (2005).
23. D. Feller and E. R. Davidson, Basis sets of *ab initio* molecular orbital calculations and intermolecular interactions, in K. B. Lipkowitz and D. B. Boyd (eds), *Reviews in Computational Chemistry*, Vol. 1, Wiley-VCH, Weinheim (1990).
24. P.-O. Lödwin, *Phys. Rev.* **97**, 1409 (1955).
25. S. R. Langhoff and E. R. Davidson, *Int. J. Quantum Chem.* **8**, 61 (1974).
26. C. Møller and M. S. Plesset, *Phys. Rev.* **46**, 618 (1934).
27. L. A. Curtiss, K. Raghavachari, G. W. Trucks, and J. A. Pople, *J. Chem. Phys.* **94**, 7221 (1991).
28. L. A. Curtiss, K. Raghavachari, P. C. Redfern, V. Rassolov, and J. A. Pople, *J. Chem. Phys.* **109**, 7764–7776 (1998).
29. L. A. Curtiss, P. C. Redfern, and K. Raghavachari, *J. Chem. Phys.* **126**, 84108 (2007).
30. R. G. Parr and W. Yang, *Density-Functional Theory of Atoms and Molecules*, Oxford University Press, New York (1989).
31. R. Stowasser and R. Hoffmann, *J. Am. Chem. Soc.* **121**, 3414 (1999).
32. M. Ernzerhof, J. P. Perdew, and K. Burke, *Top. Curr. Chem.* **180**, 1 (1996).
33. W. Kohn, A. D. Becke, and R. G. Parr, *J. Phys. Chem.* **100**, 12974 (1996).
34. A. D. Becke, *J. Chem. Phys.* **98**, 5648 (1993).
35. A. D. Becke, *Phys. Rev. A* **38**, 3098 (1988).
36. C. Lee, W. Yang, and R. G. Parr, *Phys. Rev. B* **37**, 785 (1988).
37. R. Colle and D. Salvetti, *Theor. Chim. Acta* **70**, 407 (1986).
38. S. H. Vosko, L. WIlk, and M. Nusair, *Can. J. Phys.* **58**, 1200 (1980).
39. R. H. Hertwig and W. Koch, *Chem. Phys. Lett.* **268**, 345 (1997).
40. For the most recent release of Gaussian series of programs, see M. J. Frisch, G. W. Trucks, H. B. Schlegel, G. E. Scuseria, M. A. Robb, J. R. Cheeseman, G. Scalmani, V. Barone, B. Mennucci, G. A. Petersson, H. Nakatsuji, M. Caricato, X. Li, H. P. Hratchian, A. F. Izmaylov, J. Bloino, G. Zheng, J. L. Sonnenberg, M. Hada, M. Ehara, K. Toyota, R. Fukuda, J. Hasegawa, M. Ishida, T. Nakajima, Y. Honda, O. Kitao, H. Nakai, T. Vreven, J. A. Montgomery, Jr., J. E. Peralta, F. Ogliaro, M. Bearpark, J. J. Heyd, E. Brothers, K. N. Kudin, V. N. Staroverov, R. Kobayashi, J. Normand, K. Raghavachari, A. Rendell, J. C. Burant, S. S. Iyengar, J. Tomasi, M. Cossi, N. Rega, J. M. Millam, M. Klene, J. E. Knox, J. B. Cross, V. Bakken, C. Adamo, J. Jaramillo, R. Gomperts, R. E. Stratmann, O. Yazyev, A. J. Austin, R. Cammi, C. Pomelli, J. W. Ochterski, R. L. Martin, K. Morokuma, V. G. Zakrzewski, G. A. Voth, P. Salvador, J. J. Dannenberg, S. Dapprich, A. D. Daniels, Ö. Farkas, J. B. Foresman, J. V. Ortiz, J. Cioslowski, and D. J. Fox, *Gaussian 09*, Gaussian, Inc., Wallingford, CT (2009).
41. M. D. Wodrich, C. Corminboeuf, P. R. Schreiner, A. A. Fokin, and P. v. R. Schleyer, *Org. Lett.* **9**, 1851 (2007).
42. Y. Zhao and D. G. Truhlar, *Theor. Chem. Acc.* **120**, 215 (2008).
43. Y. Zhao and D. G. Truhlar, *Acc. Chem. Res.* **41**, 157 (2008).

44. C. Reichardt, *Solvents and Solvent Effects in Organic Chemistry*, Wiley-VCH, Weinheim (2003).
45. B. Ya. Simkin and I. I. Sheikhet, *Quantum Chemical and Statistical Theory of Solutions: A Computational Approach*, Ellis Horwood, London (1995).
46. O. Acevedo and W. L. Jorgensen, *J. Am. Chem. Soc.* **128**, 6141 (2006).
47. O. Acevedo and W. L. Jorgensen, *Org. Lett.* **6**, 2881 (2004).
48. G. Vayner, K. N. Houk, W. L. Jorgensen, and J. I. Braumann, *J. Am. Chem. Soc.* **126**, 9054 (2004).
49. J. Tomasi, B. Mennucci, and R. Camini, *Chem. Rev.* **105**, 2999 (2005).
50. J. Tomasi and M. Persico, *Chem. Rev.* **94**, 2027 (1994).
51. C. J. Cramer and D. G. Truhlar, *Chem. Rev.* **99**, 2161 (1999).
52. L. Onsager, *J. Am. Chem. Soc.* **58**, 1486 (1936).
53. M. W. Wong, K. B. Wiberg, and M. J. Frisch, *J. Am. Chem. Soc.* **114**, 523 (1992).
54. J.-L. Rivail, D. Rinaldi, and M. Ruiz-López, in S. J. Formosinho, I. G. Csizmadia, and L. Arnaut (eds), *Theoretical and Computational Methods for Organic Chemistry*, Vol. 339, Kluwer Academic Publishers, Dordrecht (1991).
55. W. C. Still, A. Tempczyk, R. C. Hawley, and T. Hendrickson, *J. Am. Chem. Soc.* **112**, 6127 (1990).
56. C. J. Cramer and D. G. Truhlar, *Acc. Chem. Res.* **41**, 760 (2008).
57. S. Miertus, E. Scrocco, and J. Tomasi, *Chem. Phys.* **55**, 117 (1981).
58. M. Cossi, V. Barone, R. Cammi, and J. Tomasi, *Chem. Phys. Lett.* **255**, 327 (1996).
59. R. Cammi and J. Tomasi, *J. Comput. Chem.* **16**, 1449 (1995).
60. A. V. Marenich, R. M. Olson, C. P. Kelly, C. J. Cramer, and D. G. Truhlar, *J. Chem. Theor. Comput.* **3**, 2011 (2007).
61. A. Ben-Naim, *Solvation Thermodynamics*, Plenum Press, New York (1987).
62. C. P. Kelly, C. J. Cramer, and D. G. Truhlar, *J. Phys. Chem. B* **110**, 16066 (2006).
63. C. P. Kelly, C. J. Cramer, and D. G. Truhlar, *J. Phys. Chem. B* **111**, 408 (2007).
64. H. Staudinger and J. Meyer, *Helv. Chim. Acta* **2**, 635 (1919).
65. H. Staudinger and J. Meyer, *Helv. Chim. Acta* **4**, 861 (1921).
66. B. B. Snider and J. Zhou, *J. Org. Chem.* **70**, 1087 (2005).
67. C. Gil and S. Braese, *Chem. Eur. J.* **11**, 2680 (2005).
68. M. W. Ding, S. Z. Xu, and J. F. Zhao, *J. Org. Chem.* **69**, 8366 (2004).
69. M. A. T. Maughan, I. G. Davies, T. D. Claridge, W. S. Cournety, P. Hay, and B. G. Davis, *Angew. Chem., Int. Ed.* **42**, 3788 (2003).
70. F. Palacios, C. Alonso, D. Aparicio, G. Rubiales, and J. M. de los Santos, *Tetrahedron* **63**, 523 (2007).
71. B. Jiang, C.-G. Yang, and J. Wang, *J. Org. Chem.* **67**, 1396 (2002).
72. J. D. White, J. H. Cammack, and K. Sakuma, *J. Am. Chem. Soc.* **111**, 8970 (1989).
73. F. P. Cossío, C. Alonso, B. Lecea, M. Ayerbe, G. Rubiales, and F. Palacios, *J. Org. Chem.* **71**, 2839 (2006).
74. P. v. R. Schleyer, C. Maerker, A. Dransfeld, H. Jiao, and N. J. R. van Eikema Hommes, *J. Am. Chem. Soc.* **118**, 6317 (1999).
75. A. Arrieta, A. de Cózar, and F. P. Cossío, *Curr. Org. Chem.* **15**, 3594 (2011).

76. N. Kano, X. J. Hua, S. Kawa, and T. Kawashima, *Tetrahedron Lett.* **41**, 5237 (2000).
77. H. Staudinger, *Liebigs Ann. Chem.* **356**, 51 (1907).
78. N. Fu and T. T. Tidwell, *Tetrahedron* **64**, 10465 (2008).
79. R. B. Morin and M. Goldman, *Chemistry and Biology of β-Lactam Antibiotics*, Vols 1-3, Academic Press, New York (1982).
80. C. Palomo, J. M. Aizpurua, I. Ganboa, and M. Oiarbide, *Curr. Med. Chem.* **11**, 1837 (2004).
81. C. Palomo, A. Arrieta, F. P. Cossío, J. M. Aizpurua, A. Mielgo, and N. Aurrekoetxea, *Tetrahedron Lett.* **31**, 6429 (1990).
82. W. Jin, S. Metobo, and R. M. Williams, *Org. Lett.* **5**, 2095 (2003).
83. F. P. Cossío, A. Arrieta, and M. A. Sierra, *Acc. Chem. Res.* **41**, 925 (2008).
84. A. Arrieta, B. Lecea, and F. P. Cossío, *Top. Heterocycl. Chem.* **22**, 313 (2010).
85. Y. Liang, L. Jiao, and J. Xu, *J. Am. Chem. Soc.* **128**, 6060 (2006).
86. B. Li, Y. Wang, D.-M. Du, and J. Xu, *J. Org. Chem.* **72**, 990 (2007).
87. B. K. Banik and F. F. Becker, *Tetrahedron Lett.* **41**, 6551 (2000).
88. I. Banik, L. Hackfield, and B. K. Banik, *Heterocycles* **59**, 505 (2003).
89. I. Banik, F. F. Becker, and B. K. Banik, *J. Med. Chem.* **46**, 12 (2003).
90. B. K. Banik, I. Banik, and F. F. Becker, *Bioorg. Med. Chem.* **13**, 3611 (2005).
91. B. K. Banik, B. Lecea, A. Arrieta, A. de Cózar, and F. P. Cossío, *Angew. Chem., Int. Ed.* **46**, 3028 (2007).
92. A. Arrieta, J. M. Ugalde, F. P. Cossío, and B. Lecea, *Tetrahedron Lett.* **35**, 4465 (1994).
93. E. Eliel, S. H. Wilen, and L. N. Mander, *Stereochemistry of Organic Compounds*, John Wiley & Sons, Inc., New York (1994), pp. 550–553 and references therein.
94. M. Alajarín, A. Vidal, F. Tovar, A. Arrieta, B. Lecea, and F. P. Cossío, *Chem. Eur. J.* **5**, 1106 (1999).
95. F. P. Cossío, A. Arrieta, B. Lecea, M. Alajarín, A. Vidal, and F. Tovar, *J. Org. Chem.* **65**, 3633 (2000).
96. N. Castagnoli, Jr., *J. Org. Chem.* **34**, 3187 (1969).
97. M. Cushman and E. J. Madas, *J. Org. Chem.* **52**, 907 (1987).
98. M. González-López and J. T. Shaw, *Chem. Rev.* **109**, 164 (2009).
99. X.-Y. Xu, G.-W. Quin, R.-S. Xu, and X.-Z. Zhu, *Tetrahedron* **54**, 14179 (1998).
100. A. Morrell, S. Anthony, G. Kohlhagen, Y. Pommier, and M. Cushman, *J. Med. Chem.* **49**, 7740 (2006).
101. P. Y. Ng, Y. Tang, W. M. Knosp, H. S. Stadler, and J. T. Shaw, *Angew. Chem., Int. Ed.* **46**, 5352 (2007).
102. Y. Vara, T. Bello, E. Aldaba, A. Arrieta, J. L. Pizarro, M. I. Arriortua, X. Lopez, and F. P. Cossío, *Org. Lett.* **10**, 4759 (2008).
103. T. Bello, A. de Cózar, Y. Vara, and F. P. Cossío, Unpublished results.

CHAPTER 3

Quantum Instanton Evaluation of the Kinetic Isotope Effects and of the Temperature Dependence of the Rate Constant

JIŘÍ VANÍČEK

Laboratory of Theoretical Physical Chemistry, Institut des Sciences et Ingénierie Chimiques, École Polytechnique Fédérale de Lausanne (EPFL), Lausanne, Switzerland

3.1 INTRODUCTION

Due to the central role the reaction rate constant plays in physical chemistry, many more or less accurate approximations for this quantity have been developed over time, starting from the Arrhenius equation [1] and transition state theory (TST) [2–4]. Among the most accurate of such approximations are so-called quantum transition state theories [5–18], which treat the rate constant quantum mechanically, but, similarly to the original classical TST, still rely on some sort of a transition state assumption. A recent such approximation that can also treat general many-dimensional systems is the quantum instanton (QI) approximation of Miller et al. [17].

The main reason for introducing quantum TSTs was that simpler descriptions of the rate constant had treated the nuclei as classical particles. While this is often an appropriate point of view, it fails completely in some cases, and almost always when hydrogen is involved in bond breaking or formation. Indeed, strong nuclear quantum effects have been observed in many reactions and recently even in several enzymatic reactions [19–21]. In such situations, one should treat some or all nuclei quantum mechanically. The rate constant is a dynamic quantity and hence very difficult to compute quantum mechanically because solving the time-dependent Schrödinger equation exactly is possible only for a handful of atoms [22]. Fortunately, the long-time quantum dynamic effects on the thermal rate constant of most chemical reactions

Rate Constant Calculation for Thermal Reactions: Methods and Applications, Edited by Herbert DaCosta and Maohong Fan.

are washed out by finite temperature. The most important remaining quantum effects are thermodynamic rather than dynamic, and while the "real-time" quantum dynamics is extremely difficult, quantum thermodynamics (which is equivalent to "imaginary-time" quantum dynamics) can be computed accurately for fairly large systems using the Feynman path integral formalism [23–27].

This chapter describes how QI approximation can be combined with the path integral formalism and specific types of thermodynamic integration in order to evaluate the kinetic isotope effects and the temperature dependence of the rate constant. This chapter is organized as follows: Section 3.2 overviews the Arrhenius law, transition state theory, and the Wigner tunneling correction as theories upon which QI approximation improves. QI approximation is presented in Section 3.3. Sections 3.4 and 3.5 explain how QI approximation is combined with the thermodynamic integration to compute the kinetic isotope effects and the temperature dependence of the rate constant, respectively. In order to make the methodology useful for general polyatomic molecules, Section 3.6 describes an efficient implementation in terms of Feynman path integrals. Several examples of increasing complexity are shown in Section 3.7.

3.2 ARRHENIUS EQUATION, TRANSITION STATE THEORY, AND THE WIGNER TUNNELING CORRECTION

Probably the oldest yet still the best-known expression for the thermal rate constant $k(T)$ at temperature T is the empirical Arrhenius law [1]:

$$k_{\mathrm{A}}(T) = Ae^{-E_{\mathrm{act}}/k_{\mathrm{B}}T} \tag{3.1}$$

where E_{act} is the activation energy, k_{B} is the Boltzmann constant, and the temperature dependence is purely exponential. An improvement over the Arrhenius law was provided by TST [2–4] that expresses the rate constant as

$$k_{\mathrm{TST}}(T) = \frac{k_{\mathrm{B}}T}{h}\frac{Q^{\ddagger}}{Q_{\mathrm{r}}}e^{-E^{\ddagger}/k_{\mathrm{B}}T} \tag{3.2}$$

where h is the Planck constant, Q_{r} is the reactant partition function, $Q^{\ddagger}$ is the partition function of the activated complex (i.e., excluding the unstable motion along the reaction coordinate), and $E^{\ddagger}$ is the classical barrier height.

TST assumes the separability of the Hamiltonian, allowing the partition function to be factored into translational, rotational, and vibrational components as

$$Q = Q_{\mathrm{trans}}Q_{\mathrm{rot}}Q_{\mathrm{vib}}. \tag{3.3}$$

At high temperatures, one may use the classical forms of the three components of the partition function:

$$Q_{\mathrm{trans}} = \left(\frac{2\pi mk_{\mathrm{B}}T}{h^2}\right)^{3/2} \text{(in three dimensions)}, \tag{3.4}$$

$$Q_{\text{rot}} = 8\pi^2 \sigma_{\text{sym}}^{-1} (I_x I_y I_z)^{1/2} \left(\frac{2\pi k_B T}{h^2} \right)^{3/2} \text{ (in three dimensions)}, \tag{3.5}$$

$$Q_{\text{vib}} = \frac{k_B T}{\hbar \omega} \text{ (per degree of freedom)} \tag{3.6}$$

where σ_{sym} is the symmetry number, I_x, I_y, and I_z are the moments of inertia along the principal axes, and ω is the angular frequency of the vibration. (Note that Q_{trans} is quantum mechanically exact at all temperatures and that at room temperature, the above approximation is usually good for Q_{rot} and usually poor for Q_{vib}.)

The leading quantum correction to TST at high temperatures is the Wigner tunneling correction [28], that is, a multiplicative factor:

$$\kappa_W = 1 + \frac{1}{24} \left(\frac{\hbar |\omega_{\text{unst}}^{\ddagger}|}{k_B T} \right)^2 \tag{3.7}$$

where $\omega_{\text{unst}}^{\ddagger}$ is the imaginary frequency of the unstable asymmetric stretch along the reaction coordinate. With the Wigner correction, the rate constant becomes

$$k_{\text{TST+W}}(T) = \kappa_W k_{\text{TST}}(T).$$

The correction can be derived by treating the motion through the transition state as a vibration on an upside-down potential and expanding the corresponding quantum partition function to second order in T^{-1}. Although an improvement over TST, the Wigner tunneling correction cannot describe a truly multidimensional tunneling.

3.3 QUANTUM INSTANTON APPROXIMATION FOR THE RATE CONSTANT

QI approximation for thermal rate constants is introduced in Ref. [17] where it is derived from the exact expression for the thermal rate constant:

$$k(T) = \frac{1}{Q_r h} \int_0^{\infty} dE N(E) e^{-\beta E}. \tag{3.8}$$

In the above equation, $N(E)$ denotes the cumulative reaction probability at total energy E and $\beta \equiv 1/k_B T$ is the inverse temperature. However, the most direct derivation [29,30] starts from the Miller–Schwartz–Tromp formula for the rate constant [31]:

$$k(T) = \frac{1}{Q_r} \int_0^{\infty} dt\, C_{\text{ff}}(t) \tag{3.9}$$

where $C_{\text{ff}}(t)$ is the shorthand for the symmetrized flux–flux correlation function at time t:

$$C_{\text{ff}}(t) \equiv C_{\hat{F}_a \hat{F}_b}^{\text{sym}}(t) \tag{3.10}$$

and $\hat{F}_\gamma$ denotes the flux operator through the dividing surface γ, $\gamma \in \{a, b\}$. In general, the symmetrized correlation function between operator $\hat{A}$ at time 0 and $\hat{B}$ at time t is given by

$$C^{\text{sym}}_{\hat{A}\hat{B}}(t) = \text{Tr}(e^{-\beta\hat{H}/2}\hat{A}(0)e^{-\beta\hat{H}/2}\hat{B}(t)),$$
$$\hat{A}(t) = e^{\mathrm{i}\hat{H}t/\hbar}\hat{A}(0)e^{-\mathrm{i}\hat{H}t/\hbar}$$

where $\hat{H}$ is the Hamiltonian operator.

A stationary phase approximation at $t = 0$ applied to Eq. (3.9),

$$k(T) = Q_\text{r}^{-1}\frac{1}{2}\int_{-\infty}^{\infty} dt \frac{C_\text{ff}(t)}{C_\text{dd}(t)} C_\text{dd}(t) \approx Q_\text{r}^{-1}\frac{1}{2}\frac{C_\text{ff}(0)}{C_\text{dd}(0)}\int_{-\infty}^{\infty} dt\, C_\text{dd}(0)e^{-(\Delta H)^2 t^2/\hbar^2}, \quad (3.11)$$

yields QI approximation for the rate constant [29,30]

$$k_\text{QI}(T) = \frac{\sqrt{\pi}}{2}\frac{\hbar}{\Delta H}\frac{C_\text{ff}(0)}{Q_\text{r}} \quad (3.12)$$

where ΔH is a specific type of energy variance [32]:

$$\Delta H = \hbar\left[-\frac{\ddot{C}_\text{dd}(0)}{2C_\text{dd}(0)}\right]^{1/2}. \quad (3.13)$$

The delta–delta correlation function $C_\text{dd}(t)$ introduced above is defined as

$$C_\text{dd}(t) \equiv C^{\text{sym}}_{\hat{\Delta}_a\hat{\Delta}_b}(t). \quad (3.14)$$

This correlation function at time $t = 0$ is a kind of generalization of the partition function, but constrained to two dividing surfaces. In a system with D degrees of freedom, the generalized delta function operator $\hat{\Delta}_\gamma$ is given by

$$\hat{\Delta}_\gamma = \Delta[\xi(\hat{\mathbf{r}}) - \xi_\gamma] \equiv \delta[\xi(\hat{\mathbf{r}}) - \xi_\gamma]\|\nabla\xi\|, \quad (3.15)$$

$$\|\nabla\xi\|^2 = \sum_{i=1}^{D} m_i^{-1}|\nabla_i\xi(\mathbf{r})|^2 \quad (3.16)$$

and $\xi(\mathbf{r})$ is the reaction coordinate such that, at the transition state, $\xi(\mathbf{r}^\ddagger) = 0$. Similarly, the equation $\xi(\mathbf{r}) = \xi_\gamma$ determines the location of the dividing surface γ.

The excellent accuracy of QI approximation has already been verified in numerous applications [17,30,32–47]. The main shortcoming of QI method is the neglect of recrossing, which is, however, neglected in any quantum or classical TSTs. The recrossing effects on QI rate constant have been quantified for several collinear reactions by Ceotto and Miller [33]. Fortunately, the recrossing effects become generally less important in higher dimensions. While the original form of QI is a short-time approximation [17], it can be improved systematically by including higher order derivatives of the flux–flux correlation function in the integral (3.9) instead of just the second derivative that is used in the stationary phase approximation (3.11) [39].

However, even the original form (3.12) of QI is capable of capturing such quantum effects as tunneling, "corner cutting," and anharmonic fluctuations about the optimal semiclassical path and therefore goes far beyond the capabilities of the conventional TST. Still, it is useful to consider the rate constant in the framework of TST because this theory (i) allows a classification of various contributions to the rate constant, (ii) is still widely used for describing experimental rate constants, and (iii) gives quantitative results at least at high enough temperatures (but not too high that recrossing effects become important).

3.4 KINETIC ISOTOPE EFFECTS

As the first application of QI approximation, let us consider the calculation of the kinetic isotope effect (KIE). KIE is defined as the effect of isotopic substitution on the rate constant:

$$\mathrm{KIE} = \frac{k_{\mathrm{l}}}{k_{\mathrm{h}}} \tag{3.17}$$

where subscript "l" ("h") denotes the lighter (heavier) isotopolog. This KIE is widely used in chemical kinetics to detect nuclear tunneling and other nuclear quantum effects, and to distinguish between possible reaction mechanisms. Unlike the rate constant itself, KIE depends very little on the height of the classical reaction barrier, and so KIE can separate various effects from the simple exponential dependence on the barrier height that overwhelms the rate constant.

3.4.1 Transition State Theory Framework for KIE

Within TST (3.2), KIE becomes

$$\mathrm{KIE}_{\mathrm{TST}} = \frac{Q_{\mathrm{l}}^{\ddagger}\,Q_{\mathrm{r,h}}}{Q_{\mathrm{h}}^{\ddagger}\,Q_{\mathrm{r,l}}}. \tag{3.18}$$

TST expression (3.18) greatly simplifies [30,48–50] by employing the Teller–Redlich product theorem [51,52] that relates the product of the normal mode frequencies to the molecular mass, moments of inertia, and the product of nuclear masses. For an asymmetric top, this theorem gives

$$\left(\frac{M_{\mathrm{l}}}{M_{\mathrm{h}}}\right)^{3/2}\left(\frac{I_{x,\mathrm{l}}I_{y,\mathrm{l}}I_{z,\mathrm{l}}}{I_{x,\mathrm{h}}I_{y,\mathrm{h}}I_{z,\mathrm{h}}}\right)^{1/2} = \prod_{i=1}^{D}\frac{m_{i,\mathrm{l}}}{m_{i,\mathrm{h}}}\prod_{j=1}^{D-6}\frac{\omega_{j,\mathrm{l}}}{\omega_{j,\mathrm{h}}} \tag{3.19}$$

where $M = \sum_{i=1}^{D} m_i$ is the total mass of the molecule and ω_j is the angular frequency of the *j*th normal mode. If Eqs (3.3)–(3.6) for the partition functions are substituted into Eq. (3.18), the Teller–Redlich theorem (3.19) yields the classical (or infinite temperature) limit of KIE, which is temperature independent:

$$\mathrm{KIE}_{\mathrm{class}} = \frac{\sigma_{\mathrm{sym,h}}^{\ddagger}}{\sigma_{\mathrm{sym,l}}^{\ddagger}}\frac{\sigma_{\mathrm{sym,r,l}}}{\sigma_{\mathrm{sym,r,h}}}\frac{|\omega_{\mathrm{unst,l}}^{\ddagger}|}{|\omega_{\mathrm{unst,h}}^{\ddagger}|}. \tag{3.20}$$

Here $\sigma^{\ddagger}_{\mathrm{sym}}$ and $\sigma_{\mathrm{sym},r}$ are the symmetry numbers for the transition state and reactants, respectively.

Including the Wigner tunneling correction (3.7), the kinetic isotope ratio becomes

$$\mathrm{KIE}_{\mathrm{TST+W}} = \mathrm{KIE}_{\mathrm{TST}} \frac{\kappa_{\mathrm{W,l}}}{\kappa_{\mathrm{W,h}}}. \tag{3.21}$$

3.4.2 Quantum Instanton Approach and the Thermodynamic Integration with Respect to the Isotope Mass

Using QI expression (3.12), KIE can be expressed as [30]

$$\mathrm{KIE}_{\mathrm{QI}} = \frac{Q_{\mathrm{r,h}}}{Q_{\mathrm{r,l}}} \frac{\Delta H_{\mathrm{h}}}{\Delta H_{\mathrm{l}}} \frac{C_{\mathrm{dd,l}}(0)}{C_{\mathrm{dd,h}}(0)} \frac{[C_{\mathrm{ff,l}}(0)]/[C_{\mathrm{dd,l}}(0)]}{[C_{\mathrm{ff,h}}(0)]/[C_{\mathrm{dd,h}}(0)]} \tag{3.22}$$

where both numerator and denominator were for convenience multiplied and divided by $C_{\mathrm{dd}}(0)$. Introduction of C_{dd} into expression (3.22) simplifies the ultimate path integral evaluation: on one hand, similarly to ΔH, the flux factor $C_{\mathrm{ff}}(0)/C_{\mathrm{dd}}(0)$ can be computed directly in a single path integral simulation [32]; on the other hand, the ratio $C_{\mathrm{dd,l}}(0)/C_{\mathrm{dd,h}}(0)$ can be computed by a thermodynamic integration procedure that is described below.

It is convenient to consider the isotope change to be continuous and parameterized by a real number $\lambda \in [0, 1]$, where $\lambda = 0$ for isotopes present in isotopolog l and $\lambda = 1$ for isotopes in isotopolog h. This allows the description of KIE when not just one but possibly several atoms are replaced by their isotopes simultaneously. Within the Born–Oppenheimer approximation, the potential energy surface for the reaction does not change; the only change in the Hamiltonian is in the masses of the isotopes. The simplest choice of λ:

$$m_i(\lambda) = (1-\lambda)m_{\mathrm{l},i} + \lambda m_{\mathrm{h},i} \tag{3.23}$$

gives the mass of the ith degree of freedom as a linear interpolation between the masses $m_{\mathrm{l},i}$ and $m_{\mathrm{h},i}$ of the ith degree of freedom in the isotopologs l and h, respectively. For clarity, λ will be written as an explicit argument of all quantities. Ratios $Q_{\mathrm{r}}(\lambda = 1)/Q_{\mathrm{r}}(\lambda = 0)$ and $C_{\mathrm{dd}}(0, \lambda = 1)/C_{\mathrm{dd}}(0, \lambda = 0)$ are ratios of quantities with different values of λ and hence cannot be evaluated directly by the path integral method. In order to circumvent this obstacle, one uses a *thermodynamic integration* (also known as the *charging algorithm*) with respect to the parameter λ [53,54]. While $Q(\lambda)$ is not a normalized quantity, the λ derivative of the corresponding free energy $F(\lambda)$:

$$\frac{dF(\lambda)}{d\lambda} = -\beta^{-1}\frac{d\ln Q(\lambda)}{d\lambda} = -\beta^{-1}\frac{dQ(\lambda)/d\lambda}{Q(\lambda)}$$

is normalized and can be computed directly by the path integral method. Thus, the ratios of the partition functions and delta–delta correlation functions can be found as

$$\frac{Q_r(1)}{Q_r(0)} = \exp\left[-\beta \int_0^1 d\lambda \frac{dF_r(\lambda)}{d\lambda}\right], \tag{3.24}$$

$$\frac{C_{dd}(0,1)}{C_{dd}(0,0)} = \exp\left[-\beta \int_0^1 d\lambda \frac{dF^{\ddagger}(\lambda)}{d\lambda}\right] \tag{3.25}$$

where the free energies are defined as

$$F_r(\lambda) \equiv -\beta^{-1} \ln Q_r(\lambda) \tag{3.26}$$

and

$$F^{\ddagger}(\lambda) \equiv -\beta^{-1} \ln C_{dd}(0,\lambda). \tag{3.27}$$

Several other path integral-based approaches to compute KIE exist, which however do not use QI approximation. These include, for example, approaches using other quantum TSTs [55–60] or the quantized classical path method [7,61,62].

3.5 TEMPERATURE DEPENDENCE OF THE RATE CONSTANT

Every respectable student of physical chemistry knows the Arrhenius law (3.1) for the temperature dependence of the rate constant. While this law is only approximate, the quantity that it describes, that is, the ratio $k(T)/k(T_0)$, is extremely useful since it can be measured accurately much more easily than the rate constant $k(T)$ itself. Knowledge of the temperature dependence of the rate constant can help discern between possible mechanisms of a chemical reaction. This section will explain how the temperature dependence can be computed directly within QI approximation. For clarity, the inverse temperature β will be written as an explicit argument of all quantities.

3.5.1 Transition State Theory Framework for the Temperature Dependence of $k(T)$

At high temperatures, the rate constant is expected to behave classically and follow the Arrhenius law or the more accurate TST result. Whereas the Arrhenius law (3.1) predicts the rate constant ratio to be a simple exponential function of the inverse temperature:

$$\frac{k_A(T)}{k_A(T_0)} = e^{-E_{act}(\beta-\beta_0)}, \tag{3.28}$$

TST (3.2) gives the ratio of rate constants:

$$\frac{k_{\mathrm{TST}}(T)}{k_{\mathrm{TST}}(T_0)} = \frac{\beta_0}{\beta}\frac{Q^{\ddagger}(\beta)}{Q^{\ddagger}(\beta_0)}\frac{Q_{\mathrm{r}}(\beta_0)}{Q_{\mathrm{r}}(\beta)}e^{-(\beta-\beta_0)\Delta E^{\ddagger}}. \tag{3.29}$$

In particular, assuming the partition functions $Q^{\ddagger}$ and Q_{r} to be separable into products (3.3) of classical translational, rotational, and vibrational partition functions (Eqs (3.4)–(3.6)), the temperature dependence (3.29) of TST rate constant includes an additional fractional power law besides the exponential dependence in the Arrhenius law (3.28).

At somewhat lower temperatures, when quantum effects start to play a role, the basic TST expression (3.29) can be improved in several ways: First, classical partition functions Q_{r} and $Q^{\ddagger}$ can be replaced by their exact quantum analogues for a harmonic potential. (In fact, this is usually done automatically in TST.) Second, quantum tunneling can be included approximately via the Wigner tunneling correction (3.7):

$$\frac{k_{\mathrm{TST+W}}(T)}{k_{\mathrm{TST+W}}(T_0)} = \frac{k_{\mathrm{TST}}(T)}{k_{\mathrm{TST}}(T_0)}\frac{\kappa_{\mathrm{W}}(T)}{\kappa_{\mathrm{W}}(T_0)}. \tag{3.30}$$

3.5.2 Quantum Instanton Approach and the Thermodynamic Integration with Respect to the Inverse Temperature

In the framework of QI approximation (3.12), the T dependence can be evaluated as [33,46]

$$\frac{k_{\mathrm{QI}}(T)}{k_{\mathrm{QI}}(T_0)} = \frac{Q_{\mathrm{r}}(\beta_0)}{Q_{\mathrm{r}}(\beta)}\frac{\Delta H(\beta_0)}{\Delta H(\beta)}\frac{C_{\mathrm{dd}}(0,\beta)}{C_{\mathrm{dd}}(0,\beta_0)}\frac{[C_{\mathrm{ff}}(0,\beta)]/[C_{\mathrm{dd}}(0,\beta)]}{[C_{\mathrm{ff}}(0,\beta_0)]/[C_{\mathrm{dd}}(0,\beta_0)]} \tag{3.31}$$

where, for convenience, the numerator and denominator were again multiplied and divided by $C_{\mathrm{dd}}(0,\beta)$. The flux factor $C_{\mathrm{ff}}(0,\beta)/C_{\mathrm{dd}}(0,\beta)$ and energy variance $\Delta H(\beta)$ are computed exactly as for KIEs in a single path integral simulation. However, the ratios of the partition functions and of the delta–delta correlation functions cannot be computed directly since they involve ratios of quantities at different temperatures. The difference from the case of KIEs is that the calculation of these ratios requires a different type of thermodynamic integration, namely, a thermodynamic integration with respect to the inverse temperature β [33,46]:

$$\frac{Q_{\mathrm{r}}(\beta)}{Q_{\mathrm{r}}(\beta_0)} = \exp\left[-\int_{\beta_0}^{\beta} E_{\mathrm{r}}(\tilde{\beta})d\tilde{\beta}\right], \tag{3.32}$$

$$\frac{C_{\mathrm{dd}}(0,\beta)}{C_{\mathrm{dd}}(0,\beta_0)} = \exp\left[-\int_{\beta_0}^{\beta} E^{\ddagger}(\tilde{\beta})d\tilde{\beta}\right]. \tag{3.33}$$

In the above equations, E_r and $E^\ddagger$ are the thermal energies of the reactants and of the transition state, respectively. Such a thermodynamic integration was used within a discrete variable representation of QI approximation to compute the rate constant for several collinear triatomic reactions [33]. In Ref. [46], it is generalized and presented in a form suitable for a path integral evaluation. Unlike Q_r and C_{dd}, the energies are normalized quantities because they can be written as logarithmic derivatives:

$$E_r(\beta) = -\frac{d\ln Q_r(\beta)}{d\beta} = -\frac{dQ_r(\beta)/d\beta}{Q_r(\beta)}, \tag{3.34}$$

$$E^\ddagger(\beta) = -\frac{d\ln C_{dd}(0,\beta)}{d\beta} = -\frac{dC_{dd}(0,\beta)/d\beta}{C_{dd}(0,\beta)}. \tag{3.35}$$

Hence, they can be computed directly by the path integral procedure.

3.6 PATH INTEGRAL REPRESENTATION OF RELEVANT QUANTITIES

3.6.1 Path Integral Formalism

As should be clear from the preceding discussion, the partition function, defined as

$$Q(\beta) = \sum_n e^{-\beta\varepsilon_n} \tag{3.36}$$

is a central quantity in statistical mechanics. In the above equation, ε_n is the energy of the nth eigenstate of the system. If the partition function is known explicitly, any thermodynamic quantity can be found. For example, as already noted above, the average thermal energy can be computed from Q using the logarithmic derivative (3.34).

The strength of the path integral (PI) formalism [23–27] lies in the fact that it allows computing the partition function without finding the eigenstates of the Hamiltonian. Let us therefore consider a molecular system with D degrees of freedom with masses m_i, $i = 1, \ldots, D$. Starting from the exact expression $Q(\beta) = \mathrm{Tr}(e^{-\beta\hat{H}})$, one obtains PI representation of Q as $Q = \lim_{P\to\infty} Q_P$, with [23–27]

$$Q_P(\beta) = C\int d\mathbf{r}^{(1)}\cdots\int d\mathbf{r}^{(P)}\exp[-\beta\Phi_P(\{\mathbf{r}^{(s)}\})]. \tag{3.37}$$

In the above equation, $C \equiv (P/(2\pi\hbar^2\beta^2))^{DP/2}\prod_{i=1}^{D} m_i^{P/2}$ is a multiplicative factor, P is the number of imaginary time slices, and $\mathbf{r}^{(s)} \equiv (r_1^{(s)}, \ldots, r_D^{(s)})$ are the D coordinates of the sth imaginary time slice. Finally, $\Phi_P(\{\mathbf{r}^{(s)}\})$ is the effective potential given by

$$\Phi_P(\{\mathbf{r}^{(s)}\}) = \frac{P}{2\hbar^2\beta^2}\sum_{i=1}^{D} m_i \sum_{s=1}^{P}(r_i^{(s)} - r_i^{(s-1)})^2 + \frac{1}{P}\sum_{s=1}^{P} V(\mathbf{r}^{(s)}) \tag{3.38}$$

with $\mathbf{r}^{(0)} \equiv \mathbf{r}^{(P)}$ and $\{\mathbf{r}^{(s)}\}$ a shorthand for $\{\mathbf{r}^{(1)}, \mathbf{r}^{(2)}, \ldots, \mathbf{r}^{(P)}\}$.

From Eqs (3.37) and (3.38), it is obvious that for $P=1$ one obtains the classical partition function and hence classical thermodynamics. The exact quantum partition function and quantum thermodynamics are obtained in the limit $P \rightarrow \infty$; however, in practice, it often suffices to take quite a small value of P to obtain accurate quantum results.

To compute a thermal average $A(\beta)$ of a physical observable A such as the heat capacity or the rate constant, one starts from the exact quantum expression $A(\beta)=\mathrm{Tr}[\hat{A}\exp(-\beta\hat{H})]/Q(\beta)$ or, if possible, from an exact expression for $A(\beta)$ in terms of the partition function, such as Eq. (3.34) for energy. Using PI expression (3.37) for Q, one ends up with a PI expression for the thermal average $A(\beta)$:

$$A_P(\beta) = \langle A_P(\{\mathbf{r}^{(s)}\})\rangle_{\rho_P} = \frac{\int d\mathbf{r}^{(1)} \cdots \int d\mathbf{r}^{(P)}\, A_P(\{\mathbf{r}^{(s)}\})\rho_P(\{\mathbf{r}^{(s)}\})}{\int d\mathbf{r}^{(1)} \cdots \int d\mathbf{r}^{(P)}\, \rho_P(\{\mathbf{r}^{(s)}\})} \tag{3.39}$$

where $\langle \cdot \rangle_\rho$ denotes a weighted average over PI configurations. The reactant weight function is given by $\rho_{\mathrm{r},P} = \exp(-\beta\Phi_P)$ and the function $A_P(\{\mathbf{r}^{(s)}\})$ is called a PI "estimator" for $A(\beta)$.

For completeness, PI representation of the delta–delta correlation function is [32]

$$C_{\mathrm{dd},P}(0,\beta) = C\int d\mathbf{r}^{(1)} \cdots \int d\mathbf{r}^{(P)}\, \rho_P^{\ddagger}(\{\mathbf{r}^{(s)}\}), \tag{3.40}$$

$$\rho_P^{\ddagger}(\{\mathbf{r}^{(s)}\}) \equiv \Delta[\xi_a(\mathbf{r}^{(P/2)})]\Delta[\xi_b(\mathbf{r}^{(P)})]\exp[-\beta\Phi_P(\{\mathbf{r}^{(s)}\})] \tag{3.41}$$

where $\rho_P^{\ddagger}$ is the sampling weight for the constrained simulations.

In practice, there are two main strategies for evaluating PI average (3.39): PI molecular dynamics (PIMD) [25,63] and PI Monte Carlo (PIMC) techniques [26]. The results from Refs [30,38,42,45,46,64], shown below, were obtained either with an in-house PIMC code (using the Metropolis algorithm) or with a PIMD implementation in the molecular dynamics package AMBER 10 [65].

3.6.2 Estimators

PI estimators for all quantities needed in Eqs (3.22) for KIE and (3.31) for the temperature dependence can be obtained from PI representations of Q_r and C_{dd} substituted into the definitions of ΔH, $C_{\mathrm{ff}}/C_{\mathrm{dd}}$, E_r, $E^{\ddagger}$, and λ derivatives of $F_\mathrm{r}(\lambda)$ and $F^{\ddagger}(\lambda)$. The estimators for ΔH and $C_{\mathrm{ff}}/C_{\mathrm{dd}}$ are derived in Ref. [32].

Thus, PI procedure may seem quite straightforward. The interesting twist comes because the estimator $A_P(\{\mathbf{r}^{(s)}\})$ is not a unique function of $\{\mathbf{r}^{(s)}\}$. The art of PIs lies in finding the optimal estimator in the sense of having the smallest statistical error for a given number M of samples. As the statistical error of a simulation is proportional to $M^{-1/2}$, an estimator with a smaller statistical error can lead to a much more efficient simulation. Because of this, let us discuss several estimators and their properties in more detail.

3.6.3 Estimators for E_r

The simplest estimator for the energy E_r, the so-called Barker or thermodynamic estimator (TE) [66], can be derived directly from Eq. (3.34) and PI expression (3.37), giving

$$E_{\mathrm{r},P,\mathrm{TE}} = \frac{DP}{2\beta} - \frac{P}{2\beta^2\hbar^2}\sum_{i=1}^{D} m_i \sum_{s=1}^{P} (r_i^{(s)} - r_i^{(s-1)})^2 + \frac{1}{P}\sum_{s=1}^{P} V(\mathbf{r}^{(s)}). \quad (3.42)$$

As observed by Herman et al. [67], TE can have a large statistical error, which can be avoided with the so-called virial (VE) [67] or centroid virial (CVE) [68] estimators. Invoking the virial theorem, the kinetic energy in these two estimators is replaced by an expression involving the gradient of the potential energy [67]. This is convenient in PIMD implementations since the gradient is already available. In PIMC simulations, however, only the potential is needed for the random walk, and in order to avoid computing the gradients, alternative approaches have been proposed. One can, for example, employ the centroid thermodynamic estimator [69] or, more generally, use a procedure based on rescaling coordinates [70,71] where the gradients of the potential are replaced by a single derivative that can be evaluated by finite difference [72]. Variants of the latter approach have been applied successfully to compute thermal energies and heat capacities [72], kinetic isotope effects [38,42,45], equilibrium isotope effects [64], temperature dependence of the rate constant [45,46], or the derivatives of the flux–flux correlation function [41] needed in the generalized QI model [39].

VE for E_r can be derived most directly by the change of coordinates $\mathbf{x}^{(s)} \equiv \beta^{-1/2}\mathbf{r}^{(s)}$ in PI (3.37), yielding

$$\begin{aligned} E_{\mathrm{r},P,\mathrm{VE}} &= \frac{1}{P}\sum_{s=1}^{P}\left\{V(\mathbf{r}^{(s)}) + \beta\frac{dV[(\beta+\Delta\beta)^{1/2}\beta^{-1/2}\mathbf{r}^{(s)}]}{d\,\Delta\beta}\right\} \\ &= \frac{1}{P}\sum_{s=1}^{P}\left\{V(\mathbf{r}^{(s)}) + \frac{dV[(1+q)^{1/2}\mathbf{r}^{(s)}]}{dq}\right\} \end{aligned} \quad (3.43)$$

where q is a small dimensionless parameter and the q derivative is evaluated by finite difference at $q=0$. Similarly, CVE can be obtained by the change of variables $\mathbf{x}^{(s)} \equiv \beta^{-1/2}(\mathbf{r}^{(s)} - \mathbf{r}^{(\mathrm{C})})$, where one first subtracts the so-called centroid coordinate $\mathbf{r}^{(\mathrm{C})} \equiv P^{-1}\sum_{s=1}^{P}\mathbf{r}^{(s)}$. The resulting estimator is

$$E_{\mathrm{r},P,\mathrm{CVE}} = \frac{D}{2\beta} + \frac{1}{P}\sum_{s=1}^{P}\left\{V(\mathbf{r}^{(s)}) + \frac{dV[\mathbf{r}^{(\mathrm{C})} + (1+q)^{1/2}(\mathbf{r}^{(s)} - \mathbf{r}^{(\mathrm{C})})]}{dq}\right\} \quad (3.44)$$

$$= \frac{D}{2\beta} + \frac{1}{P}\sum_{s=1}^{P}\left[V(\mathbf{r}^{(s)}) + \frac{1}{2}(\mathbf{r}^{(s)} - \mathbf{r}^{(\mathrm{C})})\cdot\nabla V(\mathbf{r}^{(s)})\right]. \quad (3.45)$$

Equations (3.44) and (3.45) give two forms of CVE estimator: The former, based on a finite-difference q derivative, is useful for a PIMC simulation where gradients of V are not available. The latter, obtained by evaluating the q derivative analytically, requires the knowledge of gradients of V and hence is suitable for a PIMD simulation.

3.6.4 Estimators for $E^{\ddagger}$

In constrained simulations near the transition state, it is difficult to sample from the delta function constraint (3.41), especially if a Monte Carlo sampling is used. It is therefore convenient to approximate the constrained weight function (3.41) by using the Gaussian approximation of the delta function in Eq. (3.15). This Gaussian constraint corresponds to a harmonic constraint potential $V_{\text{constr},\gamma}(\mathbf{r})$ [30,32]:

$$\delta\left[\xi(\mathbf{r})-\xi_{\gamma}\right] \approx \sqrt{\frac{\beta}{2\pi\sigma^2}}e^{-\beta V_{\text{constr},\gamma}(\mathbf{r})}, \tag{3.46}$$

$$V_{\text{constr},\gamma}(\mathbf{r}) = \frac{1}{2}\left(\frac{\xi(\mathbf{r})-\xi_{\gamma}}{\sigma}\right)^2. \tag{3.47}$$

Besides a prefactor, this amounts to adding an effective constraint potential

$$\Phi_{\text{constr}}(\{\mathbf{r}^{(s)}\}) \equiv V_{\text{constr},a}(\mathbf{r}^{(P/2)}) + V_{\text{constr},b}(\mathbf{r}^{(P)})$$

to the effective potential Φ. Assuming that $V_{\text{constr},\gamma}$ is independent of temperature and following a derivation similar to that for estimators for E_{r}, one obtains the thermodynamic, virial, and centroid virial estimators for $E^{\ddagger}$ [46]:

$$E^{\ddagger}_{P,\text{TE}} = E_{\text{r},P,\text{TE}} - \frac{1}{\beta} + \Phi_{\text{constr}}(\{\mathbf{r}^{(s)}\}), \tag{3.48}$$

$$E^{\ddagger}_{P,\text{VE}} = E_{\text{r},P,\text{VE}} - \frac{1}{\beta} + \Phi_{\text{constr}}(\{\mathbf{r}^{(s)}\}) + \frac{d\Phi_{\text{constr}}[(1+q)^{1/2}\{\mathbf{r}^{(s)}\}]}{dq}, \tag{3.49}$$

$$E^{\ddagger}_{P,\text{CVE}} = E_{\text{r},P,\text{CVE}} - \frac{1}{\beta} + \Phi_{\text{constr}}(\{\mathbf{r}^{(s)}\}) + \frac{d\Phi_{\text{constr}}[\mathbf{r}^{(\text{C})} + (1+q)^{1/2}(\mathbf{r}^{(s)}-\mathbf{r}^{(\text{C})})]}{dq}. \tag{3.50}$$

Although the above estimators converge to the exact results, the statistical errors can be decreased slightly by employing an alternative set of estimators, derived using the relation

$$\langle\Phi_{\text{constr}}(\{\mathbf{r}^{(s)}\})\rangle = \beta^{-1}$$

valid for a harmonic constraint potential. The new estimators are [46]

$$E^{\ddagger}_{P,\text{TE}} = E_{\text{r},P,\text{TE}}, \tag{3.51}$$

$$E^{\ddagger}_{P,\text{VE}} = E_{\text{r},P,\text{VE}} + \frac{d\Phi_{\text{constr}}[(1+q)^{1/2}\{\mathbf{r}^{(s)}\}]}{dq}, \tag{3.52}$$

$$E^{\ddagger}_{P,\text{CVE}} = E_{\text{r},P,\text{CVE}} + \frac{d\Phi_{\text{constr}}[\{\mathbf{r}^{(\text{C})} + (1+q)^{1/2}(\mathbf{r}^{(s)}-\mathbf{r}^{(\text{C})})\}]}{dq}. \tag{3.53}$$

Estimators (3.51)–(3.53) are in a way more intuitive than estimators (3.48)–(3.50): in the limit of a sharp constraint, the constrained energy should be independent of the type of constraint. It should be stressed that the last terms in VE and CVE in Eqs (3.49), (3.50), (3.52) and (3.53) are important; without them, the agreement among TE, VE, and CVE is lost. In other words, an intuitive guess such as $E^{\ddagger}_{P,\mathrm{CVE}} = E_{\mathrm{r},P,\mathrm{CVE}}$ would not give a correct answer for the constrained energy. Note that Eqs (3.48)–(3.53) show only the estimators suitable for a PIMC evaluation. The estimators suitable for a PIMD simulation are obtained by performing the q derivative analytically, resulting in the presence of gradients of Φ_{constr}, similar to the case of E_{r} estimators.

3.6.5 Estimators for the Derivatives of F_{r} and $F^{\ddagger}$ with Respect to Mass

The simplest estimator for $dF_{\mathrm{r}}(\lambda)/d\lambda$ is again a thermodynamic estimator, obtained by substituting PI expression (3.37) into Eq. (3.26), giving [30]

$$\left(\frac{dF_{\mathrm{r}}(\lambda)}{d\lambda}\right)_{P,\mathrm{TE}} = -\frac{P}{2\beta}\sum_{i=1}^{D}\frac{dm_i/d\lambda}{m_i}\left[1-\frac{m_i}{\beta\hbar^2}\sum_{s=1}^{P}(r_i^{(s)}-r_i^{(s-1)})^2\right]. \tag{3.54}$$

By rescaling the Cartesian coordinates in PI (3.37) as $x_i^{(s)} \equiv m_i^{1/2}(r_i^{(s)}-r_i^{(C)})$, and following a procedure similar to the case of E_{r}, one obtains the centroid virial estimator [38,42]:

$$\left(\frac{dF_{\mathrm{r}}(\lambda)}{d\lambda}\right)_{P,\mathrm{CVE}} = -\frac{1}{2\beta}\sum_{i=1}^{D}\frac{dm_i/d\lambda}{m_i} + \frac{1}{P}\sum_{s=1}^{P}\frac{dV[\{r_i^{(\mathrm{C})}+(m_i(\lambda)/(m_i(\lambda+\Delta\lambda)))^{1/2}(r_i^{(s)}-r_i^{(\mathrm{C})})\}]}{d\Delta\lambda}, \tag{3.55}$$

$$\left(\frac{dF_{\mathrm{r}}(\lambda)}{d\lambda}\right)_{P,\mathrm{CVE}} = -\sum_{i=1}^{D}\frac{dm_i/d\lambda}{m_i}\left[\frac{1}{2\beta}+\frac{1}{2P}\sum_{s=1}^{P}(r_i^{(s)}-r_i^{(\mathrm{C})})\frac{\partial V}{\partial r_i}(\mathbf{r}^{(s)})\right]. \tag{3.56}$$

In the above equations, two forms of CVE are listed: the former, based on evaluating the $\Delta\lambda$ derivative by finite difference, is suitable for a PIMC simulation where gradients of V are not available. The latter, obtained by performing $\Delta\lambda$ derivatives explicitly, requires gradients of V and is suitable for a PIMD simulation.

The estimators for $dF^{\ddagger}(\lambda)/d\lambda$ are obtained in a similar fashion, although the expressions are somewhat more complicated due to terms depending on the constraint. See Refs [30,42] for details.

3.6.6 Statistical Errors and Efficiency

The reason for introducing CVEs for E_{r}, $E^{\ddagger}$, $dF_{\mathrm{r}}/d\lambda$, and $dF^{\ddagger}/d\lambda$ is simply the efficiency. The problem with the corresponding TEs is that their statistical error σ_{stat} increases with the number of slices P. If M is the number of PIMC samples (or, roughly speaking, the length of PIMD simulation) and τ is the correlation time, then in the reactant region,

$$\sigma_{\text{stat,TE}} = \mathrm{O}(P^{1/2}\tau^{1/2}M^{-1/2}).$$

If a large value of P is needed to obtain an accurate quantum result, the length of simulation will have to increase accordingly.

The advantage of CVEs is that their statistical error is approximately independent of P:

$$\sigma_{\text{stat,CVE}} = \mathrm{O}(P^{0}\tau^{1/2}M^{-1/2}).$$

This implies that one can increase P and reach the quantum limit without having to increase the number of samples.

One may object that TE for KIE does not involve the potential, and so could be faster than CVE despite the higher statistical error. However, as the correlation time in PI simulations is usually quite large, one does not have to evaluate all estimators at each step of the random walk (or PIMD), and so the marginal cost of evaluating an estimator can be made negligible compared to the cost of the sampling. On the other hand, in constrained simulations, the prefactor in the above statistical error dependences can be much larger for CVE than for TE, and so for a given value of P, TE may have a smaller statistical error [46]. As the marginal cost is zero, it is recommended to compute all types of estimators for a given quantity and use the one with the smallest statistical error. Besides helping to find the optimal estimator, comparing various estimators provides an independent check of convergence of a PI simulation.

3.6.7 Treatment of Potential Energy Surfaces for Many-Dimensional Systems

The main difficulty in applying any dynamic method to a large molecular system is the construction of an accurate potential energy surface (PES). Unfortunately, high-level *ab initio* electronic structure methods are usually too expensive to be used in PI calculation directly. On the other hand, semiempirical or force field methods do not achieve comparable accuracy. To take advantage of both the accuracy of *ab initio* methods and the rigor of PI treatment of nuclear motion, one can combine (i) the value of the energy barrier computed with the most accurate but most expensive method (V_1), (ii) the free energy obtained in the harmonic approximation (HA) using an intermediate-level method (V_2), and (iii) the anharmonicity correction, rigorously computed with PI method, using the least accurate but also the least expensive method (V_3). The anharmonicity correction is calculated as

$$\Delta\Delta F^{\text{anharm}} = \Delta F_{\text{PI}} - \Delta F_{\text{HA}} \tag{3.57}$$

where ΔF_{PI} and ΔF_{HA} are the free energy differences computed with PI method and HA, respectively.

Moreover, in the case of QI, the global PES is not needed; PES must be known accurately only in the vicinities of the reaction minimum and of the transition state.

Hence, the calculation can be sped up enormously if the three-stage approach outlined above is implemented as follows:

1. The barrier height, that is, $E^{\ddagger} \equiv V(\mathbf{r}^{\ddagger})-V(\mathbf{r}_{\text{react}})$, which requires only two single point calculations, is computed by all three methods, V_1, V_2, and V_3.
2. The Hessians of PES at the transition state and at the reactant minimum are computed by the less expensive methods V_2 and V_3.
3. Finally, PI simulation, which requires a very large amount of single point calculations, is done with the least expensive method V_3.

The results of the three different calculations are combined to obtain three contributions to KIE or the temperature dependence of the rate constant:

(a) The barrier height $E^{\ddagger}$ computed with V_1 is used to determine not only the exponential dependence of $k(T)$ in the Arrhenius law but also the leading contributions in TST and QI method. For the temperature dependence of $k(T)$, $E^{\ddagger}$ has to be very accurate due to the exponential sensitivity. The advantage of KIE is that classically the barrier height cancels out between the two isotopologs.

(b) TST and the Wigner tunneling correction can be computed from the Hessian computed by V_2 combined with the barrier height based on V_1. This gives the rate constant, assuming separability of rotations and vibrations, HA for vibrations, and rigid rotor approximation for rotations.

(c) Finally, the anharmonicity corrections, rovibrational coupling, many-dimensional tunneling, and other quantum effects are taken into account by using PI simulation based on V_3 combined with HA result based on V_2, and the barrier height based on V_1. In practice, it means that any of the free energies of interest is computed as a sum of the result of (b) and the anharmonicity correction (3.57) computed with V_3.

The potential energy V_3 can be, for example, a semiempirical method, an empirical valence bond (EVB) potential [73–76], or, more generally, a more complicated analytical fit to several points computed by a higher level electronic structure method [77].

3.7 EXAMPLES

3.7.1 Eckart Barrier

A simple model of an activated chemical reaction is provided by the Eckart barrier, a one-dimensional system described by the potential

$$V(x) = V_0[\cosh(ax)]^{-2}. \tag{3.58}$$

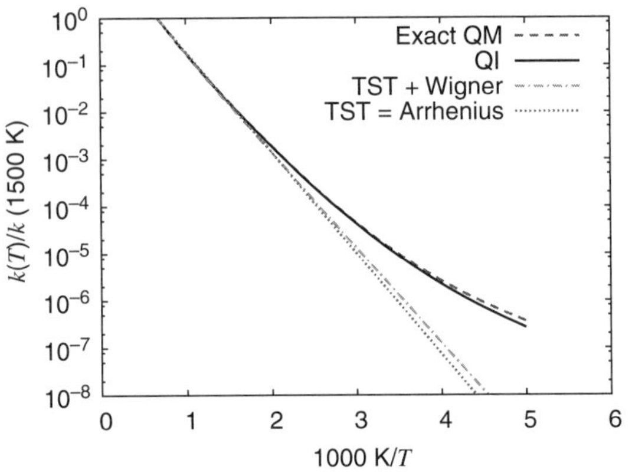

FIGURE 3.1 Eckart barrier. Temperature dependence of the rate constant.

The results given below are based on standard parameter values: $V_0 = 1.56 \times 10^{-2}$ a.u., $a = 1.36$ a.u., mass $m = 1060$ a.u., and reaction coordinate $\xi = x$. The exact quantum mechanical (QM) rate constant k_{QM} for this reaction can be obtained by integrating the exact quantum mechanical cumulative reaction probability, which is known analytically [30,78].

The temperature dependence of $k(T)$ for this system was studied in Ref. [46] by PI procedure described above. Figure 3.1 compares QI approximation with the exact QM results, TST (which is equal to the Arrhenius law here), and TST with the Wigner tunneling correction. The reference temperature is $T_0 = 1500$ K and the plot shows ratios for temperatures down to 200K. Since classical recrossing does not occur for the Eckart barrier, all TSTs should converge to the correct quantum results at high temperatures. The figure confirms this since all curves are tangent to each other at the high temperature limit. At low temperatures, one reaches the quantum regime where tunneling is important and consequently the Arrhenius plot of the exact QM result has a large curvature. While TST has a huge error, QI approximation agrees very well with QM result. Note that the Wigner tunneling correction improves over TST and captures the tunneling effect partially but still fails to recover the curvature of the exact result. QI method has an error below 3% for $T \geq 330$K and has a significant error ($\geq$10%) only at very low temperatures, below ~270K. However, this error was well understood already in the original paper by Miller et al. [17] and can be remedied by considering two separate dividing surfaces at very low temperatures. (Calculations in Ref. [46] used a single dividing surface at all temperatures for simplicity.)

The temperature dependence of the reactant and transition state energies is shown in Figure 3.2. Note that VE for E_r gives zero, but can be easily corrected with an analytical correction $1/(2\beta)$, corresponding to the thermal energy of a single free translational degree of freedom. The three different estimators (3.51)–(3.53) for the constrained energy $E^{\ddagger}$ at $T = 515.15$K are compared in Figure 3.3. Panel (a) shows that TE, VE, and CVE agree for all examined values of P and, in particular, converge to the same value for $P \to \infty$. The three estimators, however, differ in their statistical convergence. Unlike for the unconstrained result, where CVE is the optimal estimator,

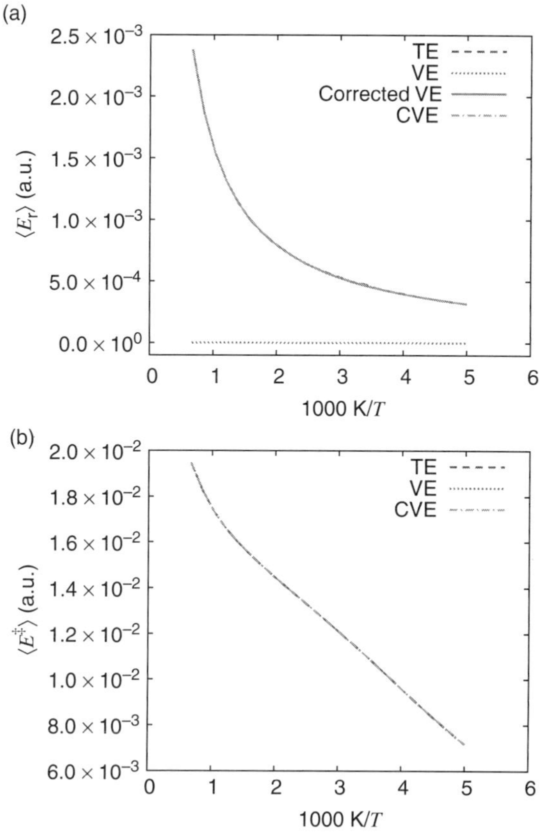

FIGURE 3.2 Eckart barrier. Temperature dependence of the (a) reactant and (b) transition state thermal energies.

for the constrained energy, the optimal estimator at this temperature turns out to be VE. This can be clearly seen in panel (b) that shows the root mean square errors (RMSEs) of the different estimators for different values of P. While RMSE of TE increases with P, RMSEs of VE and CVE remain approximately constant as a function of P, with VE having a much smaller statistical error. Assuming that the desired convergence is achieved for $P = 24$, the speedup factor achieved by using VE compared to TE and CVE is approximately $(2.9)^2 \approx 8$ and $(8.1)^2 \approx 60$, respectively. It is clear from the figure that both the speedup factor and the best estimator depend on P and hence on the temperature.

Figure 3.4 shows KIE k_H/k_D for temperatures from 100 to 2400K, calculated exactly and by various approximations in Ref. [30]. At high temperatures, the exact KIE approaches the parabola representing TST with the Wigner tunneling correction (3.21). QI result follows the exact KIE closely for all temperatures. In particular, for temperatures 300K and higher, the error of QI result is within 5%. At low temperatures (200K and lower), the error of the single-dividing-surface version

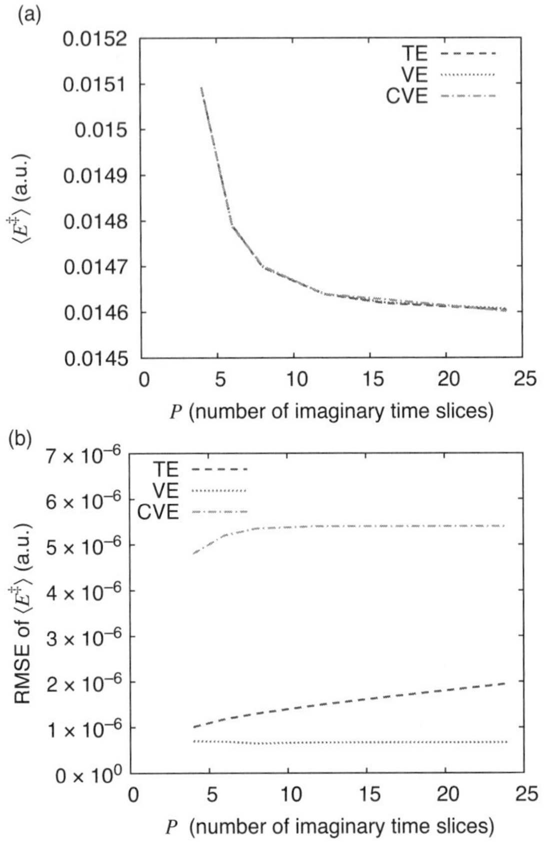

FIGURE 3.3 Eckart barrier. Dependence of (a) the transition state thermal energy and (b) the statistical RMSE of the transition state thermal energy on the number of imaginary time slices at $T = 515.15$K.

increases, which is expected since at low temperature one should correctly use two separate dividing surfaces [17,32]. One of the reasons for the excellent agreement of QI with exact results at temperatures 300K and higher is that for systems with one degree of freedom, there is strictly no classical recrossing. Every classical trajectory crosses the top of the barrier at most once. Since the classical TST yields the exact *classical* rates, the only possible errors are of quantum nature and those are well accounted for by QI approximation.

3.7.2 Full-Dimensional $H + H_2 \rightarrow H_2 + H$ Reaction

Still simple but already realistic thermally activated reaction is the hydrogen exchange reaction $H + H_2 \rightarrow H_2 + H$. Results below consider this reaction in full nine dimensions on the Boothroyd–Keogh–Martin–Peterson (BKMP2) reactive potential energy surface [79–81]. The transition state of this system is collinear with equal bond lengths

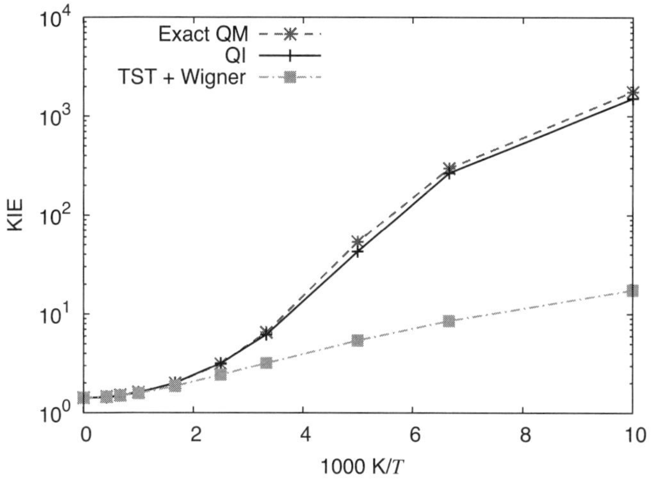

FIGURE 3.4 Eckart barrier. Temperature dependence of the kinetic isotope effect.

$d_{H_aH_b} = d_{H_bH_c}$. A suitable reaction coordinate is therefore given by the difference of the bond lengths:

$$\xi(\mathbf{r}) \equiv d_{H_aH_b} - d_{H_bH_c}. \tag{3.59}$$

Figure 3.5 shows the temperature dependence of the rate constant in the range from 200 to 1500K, computed in Ref. [46]. The exact QM results are from Ref. [30]. At high temperatures TST curve is tangent to the exact QM curve; however, at low temperatures, there is a significant discrepancy even for TST with the Wigner tunneling correction. On the other hand, QI approximation agrees very well with the exact QM result all the way to 200K.

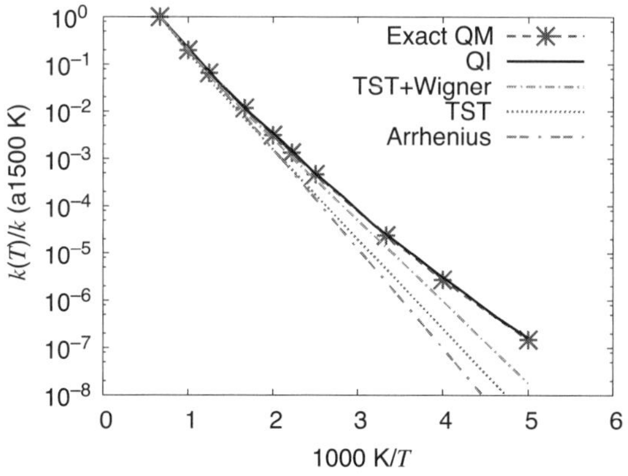

FIGURE 3.5 The $H + H_2 \rightarrow H_2 + H$ reaction. Temperature dependence of the rate constant.

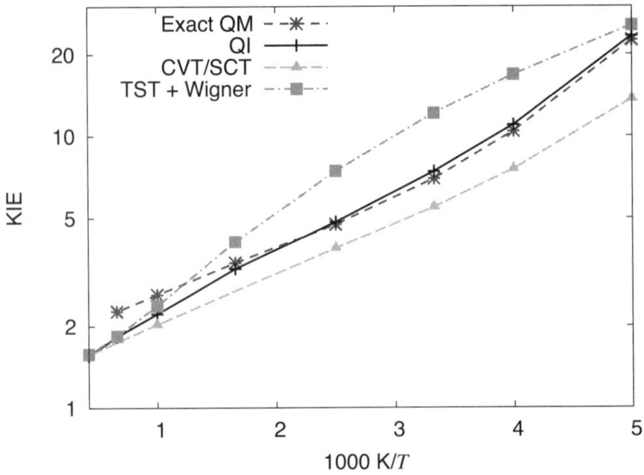

FIGURE 3.6 The $H + H_2 \rightarrow H_2 + H$ reaction. Temperature dependence of the kinetic isotope effect.

KIE $k(H + H_2)/k(D + D_2)$ for temperatures from 200 to 2400K, computed in Ref. [30], is displayed in Figure 3.6. Besides the exact QM, QI, and TST + Wigner values, this figure also displays corresponding results of the canonical variational TST with semiclassical tunneling (CVT/SCT) from Ref. [82] (this calculation was done for a slightly different—DMBE—potential [83,84]). For temperatures from 200 to 600K, the agreement between QI and QM results is excellent: the error is less than 6% in this temperature range. For high temperatures, QI and CVT/SCT results approach TST behavior predicted by Eq. (3.21), as expected, because quantum effects become small. The small deviation of TST, CVT/SCT, and QI from QM results at high temperatures is probably due to recrossing effects.

3.7.3 [1,5]-Sigmatropic Hydrogen Shift in *cis*-1,3-Pentadiene

To show the feasibility of combining QI approximation, thermodynamic integration, and PI implementation with the three-stage treatment of electronic structure in large systems, the methodology was used to study the hydrogen transfer reaction in *cis*-1,3-pentadiene. The trideuterated and dideuterated isotopologs were used by Roth and König in their experimental study of KIE in [1,5]-sigmatropic hydrogen shifts [85]. The value of KIE obtained by Roth and König was too high to be explained classically, implying that quantum effects such as tunneling are important in this reaction.

KIE and the temperature dependence of the rate constant were evaluated with QI approximation in Refs [42,45]. In the framework described above, the three electronic structure methods used in those calculations were

$$V_1 = \mathrm{CCSD/aug\text{-}cc\text{-}pVTZ},$$

$$V_2 = \mathrm{B98/6\text{-}311+(2df,p)},$$

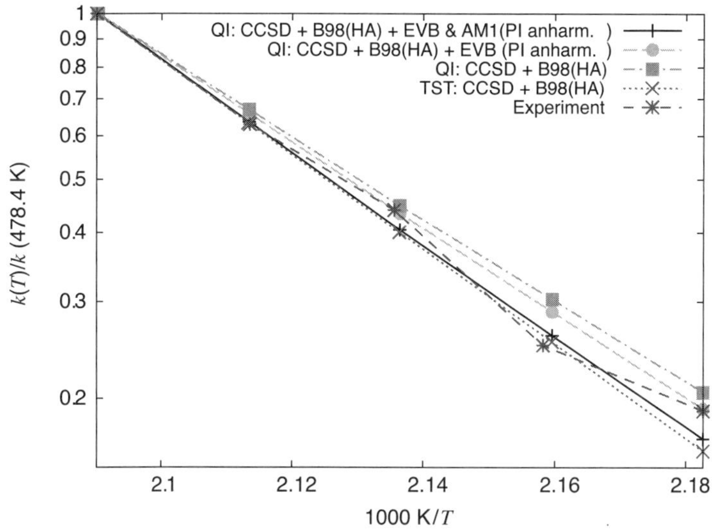

FIGURE 3.7 [1,5]-Sigmatropic hydrogen shift in *cis*-1,3-pentadiene. Temperature dependence of the rate constant for the dideuterated compound.

$$V_3 = \text{AM1 or EVB based on the GAFF force field.}$$

The temperature dependence of QI rate constant, computed in Ref. [45], is shown in Figure 3.7. "QI: CCSD + B98 (HA)" curve denotes HA result based on V_2 (i.e., B98), and "QI: CCSD + B98 (HA) + EVB (PI anharm.)" curve shows the full QI result, that is, including the anharmonicity PI correction computed with V_3 (i.e., an EVB based on GAFF force field [86]). As GAFF force field considers only harmonic stretches, the only anharmonicities included are due to nonbonded interactions. Because of this, the final and presumably the most accurate curve is "QI: CCSD + B98(HA) + EVB & AM1 (PI anharm.)" that still uses EVB potential near the transition state, but computes the reactant PI anharmonicity corrections with the semiempirical AM1 potential. It may seem more consistent to use AM1 potential also in the constrained PI simulations; however, this was computationally prohibitive. For simplicity, in the following discussion, "CCSD + B98 (HA)" is omitted as this part is common to all theoretical methods considered.

Interestingly, both the final QI: EVB & AM1 (PI anharm.) and TST results agree well the relatively broad and imprecise experimental temperature dependence. Neglecting for a moment the recrossing and anharmonicity effects, one can expect the value of k_{TST} to be smaller than HA value of k_{QI}, mainly due to the tunneling contribution to the rate that is captured by k_{QI} but ignored in k_{TST}. Moreover, as the relative importance of the tunneling increases with the decreasing temperature, the temperature dependence of the k_{QI} should be weaker than the dependence of k_{TST}. Figure 3.7 shows that this is indeed true. Another interesting feature visible in the

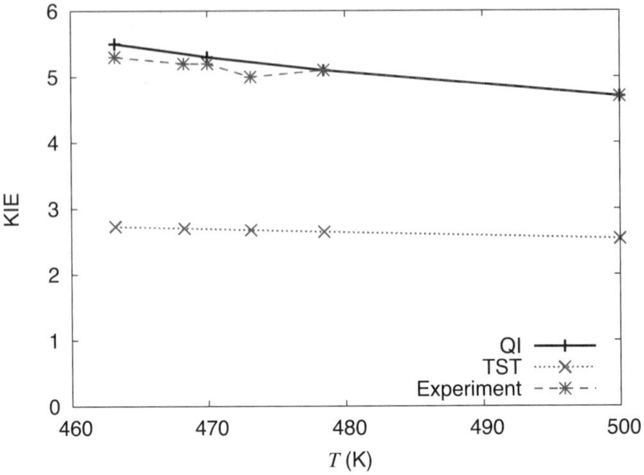

FIGURE 3.8 [1,5]-Sigmatropic hydrogen shift in *cis*-1,3-pentadiene. Temperature dependence of the kinetic isotope effect.

graph is the importance of the anharmonicity correction, which shows that the change in the temperature dependence (between k_{QI} and $k_{\mathrm{QI:EVB\&AM1(PI\ anharm.)}}$) caused by the anharmonicity can be equivalent to the change of the reaction barrier by several kcal/mol. Incidentally, the anharmonicity and tunneling corrections act in opposite directions, and the resulting cancellation explains the fortuitous accuracy of the classical TST. In order to take the anharmonicity corrections fully into account, AM1 potential had to be used in the reactant region since GAFF force field has harmonic bond stretches and hence even a PI simulation misses this part of the anharmonicity correction (compare $k_{\mathrm{QI:EVB(PI\ anharm.)}}$ and $k_{\mathrm{QI:EVB\&AM1(PI\ anharm.)}}$).

The temperature dependence of KIE, obtained in Ref. [45], is shown in Figure 3.8. QI results and experimental data agree very well; the largest relative difference is around 5%. While the displayed temperature range is relatively narrow, the breakdown of TST is clearly visible, demonstrating the importance of quantum effects on KIE.

3.8 SUMMARY

In chemical kinetics, one can often learn much by knowing only the kinetic isotope effect or the temperature dependence of the rate constant, instead of the rate constant itself. As various systematic errors can be avoided, the temperature dependence and kinetic isotope effect are easier to measure experimentally. Similarly, this chapter described a theoretical formalism based on the quantum instanton approximation to compute these two properties directly, without computing the rate constant itself. The common ingredients are the thermodynamic integration with respect to a specific parameter (inverse temperature or the isotope mass), path integral implementation, and, last but not least, the use of efficient path integral estimators.

While the quantum instanton approximation cannot describe the recrossing effect, this effect, which is predominantly classical, could be included empirically as a multiplicative factor (a "transmission coefficient") obtained in a classical molecular dynamics simulation. The main challenge for future research is to develop accurate and efficient approximations for the rate constant, its temperature dependence, and the kinetic isotope effect that include the quantum and recrossing effects simultaneously and rigorously.

APPENDIX: REACTIONS

1. Eckart Barrier (Model Reaction)
   Temperature range: 200–1500K
   Methods used: Arrhenius law, transition state theory, Wigner tunneling correction, quantum instanton approximation, and exact analytical solution
2. $H + H_2 \rightarrow H_2 + H$
   Temperature range: 200–1500K
   Methods used: Arrhenius law, transition state theory, Wigner tunneling correction, quantum instanton approximation, and exact scattering calculation
3. [1,5]-Sigmatropic Hydrogen Shift in *cis*-1,3-Pentadiene
   Temperature range: 460–500K
   Methods used: transition state theory and quantum instanton approximation

ACKNOWLEDGMENTS

This research was supported by the Swiss NSF (Grant No. 200021_124936/1) and by the EPFL. The author thanks Marcin Buchowiecki, William H. Miller, and Tomáš Zimmermann for discussions.

REFERENCES

1. S. Arrhenius, *Z. Phys. Chem.* **4**, 226 (1889).
2. H. Eyring, *J. Chem. Phys.* **3**, 107 (1935).
3. M. G. Evans and M. Polanyi, *Trans. Faraday Soc.* **31**, 875 (1935).
4. E. Wigner, *Trans. Faraday Soc.* **34**, 29 (1938).
5. J. W. Tromp and W. H. Miller, *J. Phys. Chem.* **90**, 3482 (1986).
6. G. A. Voth, D. Chandler, and W. H. Miller, *J. Phys. Chem.* **93**, 7009 (1989).
7. J. K. Hwang, Z. T. Chu, A. Yadav, and A. Warshel, *J. Phys. Chem.* **95**, 8445 (1991).
8. T. N. Truong, D. Lu, G. C. Lynch, Y.-P. Liu, V. S. Melissas, J. J. P. Stewart, R. Steckler, B. C. Garrett, A. D. Isaacson, A. Gonzalez-Lafont, S. N. Rai, G. C. Hancock, T. Joseph, and D. G. Truhlar, *Comp. Phys. Commun.* **75**, 143 (1993).
9. G. Mills and H. Jonsson, *Phys. Rev. Lett.* **72**, 1124 (1994).

10. G. K. Schenter, G. Mills, and H. Jonsson, *J. Chem. Phys.* **101**, 8964 (1994).
11. N. F. Hansen and H. C. Andersen, *J. Chem. Phys.* **101**, 6032 (1994).
12. N. F. Hansen and H. C. Andersen, *J. Phys. Chem.* **100**, 1137 (1996).
13. E. Pollak and J. L. Liao, *J. Chem. Phys.* **108**, 2733 (1998).
14. J. Shao, J. L. Liao, and E. Pollak, *J. Chem. Phys.* **108**, 9711 (1998).
15. G. Krilov, E. Sim, and B. J. Berne, *J. Chem. Phys.* **114**, 1075 (2001).
16. T. Wu, H.-J. Werner, and U. Manthe, *Science* **306**, 2227 (2004).
17. W. H. Miller, Y. Zhao, M. Ceotto, and S. Yang, *J. Chem. Phys.* **119**, 1329 (2003).
18. R. Schubert, H. Waalkens, and S. Wiggins, *Few-Body Syst.* **45**, 2003 (2009).
19. A. Kohen, R. Cannio, S. Bartolucci, and J. P. Klinman, *Nature* **399**, 496 (1999).
20. J. Basran, M. J. Sutcliffe, and N. S. Scrutton, *Biochemistry* **38**, 3218 (1999).
21. L. Masgrau, A. Roujeinikova, L. O. Johanissen, P. Hothi, J. Basran, K. E. Ranaghan, A. J. Mulholland, M. J. Sutcliffe, N. S. Scrutton, and D. Leys, *Science* **312**, 237 (2006).
22. D. C. Clary, *Science* **321**, 789 (2008).
23. R. P. Feynman and A. R. Hibbs, *Quantum Mechanics and Path Integrals*, McGraw-Hill, New York (1965).
24. H. Kleinert, *Path Integrals in Quantum Mechanics, Statistics, Polymer Physics and Financial Markets*, World Scientific, Singapore (2004).
25. B. J. Berne and D. Thirumalai, *Annu. Rev. Phys. Chem.* **37**, 401 (1986).
26. D. M. Ceperley, *Rev. Mod. Phys.* **67**, 279 (1995).
27. R. Q. Topper, *Adv. Chem. Phys.* **105**, 117 (1999).
28. E. P. Wigner, *Z. Phys. Chem.* **B19**, 203 (1932).
29. M. Ceotto and W. H. Miller, Private communication.
30. J. Vaníček, W. H. Miller, J. F. Castillo, and F. J. Aoiz, *J. Chem. Phys.* **123**, 054108 (2005).
31. W. H. Miller, S. D. Schwartz, and J. W. Tromp, *J. Chem. Phys.* **79**, 4889 (1983).
32. T. Yamamoto and W. H. Miller, *J. Chem. Phys.* **120**, 3086 (2004).
33. M. Ceotto and W. H. Miller, *J. Chem. Phys.* **120**, 6356 (2004).
34. Y. Zhao, T. Yamamoto, and W. H. Miller, *J. Chem. Phys.* **120**, 3100 (2004).
35. C. Venkataraman and W. H. Miller, *J. Phys. Chem. A* **108**, 3035 (2004).
36. Y. Li and W. H. Miller, *Mol. Phys.* **103**, 203 (2005).
37. T. Yamamoto and W. H. Miller, *J. Chem. Phys.* **122**, 044106 (2005).
38. J. Vaníček and W. H. Miller,in C. Burdik, O. Navratil, and S. Posta (eds), *Proceedings of the 8th International Conference: Path Integrals from Quantum Information to Cosmology*, JINR, Dubna (2005).
39. Ceotto, S. Yang, and W. H. Miller, *J. Chem. Phys.* **122**, 044109 (2005).
40. Y. Li and W. H. Miller, *J. Chem. Phys.* **125**, 064104 (2006).
41. S. Yang, T. Yamamoto, and W. H. Miller, *J. Chem. Phys.* **124**, 084102 (2006).
42. J. Vaníček and W. H. Miller, *J. Chem. Phys.* **127**, 114309 (2007).
43. W. Wang, S. L. Feng, and Y. Zhao, *J. Chem. Phys.* **126**, 114307 (2007).
44. W. Wang and Y. Zhao, *J. Chem. Phys.* **130**, 114708 (2009).
45. T. Zimmermann and J. Vaníček, *J. Mol. Model.* **16**, 1779 (2010).
46. M. Buchowiecki and J. Vaníček, *J. Chem. Phys.* **132**, 194106 (2010).

47. W. Wang and Y. Zhao. *J. Chem. Phys.* **132**, 064502 (2010).
48. J. Bigeleisen and M. Wolfsberg, *Adv. Chem. Phys.* **1**, 15 (1958).
49. L. C. S. Melander, *Isotope Effects on Reaction Rates*, Ronald, New York (1960).
50. C. J. Collinsand N. S. Bowman (eds), *Isotope Effects in Chemical Reactions*, Van Nostrand Reinhold, New York (1971).
51. O. Redlich, *Z. Phys. Chem.* **B28**, 371 (1935).
52. E. Teller, quoted by Angus et al., *J. Chem. Soc.* **62**, 971 (1936).
53. D. Frenkel and B. Smit, *Understanding Molecular Simulation*, Academic Press, CA (1996).
54. D. Chandler, *Introduction to Modern Statistical Mechanics*, Oxford University Press, New York (1987).
55. H. Azzouz and D. Borgis, *J. Chem. Phys.* **98**, 7361 (1993).
56. A. Calhoun and G. A. Voth, *J. Phys. Chem. B* **102**, 8563 (1998).
57. I. Feierberg, V. Luzhkov, and J. Aqvist, *J. Biol. Chem.* **275**, 22657 (2000).
58. R. Iftimie and J. Schofield, *Int. J. Quant. Chem.* **91**, 404 (2003).
59. K. Y. Wong and J. Gao, *J. Chem. Theory Comput.* **4**, 1409 (2008).
60. J. P. Bothma, J. B. Gilmore, and R. H. McKenzie, *New J. Phys.* **12**, 055002 (2010).
61. M. H. M. Olsson, P. E. M. Siegbahn, and A. Warshel, *J. Am. Chem. Soc.* **126**, 2820 (2004).
62. H. Liu and A. Warshel, *J. Phys. Chem. B* **111**, 7852 (2007).
63. M. E. Tuckerman and G. J. Martyna, *J. Phys. Chem. B* **104**, 159 (2000).
64. T. Zimmermann and J. Vaníček, *J. Chem. Phys.* **131**, 024111 (2009).
65. D. A. Case, T. A. Darden, T. E. Cheatham, C. L. Simmerling, J. Wang, R. E. Duke, R. Luo, M. Crowley, R. C. Walker, W. Zhang, K. M. Merz, B. Wang, S. Hayik, A. Roitberg, G. Seabra, I. KolossvÃiry, K. F. Wong, F. Paesani, J. Vaníček, X. Wu, S. R. Brozell, T. Steinbrecher, H. Gohlke, L. Yang, C. Tan, J. Mongan, V. Hornak, G. Cui, D. H. Mathews, M. G. Seetin, C. Sagui, V. Babin, and P. A. Kollman, *Amber 10*, University of California, CA (2008).
66. J. A. Barker, *J. Chem. Phys.* **70**, 2914 (1979).
67. M. F. Herman, E. J. Bruskin, and B. J. Berne, *J. Chem. Phys.* **76**, 5150 (1982).
68. M. Parrinello and A. Rahman, *J. Chem. Phys.* **80**, 860 (1984).
69. K. R. Glaesemann and L. E. Fried, *J. Chem. Phys.* **116**, 5951 (2002).
70. C. Predescu and J. D. Doll, *J. Chem. Phys.* **117**, 7448 (2002).
71. C. Predescu, D. Sabo, J. D. Doll, and D. L. Freeman, *J. Chem. Phys.* **119**, 10475 (2003).
72. C. Predescu, D. Sabo, J. D. Doll, and D. L. Freeman, *J. Chem. Phys.* **119**, 12119 (2003).
73. A. Warshel and R. M. Weiss, *J. Am. Chem. Soc.* **102**, 6218 (1980).
74. A. Warshel, *Computer Modeling of Chemical Reactions in Enzymes and Solutions*, John Wiley & Sons, Inc. , New York (1991).
75. Y.-T. Chang and W. H. Miller, *J. Phys. C* **94**, 5884 (1990).
76. C. Minichino and G. A. Voth, *J. Phys. Chem. B* **101**, 4544 (1997).
77. H. B. Schlegel and J. L. Sonnenberg, *J. Chem. Theory Comput.* **2**, 905 (2006).
78. H. S. Johnston, *Gas Phase Reaction Rate Theory*, Ronald, New York (1966).
79. A. I. Boothroyd, W. J. Keogh, P. G. Martin, and M. R. Peterson, *J. Chem. Phys.* **95**, 4343 (1991).

80. A. I. Boothroyd, W. J. Keogh, P. G. Martin, and M. R. Peterson, *J. Chem. Phys.* **104**, 7139 (1996).
81. http://www.cita.utoronto.ca/~boothroy/bkmp2.html.
82. P. L. Fast, J. C. Corchado, and D. G. Truhlar, *J. Chem. Phys.* **109**, 6237 (1998).
83. A. J. C. Varandas, F. B. Brown, C. A. Mead, D. G. Truhlar, and N. C. Blais, *J. Chem. Phys.* **86**, 6258 (1987).
84. B. C. Garrett, D. G. Truhlar, A. J. C. Varandas, and N. C. Blais, *Int. J. Chem. Kinet.* **18**, 1065 (1986).
85. W. R. Roth and J. König, *Justus Liebigs Ann. Chem.* **699**, 24 (1966).
86. J. Wang, R. M. Wolf, J. W. Caldwell, P. Kollman, and D. A. Case, *J. Comp. Chem.* **25**, 1157 (2004).

CHAPTER 4

Activation Energies in Computational Chemistry—A Case Study

MICHAEL BUSCH and ELISABET AHLBERG

Department of Chemistry, Electrochemistry, University of Gothenburg, Gothenburg, Sweden

ITAI PANAS

Department of Chemistry and Biotechnology, Energy and Materials, Chalmers University of Technology, Gothenburg, Sweden

4.1 INTRODUCTION

Computational chemistry program packages provide the chemist with a handy tool complementary to experimental methods for gaining information of a system of interest. However, the simplicity and user-friendliness of modern computational chemistry suites make it tempting to employ the embedded computational methods, even though their applicability to the specific problem might be questionable. Employing a formally improper method to a problem will result in numbers, which might even agree with experimental results, yet they would be right for the wrong reason and render the theoretical mechanistic analysis fundamentally flawed. More often than not, though, it is discovered that the preferred formally correct method is nevertheless not applicable due to computational limitations. In these cases the science of computational chemistry cannot be mainly focused on the numbers obtained from the calculations. Rather, we seek to convert fundamental constraints of applicability into an effective uncertainty of our computed data.

Here we address the case where the full potential energy surface (PES) is inaccessible, although some crucial regions on it are. Thus, the present study is concerned with how to extrapolate from the formally proper domains on a PES in order to learn about activation energies, where the corresponding PES sections of the latter are formally inaccessible. The purpose is to provide a formally straightforward

Rate Constant Calculation for Thermal Reactions: Methods and Applications, Edited by Herbert DaCosta and Maohong Fan.

and computationally simple tool for interpreting chemical processes, of which electrochemistry provides a particularly appealing context. This technique employs an electrode potential as chemical boundary condition, analogous to pressure or temperature, allowing chemical processes that are endergonic under ambient conditions to occur spontaneously.

Arguably one of the most well-known electrochemical reactions is the conversion of water to molecular oxygen [1–7]. Although the energy source is different, the electrochemical and biological OER bare significant similarities. In both cases the rate of conversion is controlled by catalysts. The biological photoassisted OER is employed by green plants to produce reducing equivalents for conversion of CO_2 into carbohydrates [8–12]. The apparent efficiency of the green plants to perform the photosynthesis has provided great inspiration to the field of biomimetics [5,13–19]. Yet, it is not clear whether the conditions under which biocatalysts function are also suitable for industrial application.

Our objective here is partly to contrast the conditions for working OER electrocatalysts in industry with the biomimetic catalysts based on the photocatalyst found in the green plants. When comparing the local designs of the working electrocatalysts in photosystem II (PSII) Mn_4O_4Ca nanocatalyst [20–23] to industrial catalysts such as the dimensional stable anode (DSA) based on (Ti, Ru, Ir)O_2 coating on a titanium electrode [2,24,25], there are clear similarities. For example, although the latter is a macroscopic object, it is well known from electron microscopic and electrochemical studies that the surface is nanotextured [2,26–28]. In photosystem II the OER turnover rates are very high [12]. It has been found that the O_2 release occurs in a concerted reaction step when removing the last proton/electron couple from the Mn–OH moiety by an intramolecular O–O bond formation between two Mn=O groups [8]. The understanding for this central reaction step has been improved by a recent computational study that reports on how the O–O bond formation occurs between a free Mn–O$^\bullet$ and a bridging Mn–O–Ca oxygen. This step was found to be endergonic and thus suggested to be rate limiting. The barrier was found to be 6.8 kcal/mol (=0.3 eV). The alternative nucleophilic pathway was found to inherit a reaction barrier of 28.1 kcal/mol (=1.2 eV) [11].

In contrast to the photoinduced water oxidation, the turnover rate of the electrochemical OER is decided by different kinetics and thermodynamics as determined by the applied potential. On water oxidation, surface oxy species are assumed to form independently. In the limit of high transition metal (TM) TM-oxy coverage, lateral recombination of oxy species to produce O_2 becomes a competing mechanism to said nucleophilic attack.

The price tag for the high efficiency of the manganese tetramer found in photosystem II is its chemical stability. The biocatalyst is found to be unstable toward damage of the enzyme by light or chemical stress. Thus, the enzyme must be replaced or repaired after typically 20 min [29,30]. One such channel for decomposition is dissolution of manganese ions [30]. Comparing this to the DSAs, which display a lifetime of about 10 years [2], biomimetic catalysts trade robustness for energy efficiency. One way to cure for the lack of stability might be to design a possible catalyst such that a self-repair mechanism can take place, as has been supposed to occur on a recently found homogenous cobalt catalyst [31,32].

The present expose reports on recent first-principles results for a high-current-density reaction path. It hopes to contribute to the improvement of existing processes by analyzing the relevant mechanisms involved in the OER. Here we focus on the reactivities of intermediates along a reaction path that involves the electrochemical formations of two adjacent oxy species on a binuclear catalyst. The activation energy for the rate-limiting chemical step, whereby a surface peroxy species is formed, is analyzed in terms of the normal coordinates for reactant and product. Central findings of the present study include (i) how softness of the associative force constant of the reactant is decisive and (ii) how the interoxygen distance in the reactant structure affects the activation energy. These qualities are demonstrated for a molecular binuclear catalyst and contrasted by the corresponding μ-OH bridged manganese and cobalt sites embedded in a quasi-infinite magnesium oxyhydroxide rig.

4.2 CONTEXT AND THEORETICAL BACKGROUND

4.2.1 Density Functional Theory

Contemporary applied first-principles theory is based on density functional theory (DFT). The cornerstones of this theory comprise the Hohenberg–Kohn theorems [33] that (a) employ the variational principle of quantum mechanics to demonstrate that all the properties of a nondegenerate ground state wave function are uniquely reflected in the corresponding electron density that in turn determines the external potential $v_{\text{ext}}(r)$ given by the nuclei and (b) demonstrate that there exists a variational principle for said ground state electron density. These two theorems imply the existence of a universal density functional $v_{\text{xc}}(r)$ dictating how an electron density of a nondegenerate ground state may be converted into an energy density. In the Kohn–Sham equations (see Eq. (4.1)) [34], the DFT is cast into a practical tool:

$$\left(\frac{\hbar}{2m}\nabla^2 + v_{\text{ext}}(r) + v_{\text{H}}(r) + v_{\text{xc}}(r)\right)\Psi_i = \Psi_i(r)\varepsilon_i \tag{4.1}$$

Employing a noninteracting electron reference state, the unknown functional is reformulated into a local density-dependent exchange–correlation potential, that is, $v_{\text{xc}}(r)$ is determined by the electron density at the very spot where it is acting. This is the local density approximation (LDA) for which $v_{\text{xc}}(r)$ was parameterized for the electron gas by means of quantum Monte Carlo simulations [35]. The generalized gradient approximation (GGA) [36,37] was introduced in order to account for nonuniformity in the electron density and found to improve on the results. For molecular systems the Becke gradient correction [38,39] has been very successful owing to the proper r^{-1} dependence of the exchange potential as opposed to the exponential decay in the electron gas. Significant improvements in numerical agreements with experiments are often seen on adding some portion of Hartree–Fock-type exchange interaction into the DFT framework. This is the so-called hybrid

DFT approach [40–42]. It is introduced in order to at least partly cure for some of the self-interaction errors (SIEs). This flaw of DFT is owing to the locality of the exchange interaction potential, which in Hartree–Fock theory cancels the Coulombic interaction for a single electron interacting with itself. This cancellation in Hartree–Fock theory originates from the equivalent orbital origins of the "self-Coulomb" and "self-exchange" terms for a single electron [43]. Adding some percentage of "exact exchange" is generally found to improve the agreement with experiment [44,45]. While the alleviation of some self-interaction improves the description of localized electrons, still significant self-interaction remains in the hybrid functional. Indeed, attempts to remove the self-interaction error have been found to render the numbers worse [46]. It is often said that this error to some extent compensates the lack of true multiconfigurational nature of the chemical bonding [47,48] when taking the single-determinantal wave function ansatz for building the electron density. Chemistry and indeed computational chemistry too often deal more with trends than with absolute numbers. Differences in trends between GGA and hybrid-GGA functionals are often found in systems that display weak localizations. In the solid state this effect is well understood and it is manifested in significant underestimations of band gaps. However, when the chemistry is unconditionally local, the two methods are generally found to be in qualitative agreement [49]. For a more elaborate review on DFT, the reader is referred to the papers by Kohn et al. [50,51].

4.2.2 Calculating Transition States

An inherent potentially complicating problem of DFT concerns transition states. DFT is defined for the nondegenerate case alone. This is because when an explicit multiconfigurational wave function is required, the Slater determinant may be insufficient for describing the electron density and consequently also the external potential. Luckily, this is a rare situation in case of stable molecular structures. Yet, at interstates' crossings [52], for example, between reactant and product, the validity of DFT is challenged. It turns out, though, that DFT can be found to apply also in such situations if the transition between the two states employs an avoided crossing such that the coupling between the two states is sufficiently strong. Such a situation was discussed by Lundin et al. [53].

One remaining notorious problem comprises molecular bond breaking, for example, into a biradical state, which can be generalized to include avoided crossings exhibiting small interstate couplings. In such cases, the transition state becomes truly inaccessible to contemporary DFT. Yet, for both cases, the barrier can be estimated from the properties of reactants and products, thus building on the v representability [54] of the computed electron density at the optimum structures of reactants and products. Here, we propose an approach that employs the force constants and associated normal coordinates, determined for said structures, to estimate the transition state. Moreover, if the density of vibrational states is very high, that is, $E_a \gg h\nu$, then the reaction barrier can also be estimated from the crossing of the two harmonic potentials. Based on these assumptions, an analytical expression for the activation energy E_a can be estimated from the effective harmonic oscillators of reactant and product potential energy surfaces projected on the reaction coordinate. Let

$$E_{\text{prod}} = \frac{k_{\text{prod}}}{2} x^2 \tag{4.2}$$

$$E_{\text{react}} = \frac{k_{\text{react}}}{2} (x - \Delta x)^2 + D \tag{4.3}$$

where $\Delta x = x_{\text{react}} - x_{\text{prod}}$ is the difference between the corresponding equilibrium geometries along the reaction coordinate and D is the energy difference between the product and reactant stabilities (positive D implies exothermic reaction, i.e., $D = -\Delta E$). Solving for $E_{\text{prod}} = E_{\text{react}}$ produces

$$x_{\text{TS}} = -\frac{k_{\text{react}}}{2\gamma} \Delta x \pm \sqrt{\left(\frac{k_{\text{react}}}{2\gamma} \Delta x\right)^2 + \frac{k_{\text{react}}}{2\gamma} (\Delta x)^2 + \frac{D}{\gamma}} \tag{4.4}$$

where x_{TS} is the displacement along the reaction coordinate from the product toward the reactant and

$$\gamma = \frac{k_{\text{prod}} - k_{\text{react}}}{2} \tag{4.5}$$

The corresponding activation energy E_a becomes

$$E_a = E_{\text{prod}}(x_{\text{TS}}) - D \tag{4.6}$$

Use of the above expressions is illustrated below for the rate-limiting chemical step in the electrochemical oxygen evolution reaction owing to water oxidation. Indeed, the present study comprises part of a comprehensive investigation concerning the catalytic activity of dinuclear Mn sites toward water oxidation. The reaction channel under consideration employs the two sites symmetrically, assuming the dehydrogenation of molecular water to occur on the two sites independently. Because all steps beyond the oxy-formation steps are chemical, they cannot be electrochemically driven and the regeneration of the catalyst occurs by association of the two adjacent oxo species and subsequent desorption via a superoxo intermediate. Here we produce insight into the reactivity of the first chemical step, that is, the transition from dioxo to a μ-peroxo moiety. This reaction is understood to be determined by two components: (a) the radical character of the oxy species and (b) their proximity in relation to the elastic properties of the embedding surrounding of the binuclear site. The degree of double bonding character in the Mn=O system has been subject to much debate in the literature [9,18,55,56]. The subtle distinction can be appreciated by the fact that the two different computational implementations of the same density functional DMol^3 [57,58] and CASTEP [59], employing numerical basis functions and plane waves to span the Kohn–Sham orbitals, produce slightly different Mn–O bond distances, that is, 1.67 and 1.73 Å, respectively. The purely single-bonded Mn–O$^{\bullet}$ is found at 1.85 Å. The stabilization introduced by the double bonding character is of the order of 0.9 eV.[1] In order to learn about the reactivity of the latter, the Co–O is employed as a simplified model of the single-bonded Mn–O system. The validity of this model is discussed in Ref. [49]. The ability of the catalyst support to deform in order to facilitate the Mn–O–O–Mn peroxy bond formation is the key aspect of the present study. A related issue concerns how the Mn=O bond affects the effective force constant for associative deformation in conjunction with length of the reaction coordinate toward peroxy bond formation.

Consequently, in the present study, the proposed formalism is demonstrated for the rate-limiting chemical step in the OER, that is, 2[Mn=O $\leftrightarrow$ Mn–O$^{\bullet}$]$\rightleftharpoons$ Mn–O–O–Mn. Above, the force constants of the dioxo and peroxo species comprise k_{react} and k_{prod}, respectively. By solving for x_{TS} and adding the equilibrium O–O bond lengths of the peroxo system, the interatomic O–O distance at the transition state is obtained. In what follows, the reactivity of a biomimetic binuclear molecular manganese catalyst will be contrasted by that of the binuclear Mn site supported on a magnesium oxyhydroxide rig.

4.2.3 The Tyrosine/Tyrosyl-Radical Reference Potential

One of the central problems in studying redox reactions by means of quantum chemistry is the choice of a suitable reference reaction. Similarly to electrochemical studies, where one is relatively free in the choice of reference, in principle, any redox reaction can be chosen. The most natural choice from an electrochemical point of view is the formation of molecular hydrogen from two protons and two electrons, thus connecting the results to the normal hydrogen electrode (NHE) [60]. Such an approach has successfully been employed for studying the OER on various metal oxides by Rossmeisl et al. [4]. However, in molecular quantum chemistry, major improvements in computed numbers are achieved if isodesmic reactions are employed, that is, the character and number of bonds broken being the same as those formed. Moreover, an appropriate reference system must act as both an electron and a proton sink in the case of modeling the OER, in order to avoid the nontrivial modeling of a solvated proton intermediate, which would add an additional source of errors. Here the tyrosine/tyrosyl-radical (TyrOH/TyrO$^{\bullet}$) redox couple mirrors the TM–OH/TM–O$^{\bullet}$ oxidation reaction qualitatively both in character and in any microsolvation. In addition, because the catalysts considered here can be claimed to be biomimetic, a reference system such as the recently used TyrOH/TyrO$^{\bullet}$ redox couple [49,61] can be expected to be more convenient since it provides a link to biological systems where it is known to be involved in the OER inside photosystem II [12,62,63].

In aqueous solution the redox potential of the TyrOH/TyrO$^{\bullet}$ couple has been measured to be 0.94 V versus NHE [64,65]. Compared to the 1.23 V versus NHE thermodynamically required to oxidize water, one can argue that the reference system is too weak to drive the OER. The model systems reviewed here nevertheless neglect

[1] Result taken from the CASTEP calculations.

any solvent effects. This means that the first step of removing molecular water out of bulk water and putting it into vacuum has to be added. The energy costs for this step can be approximated from the enthalpy of vaporization, which is about 40 kJ/mol [66] (0.4 eV/molecule) or 0.2 eV per redox reaction step. Thus, the resulting potential is close to the thermodynamic potential for water oxidation.

In principle, a similar argument should also apply to the OER in photosystem II if this reaction is assumed to take place in a hydrophobic cavity. Thus, in order to reach the catalyst, the water molecules subject to oxidation first have to be removed from the bulk water. While this step is artificial in the case of the reviewed systems, it would be entropy driven in photosystem II.

4.3 COMPUTATIONAL DETAILS

The force constants for the embedded transition metal oxide systems were extracted from frequency calculations using the DMol^3 code [57,58] as implemented in the Materials Studio suite [67] within periodic boundary conditions. The size of the unit cell was chosen sufficiently large as to avoid artificial interference between the reactive sites. All calculations were performed at the GGA PBE [37] level of theory using a double-ζ numerical basis set with polarization functions (DNP) and a $1 \times 1 \times 1$ k-point set. The convergence of the wave function was improved by introducing a smearing of 0.3 eV in the vicinity of the Fermi energy. The core electrons were approximated by hardness conserving semilocal pseudopotentials [68].

The force constants for the free dimer system were obtained from frequency calculations using the Gaussian 03 code [69]. All calculations were performed using the B3LYP [40,41] hybrid functional in combination with a triple-ζ 6-311++G basis set, including diffuse functions on all atoms.

For the $MgO_x(OH)_y$ embedded dinuclear transition metal system, reference plane wave calculations were also performed by means of the CASTEP [59] program package. These calculations employed the GGA PBE [37] functional in combination with an energy cutoff of 400 eV and a $1 \times 1 \times 1$ k-point set. The core electrons were approximated by an ultrasoft pseudopotential [70].

For both the periodic boundary and the free dimer calculations, any solvent effects were neglected.

4.4 RECENT ADVANCES AND NEW RESULTS

4.4.1 Homogenous OER Catalysts

A simplified model has been developed to investigate the simplest complete OER reaction cycle. It is inspired by the biomimetic manganese dimers used for studying the water oxidation reaction in biological systems. The interest in these systems is twofold since they offer candidates for possible future water oxidation catalysts [61].

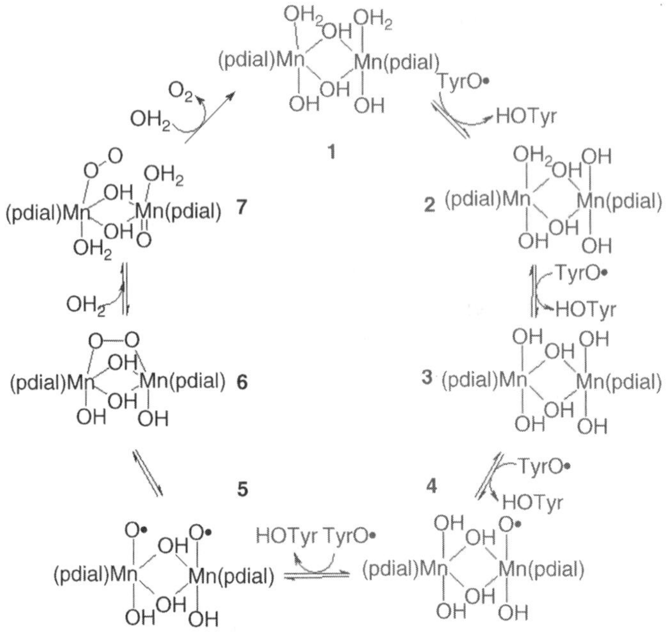

FIGURE 4.1 The full mechanism of the OER discussed in Ref. [61].

FIGURE 4.2 Overview over the energetics of the mechanism shown in Figure 4.1. Adapted from Ref. [61].

The catalyst investigated here is built up from two manganese ions bridged by two μ-OH moieties. The dimer is terminated by two hydroxides, two pdial ligands,[2] and two water molecules, which are subject to subsequent deprotonation (see Figure 4.1, structure **1**).

The suggested mechanistic OER cycle shown in Figure 4.1 can be divided into electrochemical oxidation steps and subsequent chemical steps (Figure 4.2). The four electrochemical steps comprise the removal of one electron and one proton per step. Starting from structure **1** in a nonet spin state, the first main intermediate formed after

[2] pdial is a simplified version of an acetylacetonate ligand, with two methyl groups being removed.

the first two electrochemical steps is structure **3**. It contains two $3d^3$ Mn(IV)–OH moieties. With respect to the TyrOH/TyrO$^\bullet$ redox couple, the first step is found to be strongly exothermic, while the second step is endothermic. The energetics of these two steps were found to depend strongly on the description of the SIE of contemporary DFT, as demonstrated by comparing B3LYP to BLYP results. The sensitivity is owing to the artificial delocalization due to the SIE of the unpaired electrons when subjected to weak localization. As the chemical interactions between the transition metal sites and the substrate moieties increase, the differential effect between the two methods was demonstrated to be much reduced. The subsequent two H^+/e^- oxidation steps leading to the formation of two [Mn(IV)–O$^\bullet$↔Mn(V)=O] moieties (see structure **5**) are found to be endothermic by about 2 eV in total, rendering these steps rate determining. Although the electronic structure of the Mn–O bond is still under discussion [9,18,55,56], there is evidence to suggest that the Mn–O bond displays a partial radical character on the oxygen atom, as reflected in a partial double bond. The amount of stabilization due to a Mn=O double bond can be estimated by comparing the energy of structure **5** in the quintet state Mn=O with that of the septet. In case of a septet the system is forced to exhibit six unpaired electrons with parallel spin, thus forcing the system to form one Mn(IV)–O$^\bullet$ moiety. The amount of stabilization introduced by the partial double bond nature is of the order of 0.3 eV. In the subsequent step the O–O bond is assumed to be formed between the two Mn(IV)–O$^\bullet$ moieties. This reaction is exothermic by -1.32 eV. As a consequence, this reaction step should be unproblematic from a thermodynamic point of view. Before addressing the kinetic aspects of this intermediate reaction, we conclude that the last two steps of the catalytic cycle assume the release of O_2 to be accommodated by the consecutive substitution of Mn–O bonds with water. In agreement with experiment [71], these steps were found to allow for spontaneous evolution of triplet O_2. As a possible way to allow for changing the spin state of the $O_{2(ads)}$, from singlet to triplet, it has been suggested that the manganese dimer acts as a spin catalyst [61,72].

So far, the mechanism has only been discussed from a thermodynamic viewpoint. Although this approach gives valuable information about critical steps, activation barriers are required in order to discuss rates. As described above, the transition state energies are estimated from the crossings of the harmonic oscillator potentials corresponding to the vibrational normal modes, which point toward the transition state. For the free dimer three vibrational modes are identified to be relevant for calculating the activation barrier of the O–O bond formation step. These modes are the O–O stretch vibration at 855 cm^{-1} for the μ-peroxo-bridged product structure, the symmetric bending mode around the axis through the two μ-OH ligands at 93 cm^{-1} in case of the dioxo reactant structure, and an Mn–Mn stretch-related vibration at 369 cm^{-1} for the latter structure. Owing to the low vibration frequencies found for the dioxo intermediate, the activation energy can be expected to be mainly classical in nature, that is, determined by the potential energy surfaces of reactant and product subject to the Born–Oppenheimer approximation. The soft mode at 93 cm^{-1} having a force constant of only 4.5 N/m is found to lead to an activation barrier of only 0.3 eV, which is significantly lower than what was found for the rather stiff mode at 369 cm^{-1}, displaying a force constant of 87.2 N/m that results in an activation energy of 3.4 eV.

TABLE 4.1 Summary of the O–O Bond Distances $d_{O\text{–}O}$ (in Å), Force Constants k (in N/m), and Calculated Activation Energies E_a (in eV)

Structure	Embedding	ν (cm^{-1})	$d_{O\text{–}O}$ (Å)	k (N/m)	E_a (eV)
Mn–O–O–Mn	Free dimer	855	1.46	395	
$(Mn{=}O)_2$	Free dimer	369	3.19	87.2	3.4
		93	3.19	4.5	0.3
Mn–O–O–Mn	$MgO_x(OH)_y$	813	1.44	565	
$(Mn{=}O)_2$	$MgO_x(OH)_y$	130	2.99	39.5	1.67
$(Mn\text{–}O^{\bullet})_2$	$MgO_x(OH)_y$	118	2.44	23.5	1.12[a]
Co–O–O–Co	$MgO_x(OH)_y$	723	1.46	783	
$(Co\text{–}O^{\bullet})_2$	$MgO_x(OH)_y$	127	2.47	34.3	0.64

[a] Reaction sequence: quintet Mn=O → nonet Mn–O• → Mn–O–O–Mn.

Based on these results, it is expected that the O–O bond formation proceeds without significant barrier in the free manganese dimer discussed here, owing to the flexibility of the molecular catalyst. In fact, the 0.3 eV activation barrier is similar to what has been proposed for the intramolecular O–O bond formation in photosystem II [11] (Table 4.1 and Figure 4.3).

4.4.2 Embedded Transition Metal Dimers

In order to realize a true biomimetic industrial catalyst, it is necessary to mount the molecular catalytic sites onto surfaces. This was attempted by means of theory in a quest to extend the insights gained for a free dimer shown in Figure 4.1 by embedding the μ-hydroxo-bridged dimer part onto a $MgO_x(OH)_y$ test rig [49].

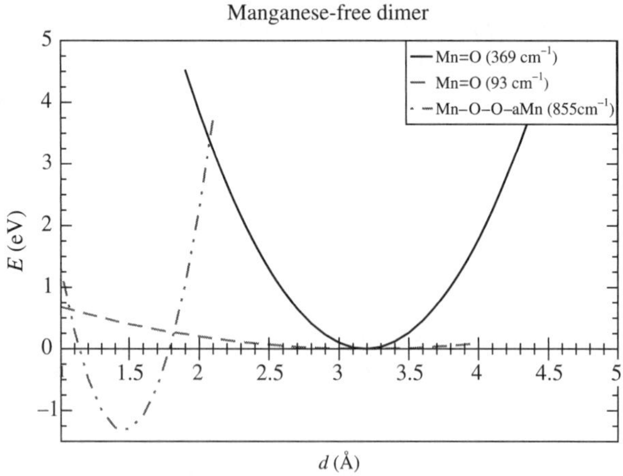

FIGURE 4.3 Plot of the relevant normal modes' PES. The activation energies can be calculated from the interception of the Mn=O and Mn–O–O–Mn PES.

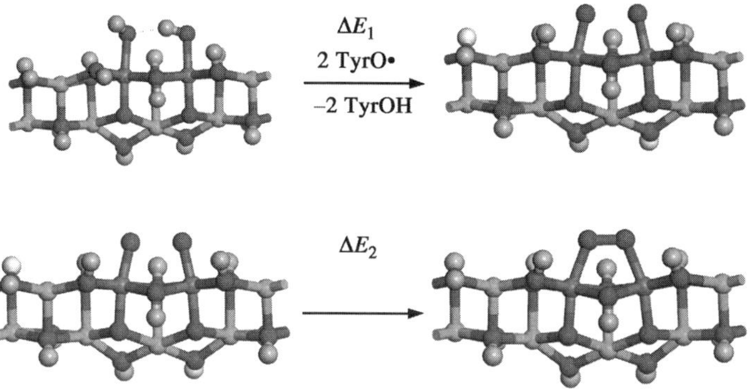

FIGURE 4.4 The reaction steps discussed for Cr, Mn, Fe, and Co embedded into a $MgO_x(OH)_y$ support. Step 1 comprises the oxidation of two TM–OH units to from two TM=O moieties. The second step considered here is the intramolecular O–O bond formation between these two TM=O groups. Adapted from Ref. [49].

The test rig used in this study was a one-dimensional $MgO_x(OH)_y$ wire containing in total six magnesium oxyhydroxide units (see Figure 4.4). This approach should allow for computationally inexpensive modeling with semiquantitative accuracy. A similar ansatz was successfully employed to address oxygen reduction [73]. The convenience in choice of the model can be appreciated from the fact that the oxidation state of the active sites could be adjusted simply by hydrogenation/dehydrogenation of oxygen atoms on the rig.

Using the described system, the OER was modeled for various transition metal oxides, such as chromium, manganese, iron, and cobalt [49]. The choice of embedded catalysts was in order to search for any possible existence of a left–right trend in catalytic performances among the transition metals in the periodic table. In this sense, the Cr(III–V) systems can be seen as a representative of a typical early transition metal, owing to their rich redox chemistry and the preference for high oxidation states. The Mn(III–V) system, which is known to be a more or less active OER catalyst in both homogenous catalysis [15,16,74] and electrochemistry [75,76], is chosen to be able to transfer the results obtained for the free dimer to the rig. The Fe(II–IV), which is known to be a less active catalyst [77], and Co(II–IV), which is an active catalyst [31,32,78], are chosen as representatives of typical late transition metals. The modeling of the OER was restricted to two steps: the oxidation of the transition metal TM–OH to TM=O modeled as a single two-electron oxidation step in conjunction with a subsequent O–O bond formation between two TM=O moieties (see Figure 4.4). For a free dimer it was shown that the rate-determining step can be expected to comprise the oxidation of the two TM–OH moieties, that is, the main electrochemical step. The subsequent O–O bond formation step was modeled because it comprises the critical first chemical step. It expresses a drive for TM=O decomposition that is unique to every transition metal ion. It is gratifying to note that a GGA, which is convenient in systems displaying periodic boundary conditions, is reliable in describing these steps. This was demonstrated in case of the manganese dimer system

TABLE 4.2 Comparison of the Reactivity of the Various Transition Metals on a $MgO_x(OH)_y$ Embedding

System	ΔE_1 (eV)	ΔE_2 (eV)
Cr(III–V)	−1.21	1.25
Mn(III–V)	1.18	−0.84
Fe(II–IV)	−0.61	0.75
Co(II–IV)	1.06	−0.75

The results are taken from Ref. [49].

by comparing the results for the pure GGA BLYP functional to those of the corresponding molecular system obtained with the hybrid B3LYP functional.

The results obtained for the various TMs are compared in Table 4.2. The main result of the study comprises the fact that the metals can be divided into two classes. The first group of metals includes Cr(III–V) and Fe(II–IV). These display oxidation steps that are exothermic with respect to the TyrO•/TyrOH reference system in conjunction with an endothermic O–O bond formation step. This is in contrast to the second group, containing Mn(III–V) and Co(II–IV), which has two consecutive endothermic single-electron oxidation steps and a subsequent strongly exothermic O–O bond formation step. The different chemical behavior of the two classes has been explained by the electronic nature of the various TM–O bonds. In case of the former class it was concluded that the reactivity is determined by the double bonding character of the TM=O, which stabilizes the intermediate. Due to the stabilization of this key intermediate, the O–O bond formation becomes endothermic. This is opposite to the behavior of the latter class of TMs that become destabilized by a partial radical nature, resulting in an activated intermediate.

Again, in what follows, the influence of the transition state energy on the overall rate for the initial chemical step toward regeneration of the dinuclear Mn catalyst is addressed. Here we build on the fact that one of the main features of the free manganese dimer leading to a very low activation energy for the O–O bond formation step is the flexibility of the complex to support O–O bond formation by bending around the axis defined by the two μ-OH ligands. This effect is expected to be reduced in case of an embedded catalytic site.

However, in case of the embedded manganese dimer, the computed surface phonons supporting the O–O bond formation are found at 130 cm^{-1} (free dimer: 93 cm^{-1}) for the dioxo species and 813 cm^{-1} (free dimer: 855 cm^{-1}) for the μ-peroxo intermediate, respectively. Hence, again the bending mode along the reaction coordinate for O–O bond formation is a low-frequency one, yet, for both vibrational modes, the force constants are found to be much higher compared to the free dimer. In case of the dioxo species the force constant is found to be 39.5 N/m, which is nearly 10 times higher than that found for the free dimer (here $k = 4.5$ N/m). A smaller effect was found for the μ-peroxo intermediate. Here the force constant is increased by a factor 1.6 from 355 N/m (free dimer) to 565 N/m (embedded manganese catalyst). As a result, the transition state energy is increased to 1.67 eV, which would render this reaction channel improbable.

In order to understand this result better, the embedded binuclear cobalt catalyst was considered. Here the force constants are of similar stiffness compared to the embedded dimanganese system. The relevant vibrational modes for this system are found at $126\,cm^{-1}$ for the dioxo species and $723\,cm^{-1}$ for the μ-peroxo species. The former frequency is similar to that found for the embedded manganese dimer, indicating a similar stiffness of this vibration, which in fact is also found ($k = 34.3$ N/m). The μ-peroxo stretching mode is shifted to much lower wave numbers, yet the force constant is found to be significantly higher than that in case of the embedded Mn dimer ($k = 783$ N/m). As a result, one could expect to obtain a similarly high activation energy compared to that found for the embedded manganese dimer. Interestingly, though, the extrapolated activation energy becomes ~0.6 eV. The dramatic difference between the Mn and Co systems despite comparable force constants can be explained by the difference in the interoxy distance found for the dioxy intermediates. In the case of the cobalt system the distance between the two TM=O moieties is found to be 2.47 Å, while this distance is 2.99 Å for the embedded manganese dimer. What adds to the complication in evaluating the latter is that the interoxy distance for this intermediate seems to be strongly computer code dependent. Hence, employing the CASTEP plane wave code, the interoxy distance is found to be only 2.13 Å, which would, assuming similar force constants, result in a much lower activation barrier. This ambiguity is contrasted by the embedded cobalt system for which the qualitative agreement between the two DFT implementations is obtained (CASTEP: 2.35 Å; $DMol^3$: 2.47 Å). These differences nevertheless provide a possible clue as to the origin of the behavior found on the embedded manganese system. The strong deviation between the CASTEP and the $DMol^3$ geometry is attributed to the well-established multiconfigurational nature on the Mn(V)=O bond consisting of a superposition of Mn(V)=O double bond and a Mn(IV)–O• single-bonded state [9,18,55,56,61]. These two states can be expected to lie relatively close in energy, but belong to two different DFT PESs. This is due to the single-determinantal wave function ansatz in DFT where the two very different geometries can be seen to reflect this problem. Hence, the $DMol^3$ result, exhibiting a very short Mn–O bond distance of only 1.67 Å, represents the Mn=O double bond character of the real wave function. The CASTEP results, on the other hand, are argued to emphasize the more Mn(IV)–O• character as can be seen from the relatively long Mn–O bond of 1.79 Å. Indeed, performing a $DMol^3$ geometry optimization for the nonet state, which effectively causes Mn–O single bonds, renders the interoxy distance 2.40 Å. Thus, it is argued that the barrier of 1.67 eV found for direct transition between the Mn(V)=O dioxo species and the peroxo species is owing to an unbalanced interpretation of the peroxy bond formation mechanism in that it does not take into account the single-bonded contribution in a proper way. A better description of this reaction should proceed in two steps. The first step comprises a transition from the Mn=O double bond PES to the Mn–O• radical PES. In fact, this reaction step should also include the transition from a quintet spin state, which is argued to be the ground state of the manganese dioxo species [61], to a nonet spin state. The energy difference between these two states is found to be of the order of 0.9 eV using the CASTEP code. Beyond this point the Mn–O system qualitatively acquires the properties of the

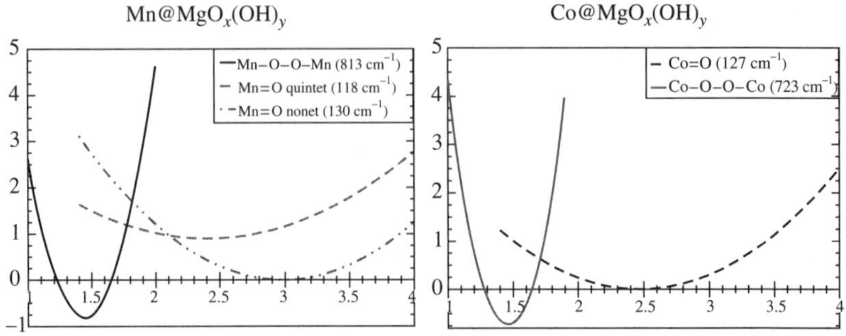

FIGURE 4.5 Plot of the relevant normal modes' PES. The activation energies can be calculated from the interception of the TM=O and TM–O–O–TM PES. While there is a direct transition found in the cobalt system, the O–O bond is formed subsequent to the Mn=O bond breaking, leading to a Mn–O$^{\bullet}$.

embedded cobalt system that displays solely radical dioxo character. The residual activation energy obtained for this nonet state is found to be approximately 0.2 eV (Figure 4.5).

Based on the above discussion, four main influences on the height of the activation barrier toward the peroxy bond formation can be identified:

1. The first factor can be attributed to the flexibility of a potential catalyst. This contribution is mainly controlled by the rigidity of the support. A more rigid support such as the $MgO_x(OH)_y$ embedding discussed in Ref. [49], which enforces a relatively high force constant, could result in a comparably high activation barrier depending on the length of the bending mode. In the limit of a free dimer, a force constant for this bending mode close to zero was obtained, which resulted in peroxy bond formation without significant barrier.
2. The second major influence is attributed to the interoxy distance. This quality is emphasized strongly here as the activation energy toward peroxy bond formation, when comparing the embedded 2 × Mn=O system to the results of the Co–O$^{\bullet}$ and Mn–O$^{\bullet}$ moieties, was shown to depend sensitively on the interoxy distance in the dioxo species. Indeed, while the force constants are found to be similar, the activation barriers differ strongly owing to the different displacements. Having said this, it should be born in mind that although the interoxy distance in the dioxo species is found to greatly influence the reactivity in case of embedded systems, in case of the free manganese dimer it was found to be negligible despite an interoxy distance of 3.19 Å. This can be attributed to the very weak bending mode resulting in a very flat potential energy surface.
3. The third factor influencing the activation barrier is the electronic nature of the TM–O bond. In case of a more double-bonding-like character, the O–O bond formation is unlikely to occur, in that this reaction would imply the simultaneous breaking of two bonds on forming the peroxy species. Instead, the

TM=O groups are expected to form TM–O˙ moieties, which allows for a subsequent structural relaxation reducing the interoxy distance, thus facilitating the subsequent peroxy bond formation. This two-step mechanism was proposed to lower the activation barrier significantly in the case of the embedded manganese system.

4. A fourth catalyst engineering aspect, not investigated here, can be envisaged. It concerns how the interoxy distance is influenced by the TM–TM distance.

While the above insights are helpful as starting point for analysis, obviously the four factors mentioned are not independent and said analysis is often complicated by the fact that all four simultaneously cooperate to decide on the resulting efficiency of a catalyst.

4.5 CONCLUDING REMARKS

The conditions for the applicability of DFT to the intermediates along the water oxidation reaction were addressed in a previous study and were reiterated here [61]. But if the applicability of DFT is already questionable for describing the intermediates, how can it be used to say anything meaningful for the transition states, which involve nonavoided crossings and thus are formally inaccessible? It was argued that, if any points along the reaction coordinate were accessible by DFT, these would comprise the stationary states' local minima on the potential energy surfaces. In general, however, the ambiguities between the various PESs render the use of traditional transition state search methods such as the synchronous transit method [79,80] or the nudged elastic band method [81] difficult. Yet, given that the stationary states are local ground states and building on the fact that all properties of a nondegenerate ground state are properties of the ground state electron density, it follows that the vibrational spectrum and the Hessian matrix taken at said optimal atomic positions are also true properties of the DFT. A straightforward way to employ this property is to estimate the reaction barriers from the crossings of the harmonic oscillator potentials for the modes in reactant and products directed toward the transition state. Such results were demonstrated in the present study.

This approach for estimating activation barriers might be employed under one central condition: the reactant mode under consideration has to be in the quasi-classical limit, given the Born–Oppenheimer approximation. This means that the density of vibrational states is high, corresponding to low wave numbers as compared to the activation energy. It must nevertheless be emphasized that even though the frequency is found to be low, the force constant might still be significant due to a large reduced mass of the oscillator. For uncomplicated cases this implies that the barrier can be traced to the ground state reactant harmonic oscillator. It was demonstrated though that absurdly high reaction barriers could be obtained if these involved multiple bond breakings. A procedure to monitor the influence of consecutive bond breaking on the barrier height was described and evaluated by comparing the partially double-bonded Mn=O system to the single-bonded Co–O. It implies evaluating force constants, normal coordinates, and optimal geometries for states with increasing spin multiplicities to assure the bond breaking. Based on this methodology, two main influences on the activation barrier height for the O–O bond formation step in the OER were identified:

1. First, the flexibility of a potential catalyst. This reduces the activation barrier in case of a flexible molecular catalyst, which can support the O–O bond formation by bending.
2. Second, the interoxy distance in the TM=O intermediate. This quality is of importance in case of rather stiff vibrational modes, which can mainly be found in embedded systems while they are found to be negligible in case of a flexible molecular catalyst.

To summarize, a handy procedure to estimate barrier heights for reactions involving avoided crossings has been described and evaluated for the dioxo to μ-peroxo step on a biomimetic binuclear manganese catalyst and a similar system embedded into a $MgO_x(OH)_y$ rig has been discussed. Its applicability is determined by the ability of DFT to describe well-defined intermediates along a preselected reaction path.

ACKNOWLEDGMENTS

This work was supported by the European Commission through the FP7 Initial Training Network "ELCAT" (Grant Agreement No. 214936-2) and the Adalbertska Stipendiestiftelsen.

REFERENCES

1. J. Bockris and T. Otagawa, *J. Phys. Chem.* **87**, 2960 (1983).
2. S. Trasatti, *Electrochim. Acta* **45**, 2377 (2000).
3. M. Yagi and M. Kaneko, *Chem. Rev.* **101**, 21 (2001).
4. J. Rossmeisl, Z.-W. Qu, H. Zhu, G.-J. Kroes, and J. Nørskov, *J. Electroanal. Chem.* **607**, 83 (2007).
5. C. Mullins and V. Pecoraro, *Coord. Chem. Rev.* **252**, 416 (2008).
6. H. Yamazaki, A. Shouji, M. Kajita, and M. Yagi, *Coord. Chem. Rev.* **254**, 2483 (2010).
7. H. Dau, C. Limberg, T. Reier, M. Risch, S. Roggan, and P. Strasser, *ChemCatChem* **2**, 724 (2010).
8. C. Hoganson, M. Pressler, D. Proshlyakov, and G. Babcock, *Biochim. Biophys. Acta* **1364**, 170 (1998).
9. P. Siegbahn and R. Crabtree, *J. Am. Chem. Soc.* **121**, 117 (1999).
10. J. Robblee, R. Cinco, and V. Yachandra, *Biochim. Biophys. Acta* **1503**, 7 (2001).

11. P. Siegbahn, *Chem. Eur. J.* **14**, 8290 (2008).
12. E. Sproviero, J. Gascn, J. McEvoy, G. Brudvig, and V. Batista, *J. Am. Chem. Soc.* **130**, 3428 (2008).
13. N. Kitajima, M. Osawa, M. Tanaka, and Y. Moro-oka, *J. Am. Chem. Soc.* **113**, 8952 (1991).
14. N. Kitajima, U. Singh, H. Amagai, M. Osawa, and Y. Moro-oka, *J. Am. Chem. Soc.* **113**, 7757 (1991).
15. J. Limburg, J. Vrettos, H. Chen, J. de Paula, R. Crabtree, and G. Brudvig, *J. Am. Chem. Soc.* **123**, 423 (2001).
16. H. Chen, R. Tagore, G. Olack, J. Vrettos, T. Weng, J. Penner-Hahn, R. Crabtree, and G. Brudvig, *Inorg. Chem.* **46**, 34 (2007).
17. R. Ruiz-Garcia, E. Pardo, M. Muñoz, and J. Cano, *Inorg. Chim. Acta* **360**, 221 (2007).
18. T. Privalov, L. Sun, B. Åkermark, J. Liu, Y. Gao, and M. Wang, *Inorg. Chem.* **46**, 7075 (2007).
19. R. Brimblecombe, D. Kolling, A. Bond, G. Dismukes, G. Swiegers, and L. Spiccia, *Inorg. Chem.* **48**, 7269 (2009).
20. R. Cinco, J. Robblee, A. Rompel, C. Fernandez, V. Yachandra, K. Sauer, and M. Klein, *J. Phys. Chem. B* **102**, 8248 (1998).
21. H. Dau, P. Liebisch, and M. Haumann, *Anal. Bioanal. Chem.* **376**, 562 (2003).
22. K. Ferreira, T. Iverson, K. Maghlaoui, J. Barber, and S. Iwata, *Science* **303** 1831 (2004).
23. B. Loll, J. Kern, W. Saenger, A. Zouni, and J. Biesiadka, *Nature* **438** 1040 (2005).
24. H. Beer,US Patent 710551 (1958).
25. H. Beer,US Patent 3711385 (1971).
26. G. Lodi, E. Sivieri, A. De Battisti, and S. Trasatti, *J. Appl. Electrochem.* **8**, 135 (1978).
27. S. Ardizzone, G. Fregonara, and S. Trasatti, *Electrochim. Acta* **35**, 263 (1990).
28. C. Malmgren, M. Hummelgård, J. Bäckström, A. Cornell, and H. Olin, *J. Phys. Conf. Ser.* **100**, 052026 (2008).
29. B. Andersson, A. Salter, I. Virgin, I. Vass, and S. Styring, *J. Photochem. Photobiol. B* **15**, 15 (1992).
30. M. Hakala, I. Tuominen, M. Keränen, T. Tyystjärvi, and E. Tyystjärvi, *Biochim. Biophys. Acta* **1706**, 68 (2005).
31. M. Kanan and D. Nocera, *Science* **321** 1072 (2008).
32. Q. Yin, J. Tan, C. Besson, Y. Geletii, D. Musaev, A. Kuznetsov, Z. Luo, K. Hardcastle, and C. Hill, *Science* **328**, 342 (2010).
33. P. Hohenberg and W. Kohn, *Phys. Rev. B* **136**, **B864** (1964).
34. W. Kohn and L. Sham, *Phys. Rev. A* **140**, A1133 (1965).
35. D. Ceperley and B. Alder, *Phys. Rev. Lett.* **45**, 566 (1980).
36. J. Perdew and W. Yue, *Phys. Rev. B* **33**, 8800 (1986).
37. J. Perdew, K. Burke, and M. Ernzerhof, *Phys. Rev. Lett.* **77**, 3865 (1996).
38. A. Becke, *J. Chem. Phys.* **84**, 4524 (1986).
39. A. Becke, *Phys. Rev. A* **38**, 3098 (1988).
40. A. Becke, *J. Chem. Phys.* **98**, 1372 (1993).
41. P. Stephens, F. Devlin, C. Chabalowski, and M. Frisch, *J. Chem. Phys.* **98**, 11623 (1994).
42. K. Raghavachari, *Theor. Chem. Acc.* **103**, 361 (2000).

43. A. Szabo and N. Ostlund, *Modern Quantum Chemistry—Introduction to Advanced Electronic Structure Theory*, McGraw-Hill Publishing Company, New York (1982).
44. M. Reiher, O. Salomon, and A. Hess, *Theor. Chem. Acc.* **107**, 48 (2001).
45. N. Schultz, Y. Zhao, and D. Truhlar, *J. Phys. Chem. A* **109**, 11127 (2005).
46. W. Koch and M. Holthausen, *A Chemist's Guide to Density Functional Theory*, Wiley-VCH, Weinheim (2001).
47. J. Gräfenstein and D. Cremer, *Mol. Phys.* **103**, 179 (2005).
48. J. Gräfenstein and D. Cremer, *Theor. Chem. Acc.* **123**, 171 (2009).
49. M. Busch, E. Ahlberg, and I. Panas, *Physical Chemistry Chemical Physics*, 13 pp. 15062–15068 (2011).
50. W. Kohn, A. Becke, and R. Parr, *J. Phys. Chem.* **100**, 12974 (1996).
51. W. Kohn, *Rev. Mod. Phys.* **71**, 1253 (1999).
52. S. Shaik and A. Shurki, *Angew. Chem. Int. Ed.* **38**, 586 (1999).
53. A. Lundin, I. Panas, and E. Ahlberg, *J. Phys. Chem. A* **111**, 9080 (2007).
54. R. van Leeuwen, *Adv. Quantum Chem.* **43**, 25 (2003).
55. M. Lundberg, M. Blomberg, and P. Siegbahn, *Inorg. Chem.* **43**, 264 (2004).
56. S. de Visser, F. Ogliaro, Z. Gross, and S. Shaik, *Chem. Eur. J.* **22**, 4954 (2001).
57. B. Delley, *J. Chem. Phys.* **92**, 508 (1990).
58. B. Delley, *J. Chem. Phys.* **113**, 7756 (2000).
59. S. Clark, M. Segall, C. Pickard, P. Hasnip, M. Probert, K. Refson, and M. Payne, *Z. Kristallogr.* **220**, 567 (2005).
60. A. Bard and L. Faulkner, *Electrochemical Methods—Fundamentals and Applications*, John Wiley & Sons, Inc., New York (2001).
61. M. Busch, E. Ahlberg, and I. Panas, *Physical Chemistry Chemical Physics*, 13 pp. 15069–15076 (2011).
62. P. Siegbahn and R. Blomberg, *Biochim. Biophys. Acta* **1655**, 45 (2004).
63. C. Goussias, A. Boussac, and A. Rutherford, *Philos. Trans. R. Soc. Lond. B* **357**, 1369 (2002).
64. A. Harriman, *J. Phys. Chem.* **91** (24), 6102 (1987).
65. M. DeFelippis, C. Murthy, M. Faraggi, and M. Klapper, *Biochemistry* **28** (11), 4847 (1989).
66. H. Kuchling, *Taschenbuch der Physik*, Fachbuchverlag Leipzig, München (2001).
67. *Materials Studio Release 5.0* (2009).
68. B. Delley, *Phys. Rev. B* **66**, 155125 (2002).
69. M. Frisch, G. Trucks, H. Schlegel, G. Scuseria, M. Robb, J. Cheeseman, J. Montgomery, Jr., T. Vreven, K. Kudin, J. Burant, J. Millam, S. Iyengar, J. Tomasi, V. Barone, B. Mennucci, M. Cossi, G. Scalmani, N. Rega, G. Petersson, H. Nakatsuji, M. Hada, M. Ehara, K. Toyota, R. Fukuda, J. Hasegawa, M. Ishida, T. Nakajima, Y. Honda, O. Kitao, H. Nakai, M. Klene, X. Li, J. Knox, H. Hratchian, J. Cross, V. Bakken, C. Adamo, J. Jaramillo, R. Gomperts, R. Stratmann, O. Yazyev, A. Austin, R. Cammi, C. Pomelli, J. Ochterski, P. Ayala, K. Morokuma, G. Voth, P. Salvador, J. Dannenberg, V. Zakrzewski, S. Dapprich, A. Daniels, M. Strain, O. Farkas, D. Malick, A. Rabuck, K. Raghavachari, J. Foresman, J. Ortiz, Q. Cui, A. Baboul, S. Clifford, J. Cioslowski, B. Stefanov, G. Liu, A. Liashenko, P. Piskorz, I. Komaromi, R. Martin, D. Fox, T. Keith, M. Al-Laham, C. Peng, A. Nanayakkara, M.

Challacombe, P. Gill, B. Johnson, W. Chen, M. Wong, C. Gonzalez, and J. Pople, *Gaussian 03 Revision E.01*, Gaussian, Inc., Wallingford, CT (2004).

70. D. Vanderbilt, *Phys. Rev. B* **41**, 7892 (1990).
71. U. Bossek, T. Weyhermueller, K. Wieghardt, B. Nuber, and J. Weiss, *J. Am. Chem. Soc.* **112**, 6387 (1990).
72. B. Minaev and H. Ågren, *Int. J. Quantum Chem.* **57**, 519 (1996).
73. J. Johnsson Wass, I. Panas, J. Åsbjörnsson, and E. Ahlberg, *J. Electroanal. Chem.* **599**, 295 (2007).
74. M. Najafpour, T. Ehrenberg, M. Wiechen, and P. Kurz, *Angew. Chem.* **122**, 2281 (2010).
75. M. El-Deab, M. Awad, A. Mohammad, and T. Ohsaka, *Electrochem. Commun.* **9**, 2082 (2007).
76. A. Mohammad, M. Awad, M. El-Deab, T. Okajima, and T. Ohsaka, *Electrochim. Acta* **53**, 4351 (2008).
77. N. Singh, S. Tiwari, K. Anitha, and R. Singh, *Faraday Trans.* **92**, 2397 (1996).
78. M. Merrill and R. Dougherty, *J. Phys. Chem. C* **112**, 3655 (2008).
79. T. Halgren and W. Lipscomb, *Chem. Phys. Lett.* **49**, 225 (1977).
80. N. Govind, M. Petersen, G. Fitzgerald, D. King-Smith, and J. Andzelm, *Comp. Mater. Sci.* **28**, 250 (2003).
81. G. Henkelmann and H. Jónsson, *J. Chem. Phys.* **113**, 9978 (2000).

CHAPTER 5

No Barrier Theory—A New Approach to Calculating Rate Constants in Solution

J. PETER GUTHRIE

Department of Chemistry, University of Western Ontario, London, Ontario, Canada

5.1 INTRODUCTION

No Barrier Theory (NBT) [1,2] is a new approach to calculating rate constants in solution that uses an experimental equilibrium constant and an assumed mechanism as the only empirical information needed in order to calculate a rate constant. What is directly calculated is the free energy of activation but conversion of this to a rate constant is trivial. The saving thing about these calculations is that relatively low-level quantum chemistry computational methods suffice; in many cases semiempirical methods are sufficient. NBT also provides a way to think qualitatively about whether a reaction is likely to be slow or fast; thus, it can be used both qualitatively to think about mechanisms and quantitatively to predict rates.

Ideally one would prefer to start with the structures of starting materials and products and proceed to rate constant; this requires being able to go from structure to equilibrium constant. Such calculations are now possible in many cases; free energies of formation in solution for small neutral molecules and sometimes for ions can be calculated but the calculations required are at relatively high levels of theory and are demanding even with powerful computer systems. Alternatively it is often possible to estimate unknown equilibrium constants by linear free energy relations. This is particularly the case for pK_a estimation, which is often needed for these calculations.

The calculation procedure is based on an assumed mechanism, and the calculations are carried out for one step of that mechanism. On the one hand, this means that one

Rate Constant Calculation for Thermal Reactions: Methods and Applications, Edited by Herbert DaCosta and Maohong Fan.

has to specify both mechanism and rate-determining step; on the other hand, the calculations provide a criterion for choosing. The mechanism that leads to the fastest reaction will be the correct one; the step with the highest free energy of activation (relative to reactants in solution) will be the rate-determining one. Thus, No Barrier Theory has the potential to provide an objective criterion for deciding on mechanism and rate-determining step, although, of course, there will be cases where alternatives differ by less than the uncertainty of the calculations.

Even more intriguing is the possibility that No Barrier Theory could provide a way to predict yields in synthetic reactions by predicting the rates of the competing reactions—provided that one thought of all the alternatives! Where the competing reactions differed in sensitivity to reactant concentrations or catalysis, it might well be possible to devise conditions to improve the yield.

All of these possible applications of No Barrier Theory depend on its having been shown to be reliable for a wide range of reactions. This testing is underway and thus far the method has proven reliable but clearly more testing is needed.

5.2 THE IDEA BEHIND NO BARRIER THEORY

The central idea of No Barrier Theory is that any real reaction can be analyzed into a small number of "simple reaction coordinates," where simple is defined as having a quadratic dependence of energy on a suitable progress variable along that coordinate. The activation barrier for a real reaction is a consequence of the need for several "simple reaction coordinates" to be traversed simultaneously even though for each simple coordinate "if only one thing happens, there is no barrier." Real reactions are thus multidimensional with the observed activation barrier the result of this multidimensional character. To represent such a reaction model, one can use two-, three-, four-dimensional, and so on, versions of the Albery–More O'Ferrall–Jencks diagram [3–5]. Such diagrams can be referred to as reaction hypercubes. Only for the two-dimensional version is it practical to show the energy as an orthogonal dimension, but in all cases it can be helpful to use the diagrams as a way to consider all the "corner intermediates," the often hypothetical species where 1, 2, 3, . . . , "simple reaction coordinates" have progressed from their initial to their final values. Up to four dimensions, it is relatively easy to draw such diagrams as illustrated in Figure 5.1.

Because No Barrier Theory is a new approach, it has been necessary to develop the language to describe it. A glossary of the terms we have found necessary is given in Table 5.1.

That all reactions can be so analyzed is an assumption of the approach, and it is already clear that there are some reactions that involve processes that cannot be so analyzed. Two such reactions are electron transfer and change of electronic state. Electron transfer processes were the basis for the development of Marcus theory [6–12]; typically the geometry changes needed to make the actual electron transfer isoenergetic were assumed to follow a quadratic dependence on reaction coordinate. Reaction involved transfer of an electron from an initial compound distorted toward the product to a final compound distorted toward the starting material, at a geometry

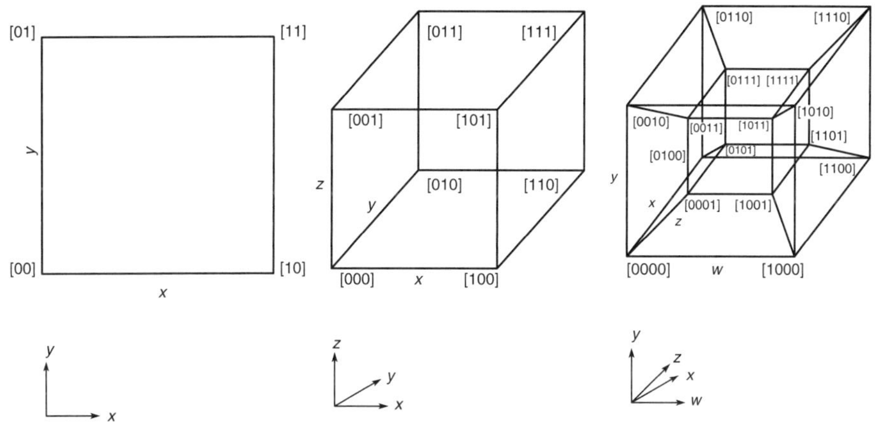

FIGURE 5.1 Two-, three-, and four-dimensional reaction coordinate diagrams. For the two-dimensional diagram, x and y are the two edge coordinates. (0, 0) is the starting point, (1, 1) is the product, and (1, 0) and (0, 1) are the corner intermediates corresponding to reaction along only one edge coordinate. For the three-dimensional diagram, x, y, and z are the three edge coordinates. (0, 0, 0) is the starting point; (1, 1, 1) is the product; (1, 0, 0), (0, 1, 0), and (0, 0, 1) are the corner intermediates corresponding to reaction along only one edge coordinate; and (1, 1, 0), (1, 0, 1), and (0, 1, 1) are the corner intermediates corresponding to reaction along two edge coordinates. For the four-dimensional diagram, w, x, y, and z are the four edge coordinates. (0, 0, 0, 0) is the starting point; (1, 1, 1, 1) is the product; (1, 0, 0, 0), (0, 1, 0, 0), (0, 0, 1, 0), and (0, 0, 0, 1) are the corner intermediates corresponding to reaction along only one edge coordinate; (1, 1, 0, 0), (1, 0, 1, 0), (1, 0, 0, 1), (0, 1, 1, 0), (0, 1, 0, 1), and (0, 0, 1, 1) are the corner intermediates corresponding to reaction along two edge coordinates; and (1, 1, 1, 0), (1, 1, 0, 1), (1, 0, 1, 1), and (0, 1, 1, 1) are the corner intermediates corresponding to reaction along three edge coordinates.

where the initial and final forms had the same energy. There are examples of reactions where the rate-determining step includes both a chemical process and a forbidden spin conversion [13]. There remain large areas of chemistry that can be so analyzed and even larger areas that might be.

The actual reaction path will avoid high-energy species as far as possible and this will include most, if not all, of the "one-thing-at-a-time" corners. In a higher dimensional process there may be a corner with two or more things changed that represents a viable intermediate, in which case the reaction may be better treated as stepwise with either formation or consumption of that intermediate as rate-determining step. There even are cases where, for a change in only one thing, the corresponding corner, though not an intermediate with a lifetime, can still be usefully regarded as the starting point for a reaction. This situation arises in hydroxide addition to carbonyls, where for reactive carbonyl compounds one must start with hydroxide separated by one water molecule from the carbonyl carbon, and include as a reaction dimension the proton transfer that converts this to hydroxide in contact with the carbonyl carbon. With unreactive carbonyl

TABLE 5.1 Glossary of Terms Used in No Barrier Theory

Term	Definition
Simple reaction coordinate	A coordinate for which there is a quadratic dependence of energy on a suitable progress variable along that coordinate
Reaction dimension	One of the edges of a reaction dimension corresponding to a simple reaction coordinate
Reaction hypercube	A generalization of the Albery–More O'Ferrall–Jencks diagram to n dimensions. The two-dimensional case is a square, the three-dimensional case is a cube, and so on. For any reaction hypercube energy could be included as an additional dimension orthogonal to the others but this is difficult to draw except in the two-reaction-dimension case
Corner species	The complex of all species (molecules, ions, solvent) involved in the reaction mechanism. For an n-dimensional hypercube there are 2^n corner species; the energies of all corner species must be known relative to the initial or final corners; the energy of the final species relative to the initial species is defined by the experimental equilibrium constant
Corner intermediate	A corner species, not the initial or final corner. It may be but generally is not a reaction intermediate
One-thing-at-a-time process	A process corresponding to progress along one of the edges of a reaction hypercube. Corner species have progressed by 0 or 1 along each of the reaction dimensions
Progress variable	Numerical measure of the extent to which a simple reaction coordinate has progressed from the initial to the final value. Normally defined so as to run from 0 (initial state) to 1 (final state)
Reaction intermediate	A species on the reaction path, neither the initial nor the final corner that has a finite lifetime under the reaction conditions
Initial species	The initial corner species with all atoms in the stereoelectronically mandated positions for the bond transformation constituting reaction to begin
Final species	The final corner species with all atoms in the stereoelectronically mandated positions for the bond transformation constituting reaction reverse reaction to begin

compounds one can start with the hydroxide in contact with the carbonyl carbon. The problem with the reactive compounds is that the total barrier to addition is comparable to the cost of transferring the proton and generating a partially desolvated hydroxide ion missing a hydrogen bond because of the contact with carbon. For unreactive compounds the overall barrier is much higher and the overall reaction can be treated as stepwise. Thus, for reactive carbonyl compounds addition of hydroxide is actually addition of water as nucleophile with general base catalysis by hydroxide while for unreactive compounds it is usefully treated as direct nucleophilic addition of water. Although the difference in time required to search the reaction hyperspace for the transition state is trivial for two-dimensional versus three-dimensional reactions, in more elaborate mechanisms, the difference between, say, four- and five-dimensional searches, is quite significant.

That reaction pathways avoid unstable intermediates but not stable intermediates is well known. This can be illustrated with some familiar organic reactions. Nucleophilic substitution at carbon can be S_N1 or S_N2. If the carbocation intermediate is relatively stable, one expects S_N1; if the carbocation is relatively unstable, then the actual reaction path will avoid both the carbocation (bond breaking without bond making) and the pentavalent species (bond making without bond breaking) and the mechanism is S_N2. The mechanism followed depends on the balance between the two possible stepwise intermediates, in this case the carbocation and the pentacoordinate species. Elimination can be E1, E2, or E1cb, depending on the relative stability of the carbocationic and carbanionic intermediates. If the carbocation is relatively stable and better than the carbanion, then the reaction is E1. If the carbanion is relatively stable and better than the carbocation, then the reaction is E1cb. If both the carbocation and carbanion are of high energy, then the reaction is E2. Although acyl transfer is most often stepwise by way of a tetrahedral intermediate, concerted acyl transfer is known [14,15]. The competing stepwise intermediates are the tetrahedral intermediate and the acylium ion. With good nucleophiles and relatively poor leaving groups, the tetrahedral intermediate is lower in energy than the acylium ion triple and the reaction proceeds in stepwise fashion by way of this intermediate. With poor nucleophiles and good leaving groups, the tetrahedral intermediate becomes increasingly less favored, and the acylium ion becomes less unfavorable, till the point is reached (as with aryl oxides of low basicity) where concerted reaction is favored.

The viewpoint of No Barrier Theory is that all real reactions are concerted. This is not really contradicting the strongly argued view that stepwise reactions are more likely than concerted ones [16–20]. No Barrier Theory leads to the view that real reactions are the resultant of more than one simple process but nonetheless the preferred path will involve as few dimensions as possible. Higher dimensional processes are inherently slower because more high-energy corner intermediates must be avoided, and each such avoided intermediate pulls the transition state to higher energy. Clearly the absolute energy of the transition state depends on the absolute energies of the avoided corners but it remains a useful generalization that multidimensional processes tend to be slower than processes involving few dimensions. If one of the corner intermediates is not of high energy, then it is likely to be an actual intermediate. Thus, the preferred reaction pathway will involve as few dimensions as possible but as many as there must be to avoid something worse.

5.3 HOW TO DEFINE THE SURFACE AND FIND THE TRANSITION STATE

NBT assumes a ground state free energy surface linking initial and final encounter complexes. This surface is defined in terms of the energies of the corner species and nothing else. NBT defines the transition state free energy successfully as shown in the examples already published, but does not define the transition state position well. In fact, since the energy is the same for the transition state and any point on an n-dimensional cross for an n-dimensional reaction diagram, there is a strong implication that in the vicinity of the saddle point the energy surface is relatively flat and the structure is not well defined. NBT does not define a reaction path and in fact defines only three points on the path: the initial and final points that have well-defined energy and geometry and the TS that has well-defined energy but only vaguely defined geometry. However, the search procedure used to find the saddle point (transition state) guarantees that it is possible to get from the saddle point to both the initial and final corners without ever going to higher energy than the saddle point.

The NBT approach is based on a hypothesis, which for the moment does not have a theoretical basis. The hypothesis is that the free energy surface for a reaction can be described in terms of the corner species (based on the simple reaction dimensions) and the assumption that for any section through the reaction diagram the energy is described by a parabola with its minimum at the low-energy end. The justification is that it has worked for hundreds of reactions.

How does one find the transition state for a reaction proceeding through such a hyperspace defined by the simple reaction coordinates? The assumption that makes it possible is that for progress along any single simple reaction coordinate the energy rises with a quadratic dependence on a suitable progress variable from a minimum at the low-energy end to the high-energy end. Then for a two-dimensional process if one projects the parabolas for one dimension and either the initial or final values of the other, then there will be a crossover point where the two energies are equal and equal to the transition state energy, or else if there is no crossover point, then there is no kinetic barrier to the overall reaction and one can proceed from low- to high-energy terminus with no need ever to go higher in energy than the high-energy terminus. Since initial and final energies for the dimension perpendicular to the parabolas are the same, the standard assumption is that the energy is the same for any value of the coordinate perpendicular to the parabolas and passes through the transition state. Unfortunately, if one examines the mathematics, the two alternative ways of calculating the transition state energy are not the same [21]. Fortunately they are not very different. A useful approximation is to use the average of the two alternative values.

To calculate the energy at any point on a higher dimensional hypersurface, one must follow and then average all possible paths to that point. For a hypersurface with n reaction dimensions (and one energy dimension), start with any of the n coordinates. The value of this coordinate specifies an $(n-1)$-dimensional hypersurface that is a section through the n-dimensional hypersurface, specified by $2^{(n-1)}$ points along each edge parallel to the axis for this first dimension. Next pick any of the $n-1$ dimensions and this specifies an $(n-2)$-dimensional section through the hypersurface, and $2^{(n-2)}$ points along each edge parallel to the axis for this second dimension. Continue this process till one has only two points left that define a parabola, and allow calculation of the energy corresponding to the point specified by all n coordinates. At each calculation of the energy of a point, one relies on the assumption that the energy is given by a parabola with its minimum at the low-energy end.

To find a transition state, one searches the reaction hyperspace for the lowest energy point from which one can proceed to either initial or final corners without ever going higher in energy. The search is carried out on a Cartesian coordinate grid. Because the number of points on such a grid goes up very rapidly with number of dimensions, the algorithm begins with a coarse search and then refines with more fine-grained searches until the transition state position is specified within 1% of each reaction dimension.

For any single simple process the energy normally rises from a minimum at the low-energy end to the high-energy end. In a few cases we have found that the energy passes through a minimum lower than the low-energy end. This happens, for example, in an analysis of the S_N2 reaction where at least three dimensions are needed: geometry change, bond making, and bond breaking. An additional feature is that the geometry changes first from the initial tetrahedral geometry to the intermediate pentacoordinate geometry (where one or two of the five ligands may not be covalently bonded to carbon), and then from pentacoordinate geometry to the final tetrahedral geometry. For the central two-dimensional section through the complete reaction diagram, all four corners have the central carbon and its three nonreacting substituents coplanar and the nucleophile and leaving group in different states of bonding. If one proceeds from the corner with leaving group bonded and nucleophile not bonded (perpendicular encounter corner) either to the corner with both nucleophile and leaving group bonded (pentacoordinate intermediate) or to the corner with neither nucleophile nor leaving group bonded (ion triple of the carbocation), then the energy passes through a minimum lower than the energy of the perpendicular encounter corner. Consequently, the transition state (which for identity reactions must fall in this reaction plane, and for nonidentity reactions is expected to fall close to it) is lower in energy than any of the corners of this planar section through the reaction diagram. The normal process for finding the transition state does not work for such reactions. Why is the behavior different for these simple processes than for the vast majority of simple processes where the minimum falls at one end of the coordinate? Here the process involves proceeding to or from a hypervalent state so that both bonding and crowding increase or decrease, leading to a minimum part way along.

The assumption that the energy follows a parabola with the minimum at the low-energy end for any section through the reaction diagram along which only one dimension changes immediately implies that the free energy surface for the reaction can be defined if the energies of all corners can be defined relative to the initial corner. This requires adding up all contributions to the energy of such a corner. Note that by using the overall equilibrium constant for the reaction in solution, we build in the energetic cost of the bond transformation and solvation, the two things that are expensive to calculate. As long as equilibrium constants are available and unknown corners can be linked to known corners with the same bonding pattern so that only geometrical distortion must be evaluated computationally, this can be done with low levels of theory. The list of contributions to the energy of a corner species that has so far proved adequate is

1. hydrogen bonding
2. electrostatic interactions
3. desolvation
4. encounter (entropic cost of bringing species into the encounter complex)
5. overall equilibrium
6. proton transfer equilibrium
7. geometrical distortion (relative to a species with the same bonding pattern)

These will now be discussed in turn. It is important to note that even though some of these contributions are evaluated using empirical relations, the parameters in these are set for the solvent or type of reagent and not adjusted to fit the rate constant.

In order to calculate the energy of the transition state, we need the energies of all the corner intermediates, which often include several species (molecules or ions) in contact. The initial corner has all atoms in place about to begin the bond transformation that will constitute the reaction, and the final corner has all atoms in the positions corresponding to having just completed that transformation. We must account for any energetic costs of bringing them together from the standard state of species free in solution, and for any interactions between species. These interactions include

1. Hydrogen bonds other than hydrogen bonds involving solvent that are presumed to be there anyway. The energy of the hydrogen bond is calculated using the Stahl–Jencks equation [22] that uses the pK_a of the hydrogen bond acid and of the conjugate acid of the hydrogen bond base and one parameter, τ, which measures the sensitivity of hydrogen bond strength to electrostatic interactions:

$$\log K_{AB} = \tau(pK_{AH} - pK_{HOH})(pK_{H_3O^+} - pK_{BH}) - 2.04$$

where $\tau = 0.013$ in water [22] and 2.04 is the logarithm of the concentration of water corrected by a statistical factor of two [22].

Although we are using the equation well outside the range of hydrogen bonds for which it was derived, it is firmly based on an electrostatic model [23] with only one empirical parameter, and thus should give a good approximate value. Since the energies of individual hydrogen bonds in solution are small [22,24–27], even a substantial relative error does not cause a major perturbation in the calculation of $\Delta G^{\ddagger}$. This treatment assumes that "hydrogen bond formation" between A–H and B is really hydrogen bond exchange:

$$\text{A—H}\cdots\text{OH}_2 + \text{B}\cdots\text{H—OH} \rightleftharpoons \text{A—H}\cdots\text{B} + \text{H}_2\text{O}\cdots\text{H—OH}$$

In a nonhydrogen bonding organic solvent one would need a different approach, such as that of Abraham [28]:

$$\log K_{\mathrm{hb}} = a\alpha_2^{\mathrm{H}}\beta_2^{\mathrm{H}} + b$$

where α_2^{H} is the hydrogen bond acidity of AH and β_2^{H} is the hydrogen bond basicity of B.

2. Electrostatic interactions between any charges or formal charges in any species in the corner intermediate. This interaction is very sensitive to solvent polarity and is relatively small in polar solvents such as water. The interaction is calculated using the standard electrostatics equation but scaled to 1 : 1 complexation of two spherical ions, for which there are experimental values. For 1 : 1 ion pairs in water at low ionic strength, the free energy of ion pair formation is taken as 0.1 kcal/mol [29]. This value applies to spherically symmetrical ions. We partition this free energy into the statistical and electrostatic contributions using the following equation [30]:

$$K_i = \frac{4\pi N a^3}{3000} e^{-(U/kT)}$$

where

$$U = \frac{z_1 z_2 e^2}{\varepsilon}\left(\frac{1}{a(1+\kappa a)}\right)$$

$$\kappa = \sqrt{\frac{8\pi N e^2 \mu}{1000\varepsilon kT}}$$

N is Avogadro's number, a is the distance of closest approach of the ions (cm), e is the electron charge, k is Boltzmann's constant, ε is the bulk solvent dielectric constant, μ is the ionic strength, and z_1 and z_2 are the charges on the ions.

The above equation was derived by Fuoss [31] and Eigen [32]. By picking an interaction distance to give $\Delta G_{\mathrm{overall}} = \Delta G_{\mathrm{encounter}} + \Delta G_{\mathrm{electrostatic}} = 0.1$ kcal/mol in water, the corresponding contributions can be calculated from the terms in this equation as $\Delta G_{\mathrm{encounter}} = 1.29$ and $\Delta G_{\mathrm{electrostatic}} = -1.19$. Simple electrostatic calculations are used to calculate the electrostatic interaction energy for all charged atoms in each corner complex, from the charges and interatomic distances, and scaling to the experimental free energy for 1 : 1 ion pair formation. The saving point is that in aqueous solution electrostatic effects are small. We are also assuming the same dielectric constant for all configurations considered. Since electrostatic effects are of modest importance in aqueous solution, the crude approximations used here are not totally unreasonable and allow some account to be taken of electrostatic effects on complex stability. For most organic solvents electrostatic effects will be larger and more important and it may be necessary to use a more elaborate treatment.

3. The desolvation cost of bringing an ion into contact with a nonhydrogen bonding species is estimated by looking at the effect on the pK_a of the species or its conjugate acid of being transferred from a hydrogen bonding to a nonhydrogen bonding solvent [33]. This transfer strips away all hydrogen bonding solvent molecules. By making an assumption about how many such hydrogen bonds there are, that is, by assuming each lone pair is involved in a hydrogen bond, one may deduce the energy cost of breaking one such hydrogen bond. This procedure can only be approximate but is tied to experimental results.
4. The entropic cost of bringing reactants into noncovalent contact is treated as the cost of increasing the concentration of one reactant from the standard state value, typically 1 M, to having the reactive atom of the reactant in one of a limited number of volumes equal to the volume of an atom and is given by

$$\Delta G_{\text{enc}} = 2.83 - 0.593\ \ln(n)$$

where 2.83 is the free energy cost in kcal/mol of bringing the nucleus of the interacting atom of a ligand to be within a sphere of radius 1.5 Å (the van der Waals radius of oxygen) from being anywhere in solution, 0.593 is RT at 25°C in kcal/mol, and n is the number of equivalent sites where the ligand might be and still satisfy the needs of the reaction.

This equation is based on a derivation by Hine [34] modified to include the number of equivalent positions where the incoming molecule could be. If there are more than two components of the complex, then ΔG_{enc} is the sum of terms for each extra species. Solvent does not contribute because it is already there.
5. The overall equilibrium constant for the observable reaction not only sets the thermodynamic cost of the reaction but also builds in the effects of solvation of the initial and final species. Where an experimental equilibrium constant is available, this avoids the two most difficult to calculate quantities governing the rate of reaction: the gas-phase equilibrium constant and the effect of solvation on this equilibrium. Nonetheless, it is helpful to be able to calculate equilibrium constants since they are not always available where wanted. Such calculations are increasingly becoming possible but still are often at the limit of what can be done. Quite often when an equilibrium constant is not available, it can be estimated with useful accuracy by the use of linear free energy relations, relative to a known value for a similar reaction. For simple neutral molecules free energies in solution can be estimated with useful accuracy by additivity schemes [35] provided there are no serious steric effects or distant polar interactions [36]. Equilibrium constants are needed for each process leading to a change in bonding pattern.
6. A special case of equilibrium constants is that of pK_a values needed for proton transfer equilibria. These are available for a great many ionizations of stable molecules [37–40] but are often needed for metastable intermediates or even for species that are too unstable to have lifetimes in their protonated or deprotonated states. Where pK_a values cannot or have not been measured, they must be estimated. Fortunately methods for estimating pK_a are well developed [41]. For

species that have no lifetime once protonated (e.g., protonated acetals) or deprotonated (e.g., tetrahedral intermediates with a good leaving group), estimates by linear free energy relations are actually appropriate and more appropriate in borderline cases where the protonated or deprotonated species might have a short lifetime but would be expected to be distorted because of an anomeric effect. Double bond–no bond resonance is expected to lead to lengthening of bonds [42] and increase in basicity [42,43] for electronegative atoms that have a neighbor with an antiperiplanar lone pair. What is estimated by the linear free energy relation is a "vertical pK_a" without geometry change. This is what is needed for the corner intermediate in a one-thing-at-a-time reaction diagram provided that it is an avoided intermediate and not a possible stepwise intermediate, and is the only well-defined pK_a one can have for such a species. The actual species, if one could generate it, would fall apart with no barrier. When the protonated or deprotonated species has a lifetime, it becomes possible to define its geometry, bearing in mind that it may only be stable with explicit solvation. Clearly there are problems here that have not been fully resolved and will need further study.

7. Distortion energies for the various corner intermediates can be calculated using relatively low levels of molecular orbital theory provided that they are calculated relative to species with the same bonding pattern and only geometry differences. This is always possible provided the equilibrium constants relating species with different bonding patterns can be supplied. Very often the starting material and/or product must adopt a reactive conformation rather than the lowest energy conformation in order to satisfy the stereoelectronic requirements of the reaction. Where this happens, it can be treated as a distortion of the initial or final corner. A distortion energy is normally calculated from the difference in energy between a partially optimized corner species, with the reaction dimensions locked at the values defining that corner, and the corresponding fully optimized equilibrium species.

   There are special problems with ionic corner intermediates because such species are often unstable to breakup in the gas phase. Computation of solvation energies remains difficult and computationally expensive, so it is not practical to use the direct approach to the problem of ionic intermediates that are stable species in solution, even if short lived, but unstable in the gas phase. We have explored various approximations to deal with this problem. Initially we tried using distortions based on the corresponding neutral species. This worked in some cases but it became clear that it was not a satisfactory general method. Then we tried locking the length of the bond that will break in unconstrained gas-phase optimization of the ionic species, using the length in the corresponding neutral species. Again this worked in some cases but was not a satisfactory general solution. The problem in both cases is that the distortion energies for the ionic corner intermediates should reflect the possible anomeric effects, with lengthening of the scissile bond and flattening of the geometry at the multivalent terminus. The current best solution we have found is the following. For each charge-bearing atom we add one water for each lone pair or hydrogen, allowing for cyclic arrays wherever geometrically feasible. This often allows the ionic species to remain bonded during optimization. During initial tests of a mechanism, the number of waters should be increased by adding layers particularly about anionic centers to ensure that the structure of the solvated ion does not change with number of waters. The goal is to have the smallest possible calculation giving reliable results. Adding more waters is computationally expensive; this becomes a strong argument for using semiempirical methods. For some ionic species, particularly but not only zwitterions, explicit solvation as described is insufficient to prevent bond breaking. For such situations we conclude that the ionic species has no lifetime even in water and then use the structure with equilibrium geometry based on the corresponding neutral as the reference for the distorted forms of this ionic species. The reference form of the ionic species is a virtual intermediate with no lifetime and the bond that breaks in the optimization will be the one involved in a bond change dimension. The energy of this species is defined by an estimated pK_a that is based on model compounds that do not break a bond on protonation, thus providing a reasonable estimate for the virtual species.

The higher the level of theory used to calculate the distortion energies, the more confidence one can have in them but the greater the computational cost; the lower the level of theory, the faster the calculations but the less confidence one can have. A workable procedure seems to be to do test calculations with higher levels of theory, perhaps on an example involving smaller molecules, and then repeat the calculation with the lower level of theory. In a recent set of calculations for aldol addition reactions we found that for all but one type of distortion, PM6 [44] (*MOPAC2009* [45]) gave results very similar to B3LYP/6-31 + G** (*Gaussian 03* [46]). The one distortion where there were serious discrepancies was the geometrical distortion of an enolate at the nucleophilic carbon. By using the B3LYP/6-31 + G**//PM6 value for this distortion and the PM6 values for the rest, one could retain most of the (very considerable) speed advantage of PM6. For other reactions it seems likely that PM6 can be used for exploratory studies, but that DFT energies should be calculated in at least some test cases for each mechanism investigated to see if PM6 distortions are good enough.

5.4 WHAT IS NEEDED FOR A CALCULATION?

The necessary input for the system of computer programs that I have written to carry out a calculation of a rate constant by NBT is provided in a set of input files:

1. structure files for starting and final compounds, in addition to any reactants, the solvent, and any other components of the initial or final corner species that must be specified
2. donbt.dat (names of the structure files and the role they play; names of structure files for the lowest energy forms of the initial and final compounds if these differ

from the reactive conformations, the equilibrium free energy change; the free energy of activation if known, for comparison with the calculated value—it is not used in the calculation and can be set to zero without affecting anything except a final comparison of calculated and observed free energies of activation)

3. mech.dat (specifies the mechanism in terms of any intermediate ionic forms that may be needed and the relative orientation of all species in the initial and final corners, in addition to the ionic forms of initial or final compounds if these are involved and not the neutral forms)
4. pka files for starting compound (and final compound), reactants (if any), and solvent
5. stj file for the solvent (giving the Stahl–Jencks parameter, and the symmetry-corrected pK_a values for the solvent)
6. information about 1:1 ion pair formation in various solvents and the symmetries of the conjugate acid and conjugate base of the common solvents coded into the program

This information is all that is needed to calculate a rate constant and the program system can carry out this calculation without human intervention once the input files are set up.

5.5 APPLICATIONS TO DATE

Numerous reactions have been treated by this method; in each case the procedure was applied to a set of reactions covering as wide a range of reactivity as possible subject to the constraints of what rate and equilibrium constants have been reported.

5.5.1 Proton Transfer Reactions

Proton transfer reactions of carbon acids were the first to be studied by NBT [47,48], but although the results showed the power of the method, proton transfers are a deceptively difficult type of reaction and a nonstandard treatment was needed. The reaction of monocarbonyl compounds was treated as having two dimensions: geometry change at carbon and proton transfer without geometry change. The problem is in the distortion energies of the enolates that as calculated in the normal way are too small to explain the observed rate constants for proton transfer. A workable solution was to estimate the vertical pK_a of the keto tautomer assuming no overlap and only the field and inductive effects of the substituents as measured by σ^*. This procedure led to calculated $\Delta G^{\ddagger}$ values in good agreement with experiment over a very wide range of compounds and reproduced the discrepancy between substituent effects on rate and equilibrium seen, for example, in acetaldehyde versus isobutyraldehyde. We are now concerned that the assumption that proton transfer from carbon to oxygen could be treated as a simple process looks unreasonable. It is a workable approximation that proton transfer between electronegative atoms is a simple process, though there is evidence that it is not and actually has a small intrinsic barrier that by NBT implies that it is the resultant of two or more simpler processes. Proton transfer between electronegative atoms involves a short movement of a proton along a preformed hydrogen bond. Proton transfer between carbon and an electronegative atom involves a very weak hydrogen bond with a very small covalent component because of the greater pK_a difference and presumably involves a longer motion of the proton. The assumption that this is also a simple process looks less reasonable. We are investigating this reaction further (Tables 5.2 and 5.3).

In Figure 5.2, as in many of the figures to follow, the reaction mechanism is shown in addition to a set of arrows showing the simple reaction dimensions into which the overall reaction was resolved. This is the set used for the publication summarized in the accompanying table and may not be quite the same as the current version.

Base-catalyzed deprotonation of dicarbonyl compounds needed a slightly more elaborate model because the cost of bringing hydroxide into "hydrogen-bonded contact" with the carbon acid (with loss of one solvating water) is too comparable to the overall kinetic barrier. This situation is analogous to that seen for hydroxide addition to reactive carbonyl compounds and the solution is analogous. The starting state is taken to have a water in contact with the C–H to be removed and hydrogen bonded to hydroxide. There is an additional reaction dimension, proton transfer from oxygen to oxygen. This more elaborate treatment led to successful calculation of $\Delta G^{\neq}$ for deprotonation of dicarbonyl compounds.

5.5.2 Addition of Water to Carbonyls

There are abundant kinetic data for addition of water to carbonyl compounds either uncatalyzed, acid catalyzed, or base catalyzed. For a considerable number of these reactions the equilibrium constant for hydration is also available and thus an extensive test of NBT is possible [73]. Over the entire range of reactivity for which data are available (from formaldehyde plus hydroxide to *N,N*-dimethylacetamide plus water), the calculated $\Delta G^{\ddagger}$ values were in good agreement with experiment. At the time this paper was published [73], the rate of uncatalyzed hydrolysis of dimethylacetamide had not been reported. Since then Wolfenden and coworkers [74] have reported a rate constant at elevated temperatures extrapolating to a $\Delta G^{\ddagger}$ of 32 kcal/mol in good agreement with our prediction of 31.11 kcal/mol.

Two models were used for uncatalyzed hydration: in one, two waters were involved (one acting as nucleophile and one as general base); in the other, three waters were involved and there was a cyclic array of waters permitting a concerted chain of proton transfers leading to the neutral hydrate. For many reactions the two mechanisms led to equally good predictions but for unreactive carbonyls the second mechanism was better. This illustrates the general phenomenon that it is safe to use a more general mechanism and let the molecules tell you if a lower dimensional stepwise version is adequate; if so, some of the possible dimensions (here the extra proton transfers) will not have happened or will have gone to completion at the transition state (Figure 5.3 and Table 5.4).

TABLE 5.2 Rate Constants for Proton Transfer Reactions Involving Monofunctional Compounds and Hydroxide or Water as Base[a]

		Hydroxide as Base			Water as Base		
Compound	pK_a	log k_{obs}	log k_{calc}	log k_{calc} − log k_{obs}	log k_{obs}	log k_{calc}	log k_{obs} − log k_{calc}
CH_3–CHO	16.7[b]	0.07[b]	0.83	0.76	−7.62[b]	−6.00	1.62
$(CH_3)_2$CH–CHO	15.49[b]	−0.85[b]	−0.81	0.04	−7.26[b]	−7.54	−0.28
$PhCH_2$–CHO	13.10[b]	1.30[b]	0.79	−0.51	−6.33[b]	−5.12	1.21
Ph_2CH–CHO	10.40[b]	2.40[b]	1.27	0.14	−4.12[b]	−3.70	0.42
CH_3–CO–CH_3	19.27[b]	−0.66[b]	−0.24	0.42	−9.34[b]	−8.57	0.77
$(CH_3)_2C\underline{H}$–CO–CH_3	20.33[c]	−2.43[d]	−3.81	−1.38			
$(CH_3)_2$CH–CO–$C\underline{H}_3$	19.3[e]	−0.98[d]	−2.78	−1.80			
$(CH_3)_2$CH–CO–$CH(CH_3)_2$	21.48[c]	−2.98[c]	−2.61	0.37			
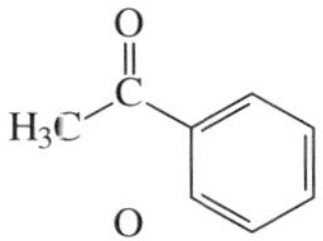	18.30[b]	−0.60[b]	0.01	0.61	−8.73[b]	−7.60	1.13
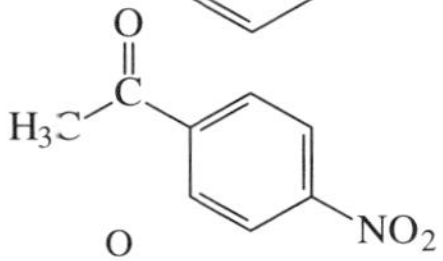	16.60[b]	−0.37[b]	0.42	0.79			
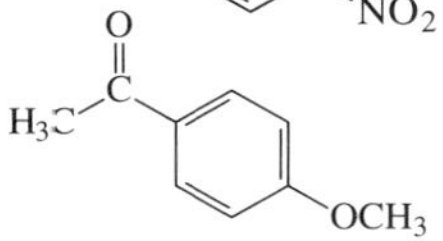	18.90[b]	−1.35[b]	−0.15	1.20			
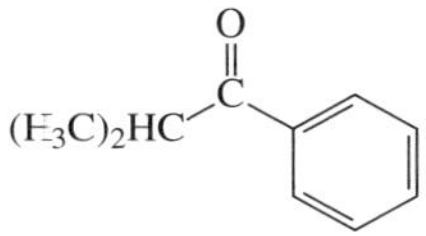	18.26[b]	−2.62[b]	−1.78	0.84	−9.78[b]	−9.37	0.41
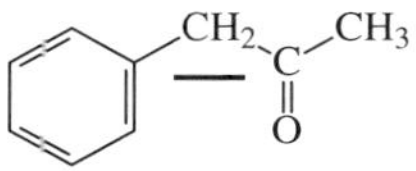	15.91[c]	0.50[d]	0.12	0.38			

(Continued)

TABLE 5.2 (*Continued*)

Compound	pK_a	Hydroxide as Base: $\log k_{obs}$	Hydroxide as Base: $\log k_{calc}$	Hydroxide as Base: $\log k_{calc} - \log k_{obs}$	Water as Base: $\log k_{obs}$	Water as Base: $\log k_{calc}$	Water as Base: $\log k_{obs} - \log k_{calc}$
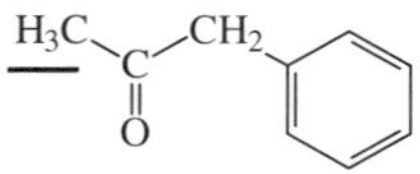	18.27[c]	−0.65[d]	−0.23	0.42			
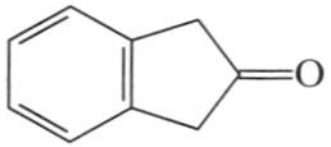	12.20[b]	2.35[b]	0.81	−1.53	−4[b]	−4.39	0.11
$ClC\underline{H}_2$–CO–CH_3	15.76[c]	0.55[c]	1.57	1.02	−7.06[f]	−6.36	0.70
$Cl_2C\underline{H}$–CO–CH_3	11.00[c]	3.24[c]	3.55	0.31	−6.36[f]	−3.42	2.94
$BrC\underline{H}_2$–CO–CH_3	13.27[c]	2.21[c]	1.98	−0.23	−5.55[f]	−4.98	0.57
$CH_3OC\underline{H}_2$–CO–CH_3	18.49[c]	−1.42[c]	−0.05	1.37	−9.86[g]	−8.24	1.62
CH_3OCH_2–CO–$C\underline{H}_3$	18.29[c]	−0.77[c]	−0.01	0.76	−9.23[g]	−7.59	1.64
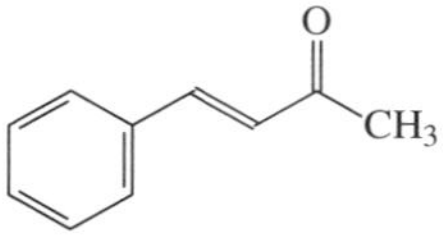	18.41[c]	−0.69[d]	0.14	0.83			
CH_3–$COOC_2H_5$	25.6[h]	−2.92[h]	−2.42	0.50			
CH_3–$COSC_2H_5$	21.0[i]	−1.70[i]	−1.00	0.70			
CH_3–CO–$COOCH_3$	13.35				−5.79[j]	−4.47	1.32

[a] All in aqueous solution at 25°C rate constants are in $M^{-1}\,s^{-1}$. Calculated rates are based on distortion energies calculated at the 3-21G level; some compounds were treated at the 3-21 + G level but this made little difference to geometries. Based on Tables 1 and 2 in Ref. [47]. Removal of a proton α to a carbonyl group; where there is ambiguity, the proton to be removed is underlined.

[b] Ref. [49].

[c] Ref. [50].

[d] Calculated from the rate constant for detritiation [51], assuming that the kinetic isotope effect is the same as for acetone [52].

[e] Since the rate constant per hydrogen is very similar to that for acetone, the pK_a is assumed to be very similar to that for acetone [50].

[f] Ref. [53], corrected for hydration as described in Ref. [54].

[g] Extrapolated from measurements of dedeuteration at higher temperatures given in Ref. [55], using a value for the isotope effect calculated using the correlation in Ref. [56].

[h] Ref. [57].

[i] Ref. [58].

[j] Ref. [59], corrected for hydration as described in Ref. [60].

TABLE 5.3 Rate Constants for Proton Transfer Reactions Involving Polyfunctional Compounds and Hydroxide or Water as Base[a]

		Hydroxide as Base			Water as Base		
Compound	pK_a	$\log k_{obs}$	$\log k_{calc}$	$\log k_{calc} - \log k_{obs}$	$\log k_{obs}$	$\log k_{calc}$	$\log k_{calc} - \log k_{obs}$
$CH_3CO–CH_2–COCH_3$	8.87[b]	4.60[b]	4.58	−0.02	−1.78[b]	−0.45	1.33
$CH_3CO–CH(CH_3)–COCH_3$	10.68[c]				−4.03[d]	−3.79	0.24
$CH_3CO–CHBr–COCH_3$	7.0[e]				−1.47[f]	−0.70	1.04
$CH_3CO–CH_2–COOC_2H_5$	10.68[c]	3.82[g]	3.44	−0.38	−2.94[f]	−3.14	−0.20
$CH_3CO–CH(CH_3)–COOC_2H_5$	12.25[g]	2.01[g]	2.32	0.31	−4.94[f]	−5.57	−0.63
$CH_3CO–CHBr–COOC_2H_5$	8.0[h]				−1.81[f]	−2.03	−0.22
$PhCO–CH_2–COCH_3$	8.53[i]				−1.96[j]	−2.85	−0.88
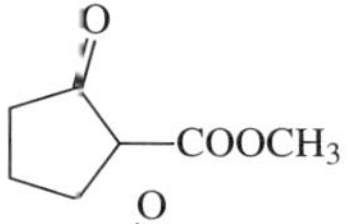	10.0[k]				−2.64[f]	−3.59	−0.95
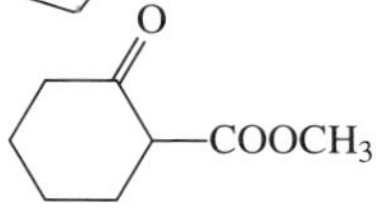	11.92[l]	2.35[l]	2.05	−0.30	−4.61[l]	−4.82	−0.21
$CH_3OCO–CH_2–COOCH_3$	13.30[m]				−4.61[f]	−5.31	−0.70

[a] All in aqueous solution at 25°C rate constants are in $M^{-1}\,s^{-1}$. Calculated rates are based on distortion energies calculated at the 3-21G level; some compounds were treated at the 3-21+G level but this made little difference to geometries. Based on Tables 3 and 4 in Ref. [47].
[b] Ref. [61].
[c] Ref. [62].
[d] Ref. [63].
[e] Ref. [64].
[f] Ref. [65].
[g] Ref. [66].
[h] Calculated from kinetic data in Ref. [67] following the procedure described in Ref. [50].
[i] Ref. [68].
[j] Ref. [69].
[k] Ref. [70].
[l] Ref. [71].
[m] Refs [68,72].

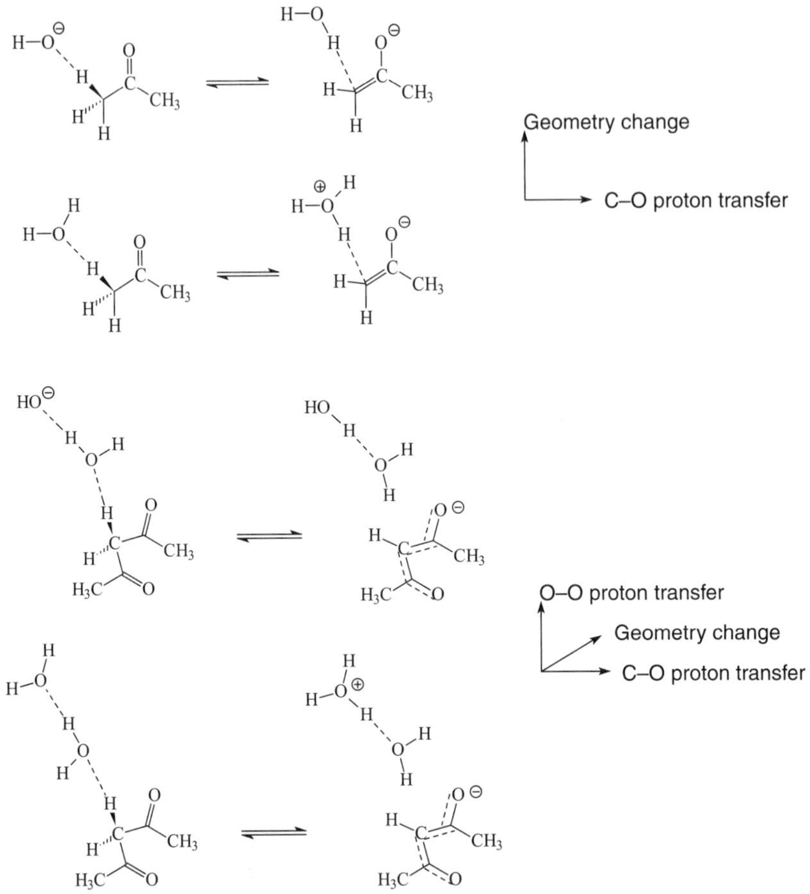

FIGURE 5.2 Mechanisms for base-catalyzed and uncatalyzed enolization of monocarbonyl and dicarbonyl compounds. Enolate formation from monocarbonyl compounds was modeled using two reaction dimensions: C–O proton transfer and geometry change. Dicarbonyl compounds required an additional dimension: O–O proton transfer. Based on structures in Ref. [47].

5.5.3 Cyanohydrin Formation

Cyanohydrin formation was reported for the range of reactions for which *k* and *K* have been reported in water (Figure 5.4). This was the first application of NBT to a carbon–carbon bond-forming reaction. The rate-determining step was taken to be the attack of cyanide on the carbonyl compound to give the cyanohydrin anion (Figure 5.4). These results are summarized in Table 5.5.

Rate constants for cyanohydrin formation have been calculated for the limited range of compounds where both rate and equilibrium constants were available. The mechanistic model involved bond making and geometry change as reaction dimensions, with allowance for desolvation of cyanide ion when it was in contact with the

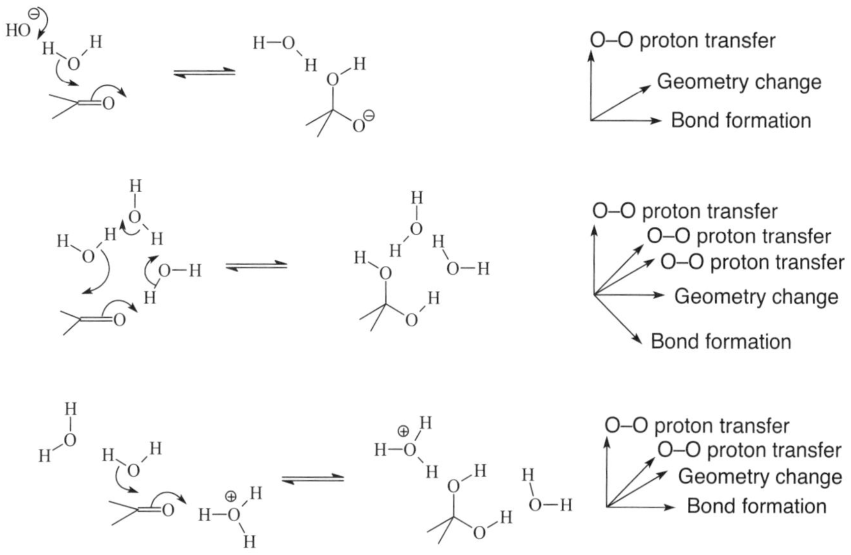

FIGURE 5.3 Mechanisms for base-catalyzed, uncatalyzed, and acid-catalyzed hydration of carbonyl compounds. The model for hydroxide-catalyzed addition involved three reaction dimensions: C–O bond formation, geometry change, and O–O proton transfer. Uncatalyzed hydration was a cyclic mechanism with three O–O proton transfer dimensions. Acid-catalyzed hydration involved two O–O proton transfers, one from hydronium ion to the carbonyl oxygen and the other from the nucleophilic water to a water acting as general base. In all three cases some compounds might only need reduced versions of these mechanisms with fewer dimensions, but these worked for all compounds. Based on Figures 3–5 and 12 of Ref. [73].

carbonyl carbon. The energetic cost of desolvation was taken from the difference in pK_a of HCN in water and DMSO.

5.5.4 The Reaction of Carbocations With Either Water or Azide Ion

The reaction of carbocations with either water or azide ion as nucleophiles has been studied. This is the second step of the S_N1 substitution reactions by way of these cations; it would be of great interest to study the S_N1 reactions of the corresponding bromides or tosylates but the equilibrium constants needed for the ionization step are not known. For the attack of water we initially used a four-dimensional model as shown in Figure 5.5, but found that in fact a simpler three-dimensional model with only one water, acting as nucleophile and leading to the protonated alcohol product, gave

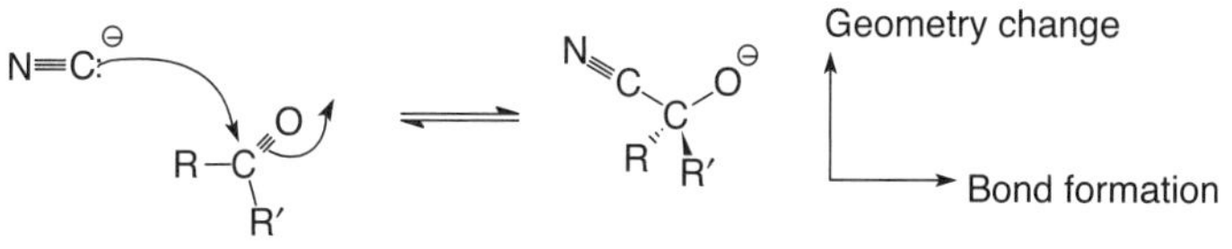

FIGURE 5.4 Mechanism for cyanohydrin formation (Ref. [21]). The model for this reaction involved two reaction dimensions, C—C bond formation and geometry change.

TABLE 5.4 $\Delta G^{\ddagger}$ for Hydration of Carbonyl Compounds by Acid-Catalyzed, Base-Catalyzed, and Uncatalyzed Mechanisms[a]

Compound	Hydroxide Catalyzed		Uncatalyzed		Acid Catalyzed	
	Observed	Calculated	Observed	Calculated	Observed	Calculated
Formaldehyde	8.56	8.36	16.08	14.65	12.35	12.07
Acetaldehyde	10.75	11.20	20.34	18.48	13.40	14.01
Propionaldehyde	11.06	11.38	20.81	19.39	13.77	14.99
Butyraldehyde	11.06	11.42	20.81	19.45	13.83	15.11
Isobutyraldehyde	13.36	11.80	21.35	19.77	13.94	15.32
Pivaldehyde	13.98	13.14	22.17	21.77	15.59	16.97
Trichloroacetaldehyde		9.26	13.83	14.69		15.39
Benzaldehyde	14.03	12.80	23.09	20.07	15.67	14.61
4-Cl-benzaldehyde	12.68	12.43	22.21	20.90	15.67	15.77
3-Cl-benzaldehyde	12.23	11.78	21.92	18.71	16.03	13.66
3,4-Dichlorobenzaldehyde	11.67	12.27	19.57	20.67	14.13	15.97
p-CF_3-benzaldehyde	11.64	11.93	21.31	20.23	15.93	15.65
m-NO_2-benzaldehyde	11.06	12.27	19.57	20.05	14.68	15.67
p-NO_2-benzaldehyde	10.73	11.81	20.54	19.91	15.93	15.67
3NO_2-4Cl-benzaldehyde	10.81	11.39	20.50	19.70	16.08	15.61
3,5-$(NO_2)_2$-benzaldehyde	8.98	11.18	18.82	18.96	16.22	15.63
5NO_2-2Cl-benzaldehyde	9.56	10.48	19.06	18.21	15.15	14.56
Acetone	14.66	14.43	24.38	23.53	15.37	17.75
Chloroacetone		12.02	19.04	20.26	16.61	16.83
1,1-Dichloroacetone		11.40		19.23		17.52
1,3-Dichloroacetone		9.51	19.94	18.73	17.45	17.25
Fluoroacetone		11.86	17.25	20.09	14.97	16.92
1,1,1-Trifluoroacetone		11.47	16.14	18.14		17.87
Cyclohexanone	14.51	14.90	23.13	22.32	14.66	16.14

Cyclopentanone	16.08	15.78		24.77	16.93	18.48
Acetophenone		15.88		25.40		19.09
α,α-Dichloroacetophenone		11.55	19.90	19.73	18.76	18.07
Trifluoroacetophenone	9.66	10.38	16.76	17.24		16.89
Methyl pyruvate[b]		12.07	18.23	19.92	16.30	17.80
2,4-Pentanedione		9.96	20.07	19.23	15.13	15.01
1,1,1-Trifluoro-2,4-pentanedione		10.33	17.21	17.41	16.58	17.77
Methyl formate	15.29	14.41	25.92	24.61	21.03	19.91
Methyl acetate	18.58	17.18	29.84	27.47	22.86	21.35
Methyl propionate	18.62	18.44		29.05	22.86	23.17
Methyl isobutyrate	18.89	19.24		29.92	23.17	23.50
Methyl difluoroacetate	12.25	12.16	22.56	21.25	22.82	19.73
Methyl trifluoroacetate	9.89	11.12	20.26	18.94		18.96
Methyl monochloroacetate	15.11	15.63	26.25	25.54	23.08	21.96
Methyl dichloroacetate	13.23	13.69	23.82	22.58	22.46	21.21
Ethyl trichloroacetate	12.57	15.44	21.67	23.59	22.09	23.44
Methyl trichloroacetate		15.73		23.42		23.23
Methyl cyanoacetate	15.25	16.25		24.78	23.97	21.96
Methyl methoxyacetate	16.84	16.90		27.67	23.20	23.14
Methyl benzoate	19.01	18.28		29.06	26.15	24.93
S-Ethyl thiolformate	14.59	13.11	25.06	22.26	18.92	18.63
S-Ethyl thiolacetate	18.71	19.46		27.54	23.67	21.94
S-Ethyl trifluoroacetate	12.3	12.95	20.64	20.37		20.25
N,*N*-Dimethylformamide	22.57	21.12		30.75	26.30	23.17
N,*N*-Dimethylacetamide	23.94	22.60		31.11	25.66	23.76
N,*N*-Dimethyltrifluoroacetamide	17.63	17.58		27.03		23.06
N-Methylformanilide	22.42	17.85		28.11	22.53	19.91

[a] All in water at 25°C. Based on Tables S1, S3, and S4 of Ref. [73]. References to origins of all observed values are found in Table 1 of Ref. [75].
[b] Reacting as a ketone.

TABLE 5.5 Equilibrium[a] and Rate[b] Constants for Cyanohydrin Formation in Aqueous Solution at 25°C[c]

Compound	log K_{obs}	log k_{obs}	log k_{calc}[d]
CH_2O	7.48	5.54	5.43[e]
CH_3CHO	1.29	2.81	2.74
CH_3COCH_3	−1.84	0.60	3.60
PhCHO	0.74[f]	1.84[f]	2.77
$PhCOCF_3$	3.98[g]	4.67[g]	3.35

[a] $K_{obs} = [RR'C(O^-)(CN)]/[RR'C{=}O][^-CN]$.
[b] $v = k_{obs}[^-CN][RR'C{=}O]$.
[c] Experimental values from Ref. [76] unless otherwise stated. Based on Table 9 of Ref. [21].
[d] Values based on distortion energies calculated at the B3LYP3-21 + G level.
[e] This reaction is predicted to have desolvation of cyanide as rate-limiting process.
[f] Ref. [77].
[g] Ref. [78].

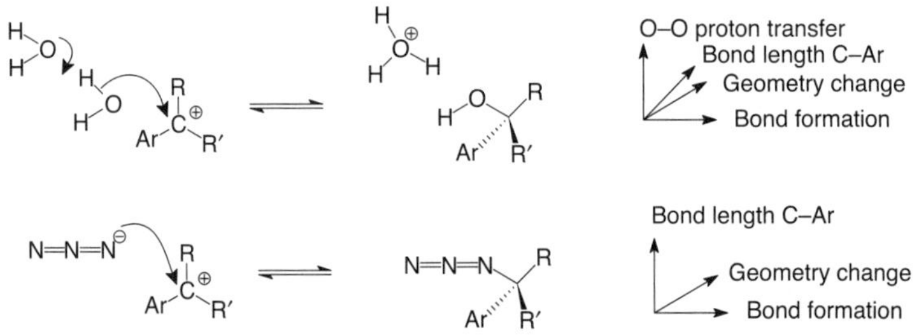

FIGURE 5.5 Mechanisms for reaction of nucleophiles with benzylic carbocations. Based on Figure 1 of Ref. [79].

comparable results. Once again, the more elaborate and more general model will give good results though in many cases a less elaborate model may work as well (Table 5.6).

Since the cations considered in this work all have a phenyl, sometimes with bulky *para* substituents, an additional factor had to be included. A geometry change at a center bearing such a substituent requires the aryl group to sweep through water displacing solvent against the internal pressure of water. This can be allowed for by calculating the volume swept out by the aryl group using the angular change and the area of the aryl group calculated by projecting the coordinates onto the plane of the ring and defining a polygon with an edge tangent to circles drawn with van der Waals radii for each outer atom of the aryl group. Inclusion of this term significantly improved the agreement between calculated and experimental values particularly for 4-dimethylaminophenyl compounds.

5.5.5 Decarboxylation

We have reported several studies of decarboxylation. The first looked at the hydration of CO_2 that is the microscopic reverse of the decarboxylation of carbonic acid. The decarboxylation was best described as a concerted process involving a hydrogen-bonded chain of water molecules. The second focused on the decarboxylation of

TABLE 5.6 Observed and Calculated Free Energies of Activation for the Reaction of Water and Azide with Carbocations[a]

	Water Reaction			Azide Reaction		
Compound	$\Delta G^{\ddagger}_{obs}$ [b]	$\Delta G^{\ddagger}_{calc}$ [c]	Error[d]	$\Delta G^{\ddagger}_{obs}$ [b]	$\Delta G^{\ddagger}_{calc}$ [c]	Error[d]
$C_6H_5–CH_2^+$	1.47	1.26	−0.21			
$CH_3O–C_6H_4–CH_2^+$	5.70	6.89	1.19	4.20	4.67	0.47
$C_6H_5–CH(CH_3)^+$	2.15	3.28	1.13			
$CH_3O–C_6H_4–CH(CH_3)^+$	6.66	7.54	0.88	4.59	6.15	1.56
$(CH_3)_2N–C_6H_4–CH(CH_3)^+$	14.85	15.61	0.75	7.61	10.40	2.79
$C_6H_5–C(CH_3)_2^+$	3.24	4.11	0.87			
$CH_3O–C_6H_4–C(CH_3)_2^+$	7.48	10.45	2.97	5.00	7.56	2.56
$C_6H_5–CH(CF_3)^+$	2.42		−0.59			
$CH_3O–C_6H_4–CH(CF_3)^+$	6.52	4.41	−2.11	4.57	5.60	1.03
$(CH_3)_2N–C_6H_4–CH(CF_3)^+$	13.08	10.61	−2.47			
$C_6H_5–C(CH_3)(CF_3)^+$	3.24	4.28	1.04			
$CH_3O–C_6H_5–C(CH_3)(CF_3)^+$	8.56	9.03	0.47	4.20	6.38	2.18
$C_6H_5–C(CF_3)_2^+$	3.25	2.09	−1.15			
$CH_3O–C_6H_5–C(CF_3)_2^+$	8.60	4.80	−3.80	4.20	6.79	2.59

[a] In aqueous solution at 25°C; all free energies are in kcal/mol. Of the entries under $\Delta G^{\ddagger}_{calc}$ and error, the first is for model 1, the four-dimensional model including a proton transfer to a second water molecule, with no consideration of $P\,\Delta V$ work; the second is for model 2, the three-dimensional model with only one water molecule considered with no consideration of $P\,\Delta V$ work; and the third is for model 3, the three-dimensional model with only one water molecule considered, but now including the effect of $P\,\Delta V$ work. Based on Tables 7 and 8 of Ref. [79].

[b] Calculated from rate constants in Table 1 of Ref. [79].

[c] Calculated as described in the text, by the application of No Barrier Theory.

[d] $\Delta G^{\ddagger}_{calc} - \Delta G^{\ddagger}_{obs}$.

acetoacetic acid and its anion and the aminoacetonitrile-catalyzed decarboxylation of acetoacetic acid. This study showed that the common textbook mechanism (an intramolecular proton transfer accompanying the decarboxylation) does not lead to the observed rate constant, while a stepwise decarboxylation by way of the zwitterion does. The concerted cyclic mechanism involving a hydrogen-bonded chain of water molecules gave the same rate constant as the zwitterion mechanism and at the transition state the proton transfers were complete.

The third looked at a wide range of thermal decarboxylations, and showed that NBT could calculate rate constants over a very wide range of reactivity. The one exception was the decarboxylation of glycine, where the calculated rate constant was too slow. It was suggested that this reaction may proceed by a more elaborate mechanism involving a concerted proton transfer to carbon. Unfortunately we do not yet know how to calculate the rate constant for such a process. Decarboxylations of neutral β-ketoacids consistently proceeded by way of the zwitterions. This contrasts with decarboxylation of carbonic acid that follows a concerted path. The difference was attributed to the energy of the zwitterions relative to the overall kinetic barrier; if the zwitterion is relatively unstable (carbonic acid case), the reaction is concerted to avoid it; if the zwitterion is not relatively unstable (β-ketoacid case), then the reaction is stepwise (Figures 5.6 and 5.7 and Table 5.7).

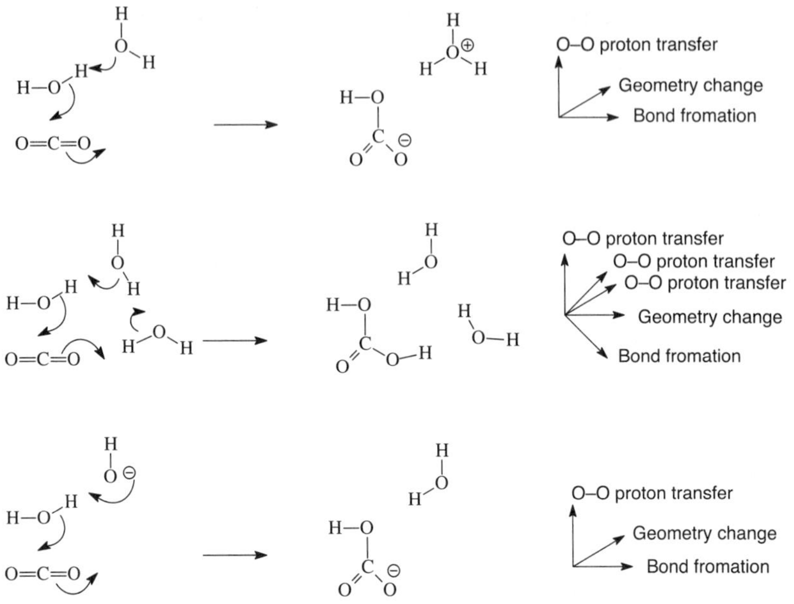

FIGURE 5.6 Mechanistic possibilities for hydration of carbon dioxide (decarboxylation of carbonic acid). Based on structures in Ref. [80].

5.5.6 The E2 Elimination Reaction

The E2 elimination reaction has been examined using NBT. The reaction involves two geometry change dimensions, two bond change dimensions (proton abstraction and loss of leaving group), and a bond length change. To match the observed rate

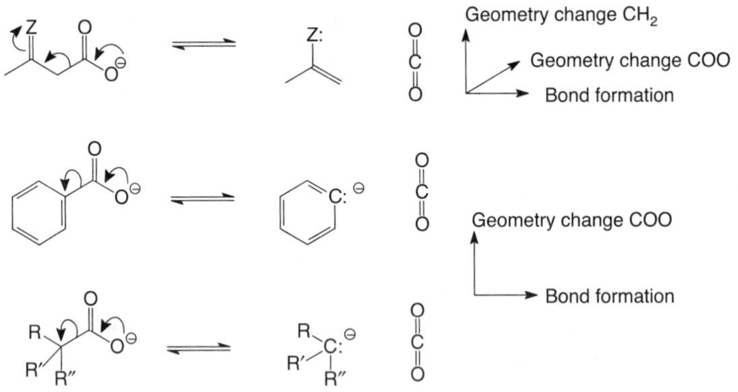

FIGURE 5.7 Mechanisms for decarboxylation. For β-keto or β-imino acids the model has three dimensions: geometry at CO_2, geometry at CH_2, and C–C bond breaking. Z could be O, OH^+, or NRR'^+. For aromatic carboxylates the model has two dimensions: geometry at CO_2 and C–C bond breaking. For carboxylates with electron-withdrawing but nonconjugating substituents, the model has two dimensions: geometry at CO_2 and C–C bond breaking. Based on Figure 1 of Ref. [81] and Figure 12 of Ref. [82].

TABLE 5.7 Observed and Calculated Free Energies of Activation for Decarboxylation Reactions[a]

Reactant	ΔG°_{obs} [b]	ΔG_{pre} or ΔG_{desolv}	ΔG_{rds}	$\Delta G^{\ddagger}_{obs}$	$\Delta G^{\ddagger}_{rds}$	G^{c}	$\Delta G^{\ddagger}_{calc}$
Acetate anion	54.39	13.40[d]	67.79	≥60.30	≥60.30	≥25.99 38.97[e]	70.91
Acetoacetic zwitterion	0.77	13.07[f]	−12.30	23.65	11.31	16.91	9.90
3-(Cyanomethylimino)-butyric acid zwitterion	1.58	2.85	−1.27	16.39	13.54	14.17	17.08
Acetoacetate anion	10.82		10.82	26.11	26.11	20.34	26.13
Benzoate anion	49.71		49.71	55.51	55.51	24.31	55.20
Carbonic acid cyclic mechanism	3.77		3.77	15.78	15.78	17.61	17.86
Cyanoacetate anion	20.06		20.06	30.80	30.80	15.92	31.55
Glycine zwitterion	45.54	11.80[d]	57.34	40.26	40.26	[g]22.09[e]	60.06
Glycine cyclic mechanism	1.30		1.30	40.26	40.26	39.60	
Malonate dianion	29.33		29.33	31.35	31.35	12.32[h] 22.99[e]	40.00
Malonate monoanion	21.26		21.26	30.80	30.80	18.66	32.67
Malonate zwitterion	15.99	13.30[f]	2.68	30.00	16.70	15.33	18.13
1-Methylorotate anion	22.72	9.21[d]	31.93	38.36	38.36	19.05	37.82
Nitroacetate	−2.25		−2.25	21.13	21.13	23.80	21.36
Pyridine-2-acetic acid	13.17		13.17		28.11	21.03	26.37
Pyridine-4-acetic acid	13.16		13.16		28.11	21.01	27.52
Trichloroacetate anion	13.54	7.00[d]	20.54	28.44	28.44	16.57	29.03
Trifluoroacetate anion	29.01	8.60[d]	37.61	40.65	40.65	19.55	42.88

[a] In aqueous solution at 25°C. Based on tables in Refs [80–82]. ΔG_{rds} is the equilibrium free energy change for the microscopic rate-determining step. $\Delta G^{\ddagger}_{rds}$ is the free energy of activation for this step.
[b] To the immediate product of decarboxylation: anion, enol, or enamine.
[c] Intrinsic barrier calculated from ΔG_{rds} and $\Delta G^{\neq}_{rds}$.
[d] ΔG_{desolv}; see Ref. [82].
[e] Calculated from ΔG_{rds} and $\Delta G^{\neq}_{calc}$.
[f] ΔG_{pre} for zwitterion formation.
[g] These data are impossible; the free energy of activation is less than the free energy change for product formation. This suggests that the mechanism is different and leads to a different product than is assumed here.
[h] As discussed in Ref. [82], it seems likely that the rate of this reaction has not yet been measured.

TABLE 5.8 Observed and Calculated Free Energies of Activation for E2 Reactions of Alkyl Bromides with Ethanolic Ethoxide Ion at 25°C, and Transition State Positions[a]

Compound	$\Delta G^{\ddagger}_{obs}$	$\Delta G^{\ddagger}_{calc}$	Error
CH_3CH_2Br	26.46	25.69	−0.76
$(CH_3)_2CHBr$	25.13	23.12	−2.01
$(CH_3)_3Br$	23.63	21.67	−1.96
$PhCH(Br)CH_3$	23.58	23.654	−0.04
$PhCH_2CH_2Br$	22.46	24.66	2.20

[a] All in ethanolic ethoxide solution at 25°C. Based on Table 6 of Ref. [83].

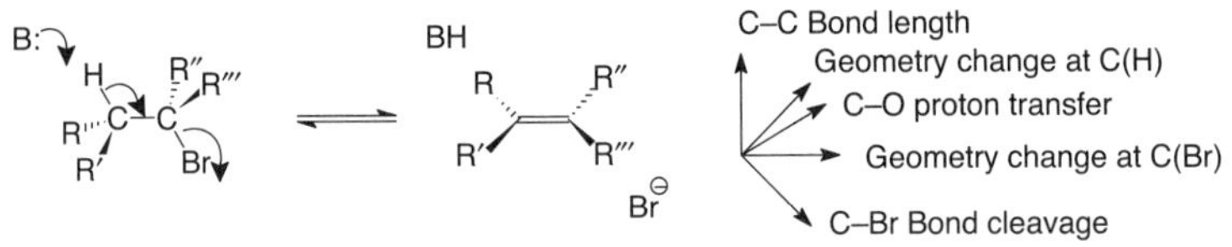

FIGURE 5.8 Mechanism for the E2 reaction. The model involved five dimensions: C–Br bond breaking, C–O proton transfer, two geometry change dimensions, and C–C bond length.

constants, it was necessary to include bond length as a reaction dimension. As in the initial studies of enolate formation, the assumption was made that proton transfer from carbon to oxygen was a simple process; this needs to be reexamined (Figure 5.8 and Table 5.8).

5.5.7 The Strecker Reaction

The Strecker reaction is a synthetically useful reaction, providing a route to amino acids and other materials, but is mechanistically complicated. First, an aldehyde or ketone must be converted to an imine, itself a multistep process, and then the iminium ion reacts with cyanide ion to give the α-aminonitrile. This final step is usually rate determining overall. All steps are reversible, which complicates mechanistic studies. Relatively few cases have been studied to the point where rate and equilibrium constants for the rate-determining step are available, but for these cases NBT led to successful predications of the rate constant. As for cyanohydrin formation, there is a desolvation to be accounted for: the cost of bringing cyanide ion into contact with the iminium carbon with loss of a hydrogen bond to solvent. For best fit this energy cost was set somewhat less for the Strecker addition than for cyanohydrin formation, though useful predicted rate constants for the Strecker addition were obtained using either the value for the cyanohydrin reaction (which was directly calculated from experimental data, though indirectly) or nothing at all (Figure 5.9 and Table 5.9).

5.5.8 The Aldol Addition

The aldol addition, the first step of the aldol condensation, is undeniably an important synthetic reaction. We have examined this reaction in its simplest, hydroxide-

TABLE 5.9 Calculated and Observed Free Energies of Activation for Strecker Reactions[a]

Reaction	ΔG°	$\Delta G^\ddagger_{obs}$	$\Delta G^{\ddagger\ b}_{calc}$	Error[b]
$C_6H_5CH{=}NH^+CH_2Ph\ +\ ^-CN$	−9.38	12.23	12.28	0.05
$C_6H_5CH{=}NH^+CH_2CH{=}CH_2 +\ ^-CN$	−9.91	11.98	11.40	−0.59
$(CH_3C)C{=}NH_2^+\ +\ ^-CN$	−11.02	21.89[c]	22.87[c]	0.98
$(CH_3C)C{=}NH^+CH_3 +\ ^-CN$	−9.28	19.99[c]	20.46[c]	0.47
$(CH_3C)C{=}N^+(CH_3)_2 +\ ^-CN$	−6.79	19.01[c]	18.62[c]	−0.39
RMSE				0.58

[a] All in aqueous solution at 25°C. Dimensions for free energy are kcal/mol. Based on Table 3 of Ref. [84]. Sources for rate and equilibrium constants given in Table 1 of Ref. [84]. Calculated for desolvation cost set to 2.6 kcal/mol; see text.
[b] Difference = $\Delta G^\ddagger_{calc} - \Delta G^\ddagger_{obs}$.[a]
[c] For the cleavage reaction of the Strecker adduct.

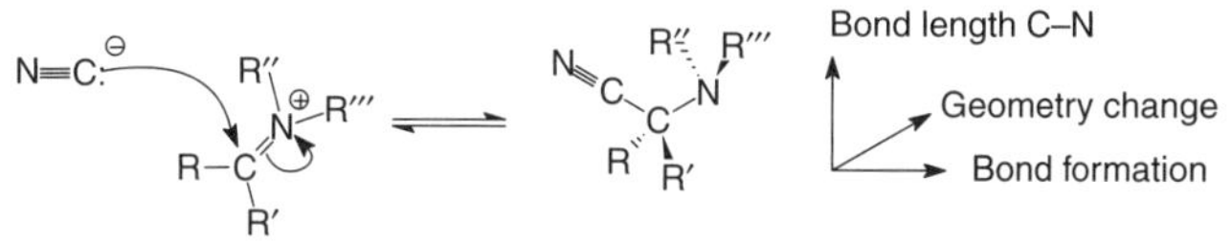

FIGURE 5.9 Mechanism for the Strecker addition. Based on Figure 1 of Ref. [84].

catalyzed form, which even though very old fashioned is also relatively inexpensive and therefore still useful. Enough data are available to permit testing NBT with a matrix of reactions: acetaldehyde, acetone, and acetophenone as nucleophiles and formaldehyde, acetaldehyde, benzaldehdye, acetone, acetophenone, and trifluoroacetophenone as electrophiles. In all cases the rate constants were calculated with useful accuracy (RMSE = 1.68). However, for this reaction it was not possible to calculate all distortions by PM6; for the geometry change dimension of the nucleophile it was necessary to use higher level calculations, in this case B3LYP/6-31 + G**. The mechanism involved three reaction dimensions: bond formation, geometry change at the nucleophilic carbon, and geometry change at the carbonyl carbon. The application of NBT to the second, dehydration, step of the aldol condensation is now being investigated (Figure 5.10 and Table 5.10).

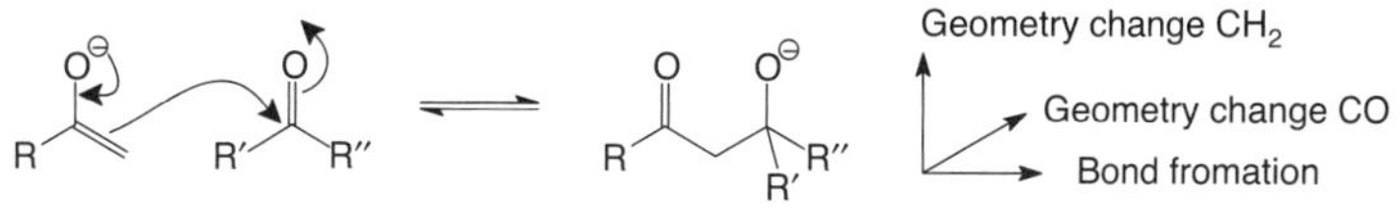

FIGURE 5.10 Mechanism for the aldol addition reaction.

TABLE 5.10 Comparison of Observed and Calculated Free Energies of Activation[a] for the Aldol Addition Reaction

	Nucleophile								
	$CH_2{=}CHO^-$			$CH_2{=}C(CH_3)O^-$			$CH_2{=}C(Ph)O^-$		
Electrophile	$\Delta G^{\ddagger}_{obs}$	$\Delta G^{\ddagger}_{calc}$	ERROR[b]	$\Delta G^{\ddagger}_{obs}$	$\Delta G^{\ddagger}_{calc}$	ERROR[b]	$\Delta G^{\ddagger}_{obs}$	$\Delta G^{\ddagger}_{calc}$	ERROR[b]
CH_2O	8.79	6.33	−2.46	(9.02)	6.46	(−2.56)	(9.75)	7.43	(−2.32)
CH_3CHO	12.81	10.14	−2.67	12.97	11.28	−1.69	(12.87)	10.78	(−2.09)
PhCHO	13.56	11.74	−1.82	11.47	10.58	−0.89	11.77	11.04	−0.73
$(CH_3)_2CO$	15.59	14.28	−1.31	12.78	13.1	0.32	16.40	14.6	−1.80
CH_3COPh	(16.05)	14.72	(−1.33)	15.16	14.11	−1.05	(15.69)	12.63	(−3.06)
CF_3COPh	(10.95)	12.2	(1.29)	9.75	11.36	1.61	(9.96)	10.33	(0.37)

[a] All in aqueous solution at 25°C. Based on Table 3 of Ref. [85]. Observed free energies of activation were calculated from the rate constants in Table 2 of Ref. [85]. Those in parentheses were estimated using Marcus theory [86].

[b] Difference = $\Delta G^{\ddagger}_{calc} - \Delta G^{\ddagger}_{obs}$.

5.6 FUTURE PROSPECTS FOR NBT

When the NBT approach to calculating rate constants was first developed, all the steps in setting up the MO calculations and then proceeding to the calculation of a rate constant had to be done "by hand." As will be apparent, this involved paying attention to a great many details. Even setting up a spreadsheet to deal with the calculation of corner energies was a welcome simplification. We are now developing computer programs to automate as much as possible of the process, with the ultimate goal being to ask the chemist applying the method only to supply structures for starting materials and products, any reactants used or produced, and the solvent, and to specify the mechanism to be followed. Considerable progress has been made toward these goals but much remains to be done.

An important ongoing task is the extension of the NBT approach to a wider range of chemistry. Extension to intramolecular reactions is an obvious goal and will permit study of many important synthetic reactions. This will require new definitions of reaction dimensions, added to those already used. Extension to organometallic reactions is important but conceptually challenging: what are the "simple reaction coordinates" appropriate to the analysis of these reactions? Proton transfer and atom transfers in general require more study. For proton transfers between electronegative atoms the approximation that this is a simple process (with no barrier) is useful for reactions involving other processes (heavy atom bond making, geometry change, etc.) but it fails for reactions involving only proton transfers between electronegative atoms such as the water-mediated proton switch. It is known that there is a small intrinsic barrier associated with proton transfers between electronegative atoms [19]. An intriguing possible solution is to treat the proton transfer as the resultant of two simpler processes, bond making and bond breaking that introduces two corner intermediates—one (bond breaking only) with a proton between two basic atoms and one (bond making only) with a hypervalent hydrogen. The problem is to deal adequately with the solvation of such bizarre species. There are similar conceptual problems with other atom transfers, such as halogenations. Carbocation rearrangements pose an interesting challenge: these reactions are fast if thermodynamically favored, yet they look elaborate, with geometry changes at the end carbons, one bond made and one broken, and possible transient geometry change at the migrating center; if at the transition state the migrating alkyl group resembles a pentacoordinate carbon, then there should be some geometry change in addition to the extra bond. Despite all these possible contributions to the intrinsic barrier, the process should be fast. Clearly there is much to be done to extend NBT.

The question of whether bond length should be treated as a dimension, for example, in addition reactions, has been troublesome. The alternatives were to treat bond length as an additional reaction dimension or to allow bond length to alter during partial optimization of corner species while the dimensions involved in geometry change, bond formation, and so on, were locked. Which seemed to be necessary was linked to how the geometry of anionic tetrahedral intermediates was determined by optimization: initially we used the geometry of the corresponding neutral intermediate, and the inclusion of bond length either had little effect or led to poorer agreement with experiment. As more reactions were studied the inadequacy of basing distortion energies of anionic intermediates on geometries of neutral species became clear and a better approximation was sought. Initially (because geometries were based on DFT calculations) there was a strong pressure to use limited numbers of explicit water molecules to solvate anions because of the rapidly increased computational cost of adding more waters. With one water per lone pair, good results were obtained provided bond length was treated as a dimension, but only for intermediates with a relatively good leaving group, as opposed to aldol addition with a relatively poor leaving group and for which inclusion of bond length as a dimension was neither necessary nor desirable. This situation was unsatisfactory and the answer may now have been found: with much larger numbers of explicit water molecules solvating an anionic center (12 rather than 3, quite practical with semiempirical methods though slow with DFT), to define the geometry of the anionic intermediate, it was again neither necessary nor desirable to include bond length as a reaction dimension. This implied that bond length was never an essential reaction dimension, but sometimes served as a proxy, compensating for inadequacies elsewhere. NBT is an evolving story, but the goal is always to find a computationally inexpensive, general, and logically consistent approach to calculating rates in solution.

5.7 SUMMARY

NBT is based on the following assumptions. Given structures of the species involved in the chemical reaction, a specification of the arrangement of all species involved in the initial and final complexes, all pK_a values for these species, desolvation energies if needed, and the overall equilibrium constant, the calculated rate constant follows from the following assumptions with no other input or adjustment of parameters:

1. All species involved in the reaction are in equilibrium at each point along each reaction coordinate.
2. For any process where only one elementary reaction coordinate changes, the energy will be a quadratic function of the corresponding reaction coordinate with the minimum at the lower energy end of the coordinate.
3. Heterolytic bond cleavage constitutes an elementary reaction coordinate.
4. Proton transfer between two atoms constitutes an elementary reaction coordinate.
5. Geometry change (e.g., from trigonal to tetrahedral) can be described by a single elementary reaction coordinate controlling all the bond angle and dihedral changes associated with the geometry change.

The long-term goal for NBT is to develop a set of consistent mechanistic models that will allow satisfactory calculation of rate constants for the widest possible range of reactions. This, of course, assumes that it will be found possible to describe that wide range of reactions in terms of NBT; thus far, it has been possible for a considerable range of mechanistic types but it remains to be seen how wide the range of applicability will be. An important part of this goal is that the mechanistic models be consistent so that for a new reaction falling within the range of applicability of NBT it will already be clear how to carry out the calculation. It is also desirable that the calculation process be as computationally inexpensive as possible so that it can be used readily as a standard tool for thinking about chemistry. Because the method is still evolving, it is from time to time necessary to go back and confirm that the latest variants still work for families of reactions treated previously. This is, for example, the case for carbonyl hydration reactions where the approach to defining the geometry of anionic tetrahedral intermediates has altered and with it the question of whether bond length is treated as a reaction dimension. The latest version of the approach, with 12 water molecules solvating O^- and bond length not included, works for these reactions. Currently 25 mechanisms have been shown to be successfully treated by NBT (i.e., rms errors of less than 2 kcal/mol in DG*), and another 25 have been defined but not yet confirmed; in some cases new capabilities have to be added to the program system. Extending and developing NBT as a tool for chemistry is now the major focus of my research.

REFERENCES

1. J. P. Guthrie, *Can. J. Chem.* **83**, 1–8 (2005).
2. J. P. Guthrie, *ChemPhysChem* **4**, 809–816 (2003).
3. W. J. Albery, *Prog. React. Kinet.* **4**, 353–398 (1967).
4. R. A. More O'Ferrall, *J. Chem. Soc. B* 274–277 (1970).

5. W. P. Jencks, *Chem. Rev.* **72**, 705–713 (1972).
6. R. A. Marcus, *J. Chem. Phys.* **24**, 966–978 (1956).
7. R. A. Marcus, *Discuss. Faraday Soc.* **29**, 21–31 (1960).
8. R. A. Marcus, *Annu. Rev. Phys. Chem.* **15**, 155–196 (1964).
9. R. A. Marcus, *J. Chem. Phys.* **43**, 679–701 (1965).
10. A. O. Cohen and R. A. Marcus, *J. Phys. Chem.* **72**, 4249–4256 (1968).
11. R. A. Marcus, *J. Phys. Chem.* **72**, 891–899 (1968).
12. R. A. Marcus, *J. Am. Chem. Soc.* **91**, 7224–7225 (1969).
13. V. Shafirovich and S. Lymar, *J. Am. Chem. Soc.* **125**, 6547–6552 (2003).
14. A. Williams, *Acc. Chem. Res.* **22**, 387–392 (1989).
15. J. P. Guthrie, *J. Am. Chem. Soc.* **113**, 3941–3949 (1991).
16. F. G. Bordwell, *Acc. Chem. Res.* **3**, 281–290 (1970).
17. F. G. Bordwell, *Acc. Chem. Res.* **5**, 374–381 (1972).
18. M. J. S. Dewar, *J. Am. Chem. Soc.* **106**, 209–219 (1984).
19. J. P. Guthrie, *J. Am. Chem. Soc.* **118**, 12886–12890 (1996).
20. J. P. Guthrie, *J. Am. Chem. Soc.* **118**, 12878–12885 (1996).
21. J. P. Guthrie, *J. Am. Chem. Soc.* **120**, 1688–1694 (1997).
22. N. Stahl and W. P. Jencks, *J. Am. Chem. Soc.* **108**, 4196–4205 (1986).
23. J. Hine, *J. Am. Chem. Soc.* **94**, 5766–5771 (1972).
24. H. N. Farrer and F. J. C. Rossotti, *Acta Chem. Scand.* **17**, 1824–1825 (1963).
25. H. Persson, *Acta Chem. Scand.* **25**, 1775–1789 (1971).
26. E. S. Hand and W. P. Jencks, *J. Am. Chem. Soc.* **97**, 6221–6230 (1975).
27. R. L. Scott, *J. Phys. Chem.* **75**, 3843–3845 (1971).
28. M. H. Abraham, *Chem. Soc. Rev.* **22**, 73–83 (1993).
29. M. H. Abraham, *J. Chem. Soc. Perkin Trans.* **2** 1893–1899 (1973).
30. R. B. Jordan, *Reaction Mechanisms of Inorganic and Organometallic Systems*, Oxford University Press, New York (1991), p. 33.
31. R. M. Fuoss, *J. Am. Chem. Soc.* **80**, 5059–5061 (1958).
32. M. Eigen, *Z. Elektrochem.* **64**, 115–123 (1960).
33. J. P. Guthrie, *Can. J. Chem.* **68**, 1643–1652 (1990).
34. J. Hine, *J. Am. Chem. Soc.* **93**, 3701–3708 (1971).
35. J. P. Guthrie, *Can. J. Chem.* **70**, 1042–1054 (1992)
36. J. P. Guthrie, *Can. J. Chem.* **69**, 1893–1903 (1991)
37. W. P. Jencks and J. Regenstein, Physical and chemical data, in G. Fassman (ed.), *Handbook of Biochemistry and Molecular Biology*, Vol. 1, 3rd edition, Chemical Rubber Company, Cleveland (1976), pp. 305–451.
38. D. D. Perrin, *Dissociation Constants of Organic Bases in Aqueous Solution*, Butterworth, London (1965).
39. D. D. Perrin, *Ionization Constants of Inorganic Acids and Bases in Aqueous Solution*, 2nd edition, Pergamon Press, Oxford (1982).
40. E. P. Serjeant and B. Dempsey, *Ionization Constants of Organic Acids in Aqueous Solution*, Pergamon Press, Oxford (1979).

41. D. D. Perrin, B. Dempsey, and E. P. Serjeant, *pKa Prediction for Organic Acids and Bases*, Chapman and Hall, London (1981).
42. E. Juaristi and G. Cuevas, *The Anomeric Effect*, CRC Press, Boca Raton (1995).
43. P. Deslongchamps, *Stereoelectronic Effects in Organic Chemistry*, Pergamon Press, Oxford (1983).
44. J. J. P. Stewart, *J. Mol. Model.* **13**, 1173–1213 (2007).
45. J. J. P. Stewart, *MOPAC2009, Version 9.189W,* Stewart Computational Chemistry, CO (2009), available at http://openmopac.net/.
46. M. J. Frisch, G. W. Trucks, H. B. Schlegel, G. E. Scuseria, M. A. Robb, J. R. Cheeseman, J. J. A. Montgomery, T. Vreven, K. N. Kudin, J. C. Burant, J. M. Millam, S. S. Iyengar, J. Tomasi, V. Barone, B. Mennucci, M. Cossi, G. Scalmani, N. Rega, G. A. Petersson, H. Nakatsuji, M. Hada, M. Ehara, K. Toyota, R. Fukuda, J. Hasegawa, M. Ishida, T. Nakajima, Y. Honda, O. Kitao, H. Nakai, M. Klene, X. Li, J. E. Knox, H. P. Hratchian, J. B. Cross, C. Adamo, J. J. Aramillo, R. Gomperts, R. E. Stratmann, O. Yazyev, A. J. Austin, R. Cammi, C. Pomelli, J. W. Ochterski, P. Y. Ayala, K. Morokuma, G. A. Voth, P. Salvador, J. J. Dannenberg, V. G. Zakrzewski, S. Dapprich, A. D. Daniels, M. C. Strain, O. Farkas, D. K. Malick, A. D. Rabuck, K. Raghavachari, J. B. Foresman, J. V. Ortiz, Q. Cui, A. G. Baboul, S. Clifford, J. Cioslowski, B. B. Stefanov, G. Liu, A. Liashenko, P. Piskorz, I. Komaromi, R. L. Martin, D. J. Fox, T. Keith, M. A. Al-Laham, C. Y. Peng, A. Nanayakkara, M. Challacombe, P. M. W. Gill, B. Johnson, W. Chen, M. W. Wong, C. Gonzalez, and J. A. Pople, *Gaussian 03, Revision B.04*, Gaussian, Inc., Pittsburgh, PA (2003).
47. J. P. Guthrie, *J. Phys. Org. Chem.* **11**, 632–641 (1998).
48. J. P. Guthrie, *J. Am. Chem. Soc.* **119**, 1151–1152 (1997).
49. J. R. Keeffe and A. J. Kresge, Kinetics and mechanism of enolization and ketonization, in Z. Rappoport (ed.), *The Chemistry of Enols*, John Wiley & Sons, Inc., New York (1990). pp. 399–480.
50. J. P. Guthrie, *Can. J. Chem.* **57**, 1177–1185 (1979).
51. A. Kankaanpera, L. Oinonen, and P. Salomaa, *Acta Chem. Scand.* **A31**, 551–556 (1977).
52. Y. Chiang, A. J. Kresge, Y. S. Tang, and J. Wirz, *J. Am. Chem. Soc.* **106**, 460–462 (1984).
53. R. P. Bell and O. M. Lidwell, *Proc. R. Soc. Lond. Ser. A* **176**, 88–113 (1940).
54. R. P. Bell and J. Hansson, *Proc. R. Soc. Lond. Ser. A* **255**, 214–217 (1960).
55. J. Hine, K. G. Hampton, and B. C. Menon, *J. Org. Chem.* **89**, 2664–2668 (1967).
56. A. J. Kresge, D. S. Sagatys, and H. L. Chen, *J. Am. Chem. Soc.* **99**, 7228–7233 (1977).
57. T. L. Amyes and J. P. Richard, *J. Am. Chem. Soc.* **118**, 3129–3141 (1996).
58. T. L. Amyes and J. P. Richard, *J. Am. Chem. Soc.* **114**, 10297–10302 (1992).
59. R. P. Bell and H. F. F. Ridgewell, *Proc. R. Soc. Lond. Ser. A* **298**, 178–183 (1967).
60. Y. Pocker, J. E. Meany, and C. Zadorojny, *J. Phys. Chem.* **75**, 792–799 (1971).
61. M. L. Ahrens, M. Eigen, W. Kruse, and G. Maas, *Ber. Bunsenges. Phys. Chem.* **74**, 380–385 (1970).
62. M. L. Eidinoff, *J. Am. Chem. Soc.* **67**, 2072–2073 (1945).
63. D. B. Dahlberg and F. A. Long, *J. Am. Chem. Soc.* **95**, 3825–3831 (1973).
64. G. Schwartzenbach and E. Felder, *Helv. Chim. Acta* **27**, 1701–1711 (1944).
65. R. P. Bell, *Discuss. Faraday Soc.* **39**, 16–23 (1965).
66. R. Brouillard and J. E. Dubois, *J. Org. Chem.* **39**, 1137–1142 (1974).

67. K. J. Pedersen, *J. Phys. Chem.* **38**, 601–621 (1934).
68. J. W. Bunting, J. P. Kanter, R. Nelander, and Z. Wu, *Can. J. Chem.* **73**, 1305–1311 (1994).
69. R. P. Bell, E. Gelles, and E. Moller, *Proc. R. Soc. Lond. Ser. A* **310**, 308–321 (1949).
70. R. P. Bell and E. Gelles, *Proc. R. Soc. Lond. Ser. A* **210**, 310–321 (1952).
71. H. Groth-Andersen and P. E. Sorensen, *Acta Chem. Scand.* **43**, 32–38 (1989).
72. R. G. Pearson and R. L. Dillon, *J. Am. Chem. Soc.* **75**, 2439–2443 (1953).
73. J. P. Guthrie and V. Pitchko, *J. Am. Chem. Soc.* **122**, 5520–5528 (2000).
74. B. P. Callahan, Y., Yuan, and R. Wolfenden, *J. Am. Chem. Soc.* **127**, 10828–10829 (2005).
75. J. P. Guthrie, *J. Am. Chem. Soc.* **122**, 5529–5538 (2000).
76. G. Schlesinger and S. L. Miller, *J. Am. Chem. Soc.* **95**, 3729–3735 (1973).
77. W. M. Ching and R. G. Kallen, *J. Am. Chem. Soc.* **100**, 6119–6124 (1978).
78. C. D. Ritchie, *J. Am. Chem. Soc.* **106**, 7187–7194 (1984).
79. J. P. Guthrie and V. Pitchko, *J. Phys. Org. Chem.* **17**, 548–559 (2004).
80. J. P. Guthrie, *Can. J. Chem.* **77**, 934–942 (1999).
81. J. Guthrie, *Bioorg. Chem.* **30**, 32–52 (2002).
82. J. P. Guthrie, M. Peiris, S. Simkin, and Y. Wang, *Can. J. Chem.* **88**, 79–98 (2010).
83. J. P. Guthrie, L. Leandro, and V. Pitchko, *Can. J. Chem.* **83**, 1654–1666 (2005).
84. J. P. Guthrie and G. Bhaskar, *Can. J. Chem.* **86**, 285–289 (2008).
85. J. P. Guthrie, A. R. Bannister, and S. Peiris, *Faraday Discuss.* **145**, 315–325 (2010).
86. J. P. Guthrie, *J. Am. Chem. Soc.* **113**, 7249–7255 (1991).

PART II

MINIREVIEWS AND APPLICATIONS

CHAPTER 6

Quantum Chemical and Rate Constant Calculations of Thermal Isomerizations, Decompositions, and Ring Expansions of Organic Ring Compounds, Its Significance to Cohbusion Kinetics

FAINA DUBNIKOVA and ASSA LIFSHITZ

Institute of Chemistry, The Hebrew University of Jerusalem, Jerusalem, Israel

6.1 PROLOGUE

6.1.1 Introduction

This chapter represents a long-time effort in performing quantum chemical calculations on various thermal isomerizations, ring expansion, and decomposition reactions of cyclic organic polyatomic molecules. The results of these calculations—potential energy surfaces, transition states and intermediates, their energetics, their vibrational frequencies, and imaginary frequencies—were used to determine reaction mechanisms and calculate rate constants. Since many of the isomerizations and decomposition reactions may have rather complex potential energy surfaces with a number of transition states and intermediates, each such reaction behaves like a complete kinetics scheme that requires a computer modeling in order to obtain the final rate constant. The available structure of the potential energy surfaces, with its transition states and intermediates, enables us to examine in very detail as to how the reaction proceeds, what bonds are broken, and how atom migrations take place toward the formation of the final product(s).

Rate Constant Calculation for Thermal Reactions: Methods and Applications, Edited by Herbert DaCosta and Maohong Fan.

The reaction systems on which the quantum chemical calculations were done and their results that are reported in this chapter are compared with existing experimental data that were obtained mainly in our laboratory. An important issue that should be mentioned at this point is the necessity to compare calculated rate parameters to those reduced in the experiment. One can calculate a very convincing potential energy surface, transition states, and intermediates on a surface and evaluate from them rate parameters. If, however, the calculated values cannot be anchored into experimental results, no one can be sure whether this particular surface is really the relevant one.

This section is a brief description of the various methods of calculations, quantum chemistry, transition state theory, tunneling, intersystem crossing, and so on. We will then very briefly describe the technique with which the experimental results were obtained.

6.1.2 Quantum Chemical Calculations

The quantum chemical calculations were carried out using the density functional theory (DFT) method using Gaussian code package [1] with an appropriate basis set. The three-parameter hybrid density functional method was used with the Lee–Yang–Parr correlation functional approximation (B3LYP) [2,3]. In the majority of cases, the Dunning correlation-consistent polarized valence double ζ (cc-pVDZ) basis set was employed [4]. In several cases where triple ζ (cc-pVTZ) and augmented (aug-cc-pVDZ) basis sets [5,6] were used when radicals were involved, no significant difference was obvious. Structures that had biradical character were located using the unrestricted uB3LYP method including wave functions with the destructive α–β and spatial symmetry. This spin-unrestricted approach has been demonstrated in the past to describe very reliably biradical intermediates and transition states when multireference problems were present [7]. Each optimized B3LYP structure was recalculated at a single point using coupled cluster or quadratic CI including single and double substitutions with a triple contribution to the energy—CCSD(T) [8,9] or QCISD(T) [10].

In some studies the biradical structures and energetics were also optimized by using a complete active space self-consistent field (CASSCF) for comparison. Such calculations were carried out using the GAMESS-USA program [11]. Also, in some cases MP2 calculations were done to verify the B3LYP results.

Structure optimization of the reactants, products, and some transition states was performed by the Berny geometry optimization algorithm without symmetry constraints [12]. In cases where identification of transition states was rather complicated, the relaxed potential energy surface scan and/or the combined synchronous transit-guided quasi-Newton (STQN) method was employed [13].

Vibrational analyses were performed in order to characterize the optimized structures as local minima or transition states. All the calculated frequencies, the zero point, and the thermal energies correspond to harmonic oscillators. The calculations of the intrinsic reaction coordinates (IRCs) for each transition state with mass-weighted internal coordinates were performed in order to check whether the transition states under consideration connect the expected reactants and products [14].

6.1.3 Rate Constant Calculations

In order to evaluate the high-pressure limit first-order rate constants from the quantum chemical calculations, the relation

$$k_\infty = \sigma\Gamma(T)(kT/h)\exp(\Delta S^\#/R)\exp(-\Delta H^\#/RT) \tag{6.1}$$

was used [15,16], where σ is the degeneracy of the reaction coordinate and $\Gamma(T)$ is a possible tunneling correction. Since we deal with unimolecular reactions, $\Delta H^\# = \Delta E^\#$, where $\Delta E^\#$ is the energy difference between the transition state and the reactant.

The tunneling effect $\Gamma(T)$ was estimated using Wigner's inverted harmonic model [17], and it is given by

$$\Gamma(T) = 1 + \frac{1}{24} \times \left(\frac{hc\bar{\lambda}^\#}{kT}\right)^2 \tag{6.2}$$

where $\bar{\lambda}^\#$ is the imaginary frequency of the reaction coordinate in cm^{-1} [18,19]. However, in most cases that are presented in this chapter, this effect was negligible, depending on $\bar{\lambda}^\#$.

Each rate constant was calculated using Eq. (6.1) at several temperatures in order to evaluate the rate constant in terms of an Arrhenius equation. RRKM calculations corresponding to the pressure and temperature range of the experiments were done, particularly when small molecules and high temperatures were involved.

When triplet pathways were significant in product formations, the multiple crossing probability at the minimum energy crossing point (MECP) was estimated with multidimensional semiclassical Landau–Zener (LZ) approximation [20,21]. This probability depends on the spin–orbit coupling (SOC) and on the local topology of the PES (through the energy difference gradient)

$$P_{LZ} = [1-\exp(-2\pi\gamma)]/[1-0.5\exp(-2\pi\gamma) \tag{6.3}$$

where $\gamma = 2\pi^2(SOC)^2/hv\ \Delta F$. v is the effective velocity of passing through the crossing point and ΔF is the difference in the slopes of the two crossing potential curves in the direction orthogonal to the intersection seam at the MECP. The effective velocity was approximated by the average velocity in a one-dimensional Maxwell–Boltzmann distribution [22]

$$v = (k_B T/2\pi\mu)^{1/2}, \tag{6.4}$$

where μ is the reduced mass of the movement along the reaction coordinate. To calculate the SOC at the minimum energy crossing point, the one- and two-electron full Pauli–Breit operator method was used [23]. The wave functions for the SOC were optimized with the complete active space self-consistent field theory in the region of the MECP that was found by using B3LYP method.

A list of all the calculated rate constants and the methods of calculation is given in Appendix.

6.1.4 Experimental Methods

The experimental studies were all done in the gas phase, using the single pulse shock tube technique. In this technique, a reaction mixture is elevated to a high temperature by a reflected shock wave. It stays and reacts at that temperature for a measurable period of time, of the order of 2 ms, and is then very rapidly cooled at a rate of $(0.5 - 1) \times 10^6$ K/s with a set of expansion waves that are generated in the tube. Due to the fast cooling, the reaction is frozen and portions of the shock-heated mixture are then transferred to a variety of analytical devises for analysis. The reactant, in most cases, is highly diluted in an inert gas—in most cases argon. In many cases high-molecular-weight molecules, either solids at room temperature or high boiling point liquids, are studied. In this case the shock tube and all its auxiliary periphery had to be heated, in many cases up to ~200°C, in order to elevate the vapor pressure of the reactants so that the reaction can be studied in the gas phase [24].

The knowledge of the exact temperature is a very important element in evaluating the rate parameters of a chemical reaction. In view of the short reaction times, the latter cannot be measured. There are thus two ways to evaluate the temperature. One is based on the solution of the conservation equations in front and behind the shock wave [25]. Ideal equations are being used and owing to nonidealities in the system, the scatter is relatively high. The second method uses standard reactions whose rate parameters are known and can thus serve as chemical thermometers [26]. For example, the rate parameters of the HF elimination from 1,1,1-trifluoroethane are very well known [27]. The reaction mixture is thus seeded with a minute concentration of the latter (~0.01%) and from the ratio [1,1,-difluoroethylene]/[1,1,1-trifluoroethane] in the shocked sample the temperature can be evaluated. There are several such thermometers depending upon the temperature range involved [28,29]. The scatter in the experimental data is very much reduced as the nonidealities that cause the scatter affect both the reaction under investigation and the standard reaction to the same extent.

Almost all the reactions that were studied both theoretically and experimentally, which will be described in this chapter, simulate in one way or another the processes that prevail in combustion. When a fuel-related molecule is elevated to high temperatures, it decomposes (or isomerizes) regardless of whether oxygen is present in the environment or not. If oxygen is present, the decomposition products react with the oxygen and combustion then takes place. In order to understand the overall combustion mechanism, one must first construct a kinetics subscheme, without oxygen, that will explain the behavior of the fuel in the absence of oxygen. Only then, this subscheme will be combined with the scheme that contains the reactions of the various fuel fragmentation species with oxygen. Let us add that the single pulse shock tube simulates perfectly the conditions that prevail in combustion environments.

6.2 SMALL ORGANIC RING COMPOUNDS

6.2.1 Cyclopropane

Cyclopropane, C_3H_6, is the smallest organic ring compound. It is a strained three-member ring. When elevated to high temperatures, it undergoes three types of

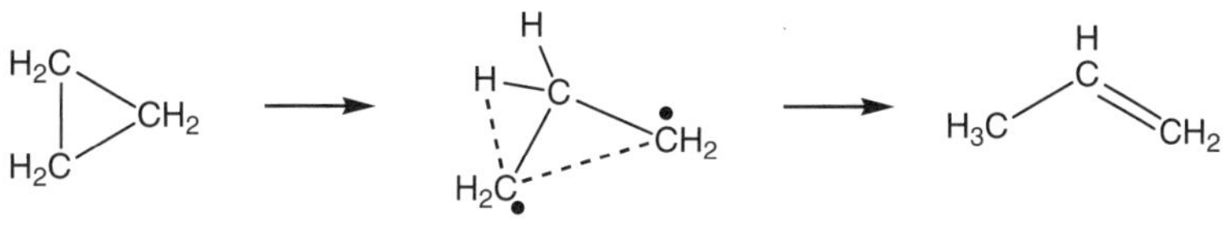

Cyclopropane (**0.0**) TS (**65.1**) Propylene (–**9.3**)

FIGURE 6.1 Reaction scheme for cyclopropane → propylene isomerization.

isomerization processes: optical, geometrical, and structural. Doubleday et al. [30,31] found a trimethylene intermediate as part of structural isomerization surface. Its formation followed by H-atom shift from the center to a terminal carbon atom yields propylene. They could not find, however, a route in which the C–C bond cleavage and the H-atom shift occurred simultaneously. We were able to locate such a transition state as shown in Figure 6.1 [32].

The energy level of the transition state connecting the trimethylene with propylene was very similar to the one going directly from cyclopropane to propylene [30,31,33,34]. We have calculated a rate constant using Eq. (6.1) and performed RRKM calculations to adapt the calculated high-pressure limit first-order rate constant to the pressure where the experiment was done. The comparison between our calculated rate constants and the experimental values of Waage and Rabinovitch [35], for example, was very good [32,36].

The reaction has a factor of 12 degeneracy of the reaction coordinate. The high preexponential factor of $8 \times 10^{15}\ s^{-1}$ is due to the high degeneracy rather than to a high, abnormal entropy of activation. The calculated $\Delta S^{\#}$ was found to be only ~3 cal/(K mol). An additional two-step pathway via propylidene ($H_3C{-}CH_2{-}HC{:}$) formation as an intermediate that was proposed by Bettinger et al. has ~2 kcal/mol higher barrier (of ~67 kcal/mol) [37].

Trimethylene as an intermediate for the geometrical and optical isomerizations has been calculated by a large number of investigators [38–46]. Our results on the geometrical and the optical isomerizations were similar to the other calculations.

6.2.2 Cyclopropane Carbonitrile

When one hydrogen atom is replaced by a functional group, four isomers are formed in the process of the structural isomerization. Calculations [47] and experimental study [48,49] were performed on cyclopropane carbonitrile (C_3H_5CN), where the functional group is a –C≡N group.

The production of three isomers is initiated by a C–C bond breaking adjacent to the carbon connected to the functional group, channel 1 (Figure 6.2). The three isomers in channel 1 are *cis*-crotonitrile, *trans*-crotonitrile, and vinyl acetonitrile. The fourth isomer, methacrylonitrile, is obtained via channel 2. The production of the last is initiated by the C–C bond breaking opposite to the C–R bond, channel 2 (Figure 6.3).

The quantum chemical calculations show that the first three isomers are produced via two transition states and one intermediate each. Methacrylonitrile is obtained directly from the cyclopropane carbonitrile via one transition state with no intermediates.

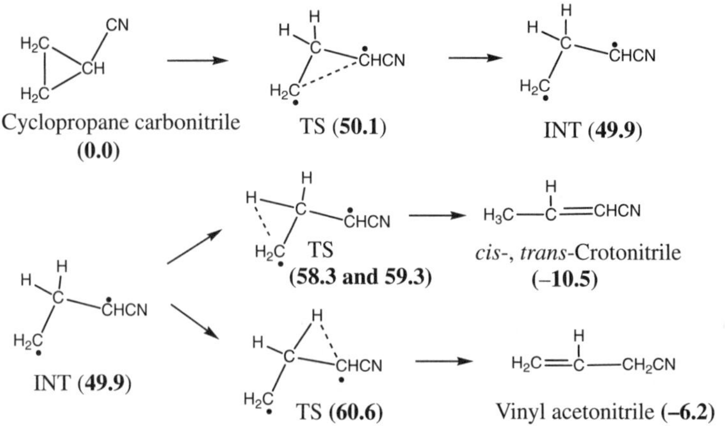

FIGURE 6.2 Reaction schemes for cyclopropane carbonitrile isomerizations to *cis*-, *trans*-crotonitrile, and vinyl acetonitrile (channel 1). All the energy values are calculated relative to the reactant.

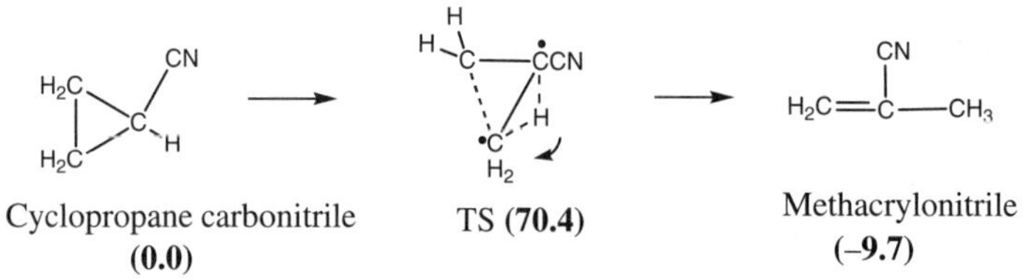

FIGURE 6.3 Reaction schemes for cyclopropane carbonitrile isomerizations to methacrylonitrile (channel 2).

The reaction scheme and the energetics of all the four isomerizations are shown in Figures 6.2 and 6.3.

Rate constants were calculated using Eq. (6.1) and were compared with our past experimental results [49]. They are shown in Figure 6.4. The comparison between the experimental and calculated rate constants is excellent.

6.2.3 The Epoxy Family of Molecules

The epoxy family of molecules is another group of three-member ring compounds where one CH_2 group in the ring is replaced by an oxygen atom. The simplest ring compound is ethylene oxide, C_2H_4O. It has a number of derivatives. For example, when one hydrogen atom is replaced by a methyl group, propylene oxide is obtained. When two hydrogen atoms are replaced by methyl groups, 1,1- and 1,2-butylene oxide are formed. In propylene oxide, when elevated to high temperatures, four isomerization products are formed: propyl aldehyde (propanal), allyl alcohol, acetone, and methyl vinyl ether [50–55].

The first three isomerizations take place without any intermediates and with a single transition state on their potential energy surfaces [52]. All that is required for the isomerizations to take place is a C–O bond cleavage and an H-atom shift that

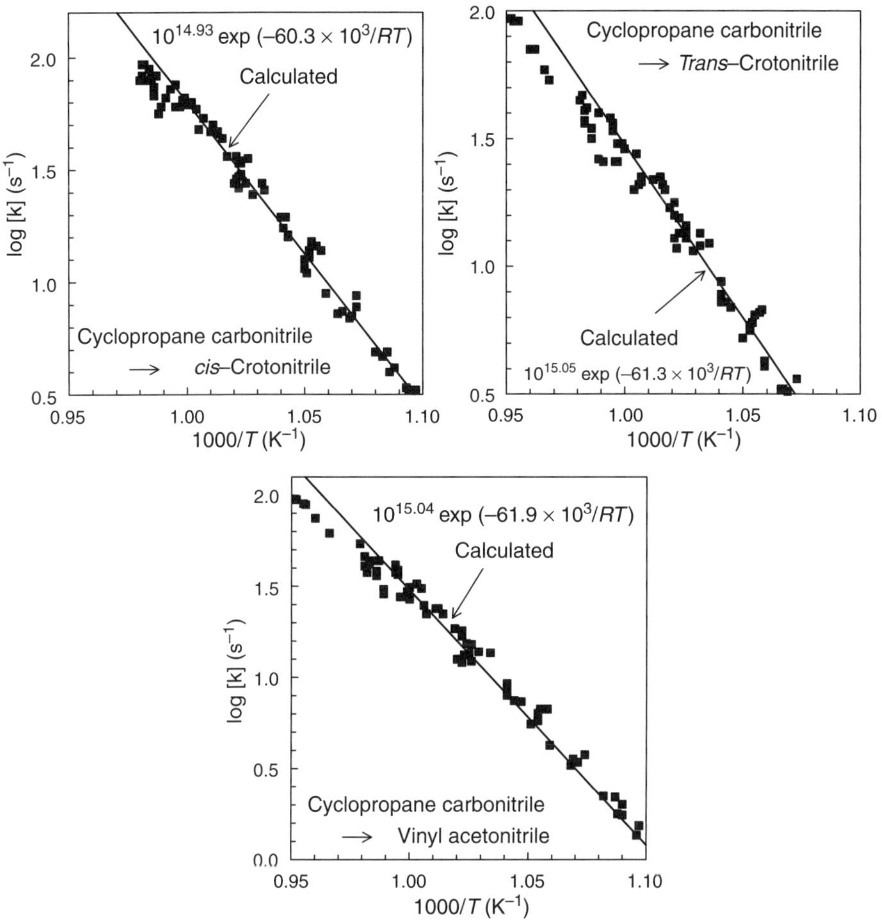

FIGURE 6.4 Comparison between the calculated [47] and experimental [48,49] rate constants of the cyclopropane carbonitrile isomerizations.

occur simultaneously. The production of propanal is quite similar to that of crotonitrile in the isomerization of cyclopropane carbonitrile. The fourth isomerization product is methyl vinyl ether. Here, a C—C bond breaks rather than a C—O bond. It is then followed by a rotation of the CH_3 group around the existing C—O bond and ends by H-atom shift from the methyl group to a CH_2 group. This pathway has two intermediates and three transition states. There is also a concerted mechanism for the methyl vinyl ether production via a carbene intermediate, but its barrier is unrealistically high. A diagram of the four reaction schemes is shown in Figure 6.5.

Rate constants for the four processes were calculated [52] and compared with the experimental values [50,51]. As can be seen in Figure 6.6, the agreement is quite good.

There are some investigations on the influence of a proton and water molecules on the ring-opening process of propylene oxide. It was found that the ring-opening

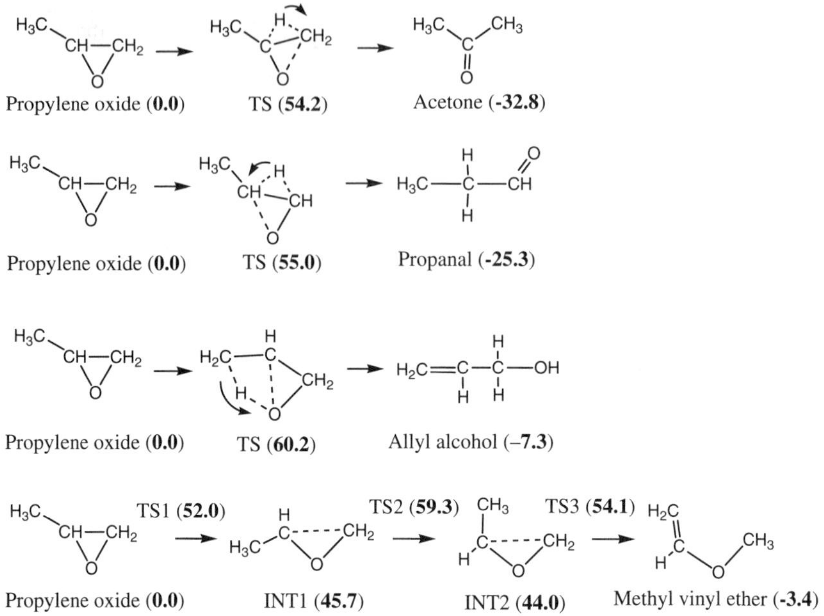

FIGURE 6.5 Reaction schemes for propylene oxide isomerizations. All the energy values are calculated relative to the reactant.

process occurs in a protonated system that is easier than in the isolated monomer (<30 kcal/mol) [56–58].

6.3 PYRROLE AND INDOLE

6.3.1 Pyrrole

Pyrrole, C_4H_5N, is an isomer of cyclopropane carbonitrile. It is a five-member ring compound where the nitrogen atom is part of the ring. It isomerizes to *cis*-crotonitrile and vinyl acetonitrile, both of which are also isomerization products of cyclopropane carbonitrile [59–62]. Methacrylonitrile and *trans*-crotonitrile, which are also isomers of cyclopropane carbonitrile, are not formed in pyrrole isomerization. *trans*-Crotonitrile, however, can be obtained by *cis* → *trans* isomerization. The first step in the isomerization is an H-atom shift from the nitrogen atom to an adjacent sp^2 carbon to form an sp^3 carbon atom. This step produces pyrrolenine that is then followed by N–C bond cleavage [59–63]. The biradical intermediate that is formed (INT1, Figure 6.7) further isomerizes via two pathways, path (A) and path (B). Path (A) produces vinyl acetonitrile and path (B) produces *cis*-crotonitrile, as can be seen in Figure 6.7.

Pyrrolenine [64–66] formation was shown to be the lowest energy pathway also in pyrrole pyrolysis (HCN formation) [59,63], starting from both pyrrole [59–63,67,68] and pyrrolyl [69].

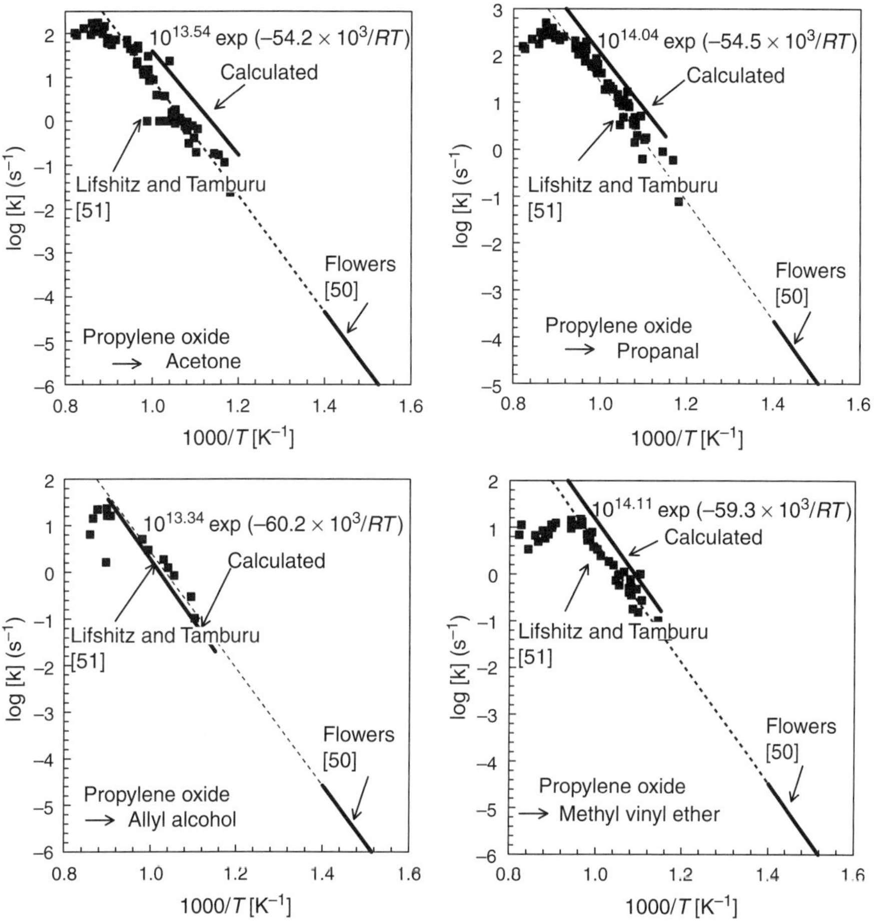

FIGURE 6.6 Comparison between the calculated [52] and experimental [50,51] rate constants of propylene oxide isomerizations.

Figure 6.8 shows a comparison between the experimental [59] and the calculated isomerization rate constants [61]. The agreement is reasonable.

6.3.2 Indole

Indole is a compound composed of a pyrrole ring fused to benzene. It isomerizes to *o*-toluisocyanide and phenyl acetonitrile [70,71]. *o*-Toluisocyanide then further isomerizes to *o*-tolunitrile (Figure 6.10). Over the temperature range of the present experiments (1050–1540 K), $-N{\equiv}C \rightarrow -C{\equiv}N$ isomerization is very fast [72,73] so that the isocyanide is not observed experimentally and the final product is the *o*-tolunitrile.

Similar to pyrrole, the isomerization of indole to produce phenyl acetonitrile begins with an H-atom shift from the nitrogen atom to an adjacent sp^2 carbon in the

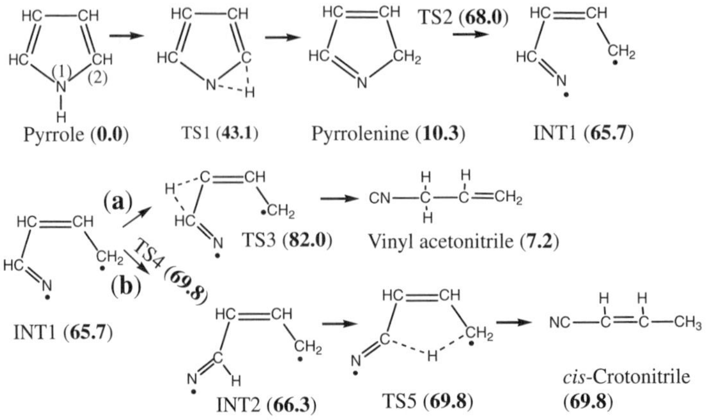

FIGURE 6.7 Reaction schemes for pyrrole isomerizations. All the energy values are calculated relative to the reactant.

benzene ring, to form an sp^3 carbon atom with a considerable loss of resonance of the benzene ring. The N(1)–C(9) bond is then cleaved and with an additional H-atom shift phenyl acetonitrile is formed. The pathway can be seen in Figure 6.9.

The isomerization pathway that leads to the formation of *o*-tolunitrile has also three transition states and two intermediates. First, 2*H*-indole is formed, and then 3*H*-indole [73–75] followed by cleavage of the C(2)–C(3) bond together with an H-atom shift from C(2) to C(3). The reaction scheme and the energetics are shown in Figure 6.10.

Figure 6.11 shows a comparison between the calculated [71] and the experimental [70] rate constants of the two final products. As can be seen, the agreement for phenyl acetonitrile is excellent, whereas for *o*-tolunitrile the deviation between the

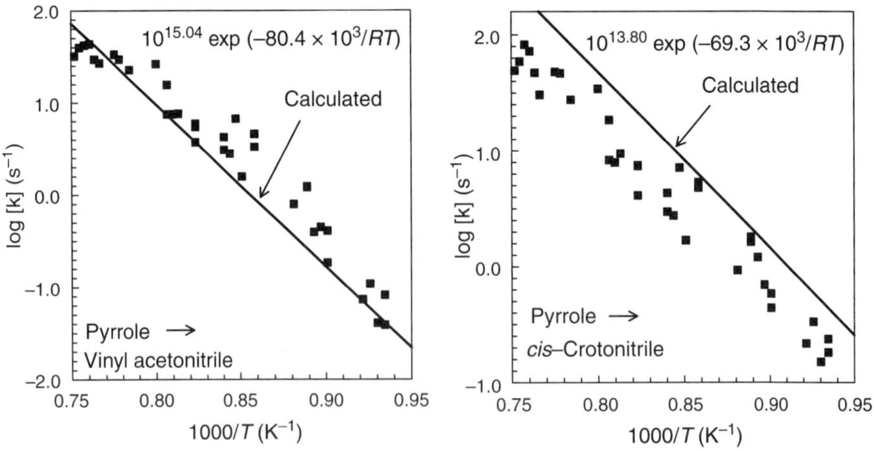

FIGURE 6.8 Comparison between the calculated [61] and experimental [59] rate constants of pyrrole isomerization.

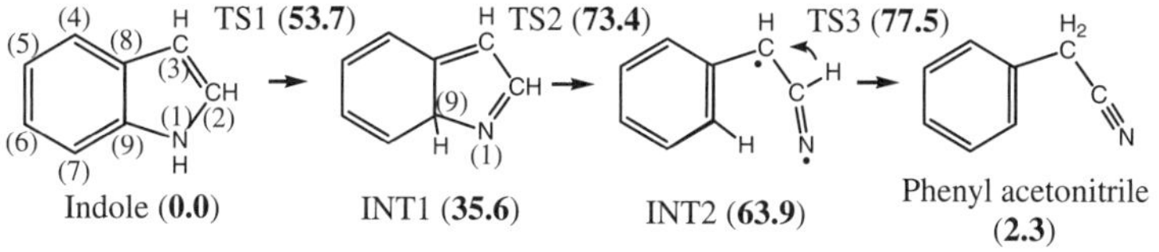

FIGURE 6.9 Reaction scheme for indole → phenyl acetonitrile isomerization. All the energy values are calculated relative to the reactant.

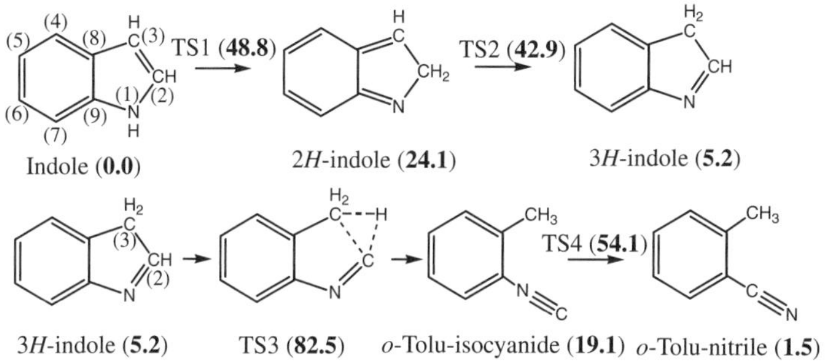

FIGURE 6.10 Reaction schemes for indole → *o*-tolunitrile isomerization. All the energy values are calculated relative to the reactant.

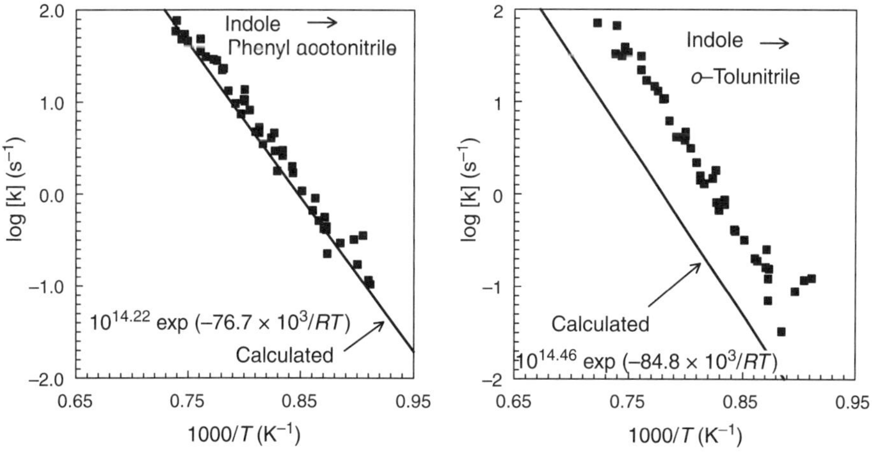

FIGURE 6.11 Comparison between the calculated [71] and experimental [70] rate constants of indole isomerization.

calculated and the experimental value is a factor of ~5 that is equivalent to approximately 4 kcal/mol.

6.4 DIHYDROFURANS AND DIHYDROBENZOFURANS

Another group of molecules that undergo isomerization and decomposition processes is five-member ring compounds, where oxygen is part of the ring. These are the furan–hydrofuran family of molecules, with and without fused benzene ring. Five of these compounds will be described in this chapter. They are 2,3-dihydrofuran, 5-methyl-2,3-dihydrofuran, 2,5-dihydrofuran, dihydrobenzofuran, and *iso*-dihydrobenzofuran. Except for 2,5-dihydrofuran that undergoes 2,5-H_2 elimination to form furan, and does not isomerize, all the other compounds undergo isomerization processes.

6.4.1 2,3-Dihydrofuran

The main isomerization products of 2,3-dihydrofuran are cyclopropane carboxaldehyde (reaction (1)) and propenyl aldehyde (reaction (2)) [76,77]. Propenyl aldehyde can be obtained also via isomerization of the product—cyclopropane carboxaldehyde (reaction (3)). Reactions (1) and (2) go directly from the reactant to the product via one transition state. Reaction (3) has one intermediate and two transition states. The transition state for the production of cyclopropane carboxaldehyde (reaction (1)) involves a C(sp^3)–O bond cleavage and formation of a new C=O double bond with the second adjacent carbon atom, together with a C–C bond formation (Figure 6.12).

The isomerization of cyclopropane carboxaldehyde to propenyl aldehyde involves an opening of the cyclopropane ring and H-atom shift to form a methyl group. The pathway for this process is very similar to the one that has been found for cyclopropane carbonitrile that has been described in Section 6.2.2.

Rate constants for the three processes were calculated using the methods described in Section 6.1.3, but experimental data for comparison are unavailable.

6.4.2 5-Methyl-2,3-Dihydrofuran

The pathways of 5-methyl-2,3-dihydrofuran isomerization are exactly identical to those of the 2,3-dihydrofuran except that the energetics is slightly higher. Also, instead of the aldehyde, the product is ketone. As experimental rate parameters for the isomerization of this reactant were available [78,79], a comparison between the experimental and calculated [77] rate parameters was thus made.

The rate constants were evaluated using the calculated $\Delta S^{\#}$ and $\Delta E^{\#}$ at several temperatures over the temperature range covered in the experiment. They were then transferred to Arrhenius-type rate equations in order to compare the calculated and the experimental rate constants. The comparisons are shown in Figure 6.13. As can be seen, the agreement between the experimental and the calculated rate constants is

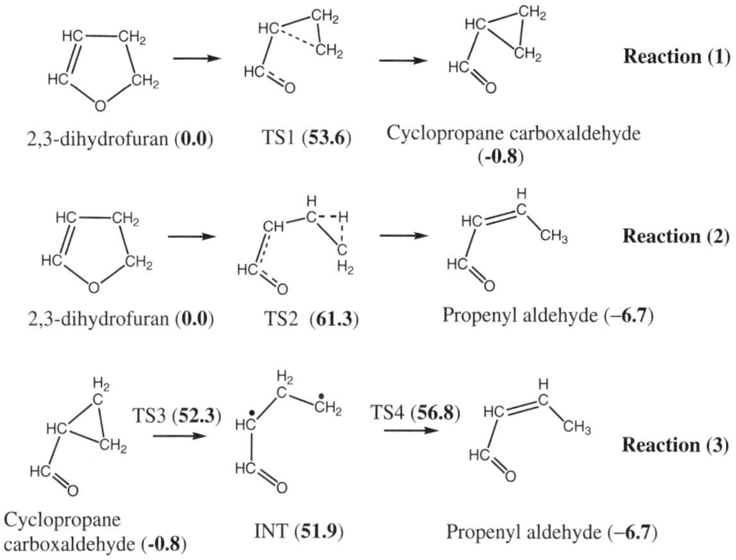

FIGURE 6.12 Reaction schemes for 2,3-dihydrofuran isomerizations and its products' interconversion. All the energy values are calculated relative to the reactant—2,3-di-hydrofuran.

good for reactions (1) and (3). For reaction (2) the calculated value is lower than the experimental one by the equivalence of ~4.5 kcal/mol.

6.4.3 Van der Waals Interactions in H_2 Elimination: 2,5-Dihydrofuran

An additional reaction in 2,3-dihydrofuran, and the only reaction in 2,5-dihydrofuran, is H_2 elimination to yield furan [80–84]. In 2,5-dihydrofuran the mechanism is rather simple, one transition state and no intermediates. The two hydrogen atoms come close to one another to form a six-center transition state and produce furan and hydrogen molecule. The barrier is approximately 49 kcal/mol. This is shown in Figure 6.14.

In 2,3-dihydrofuran the process is more complicated; the two hydrogen atoms are eliminated from a single carbon atom to form a hydrogen molecule. The barrier here is considerably higher than the one in 2,5-dihydrofuran, namely, 82 kcal/mol compared to 49 kcal/mol. It will thus isomerize before H_2 elimination can take place. The mechanism is shown in Figure 6.15.

The interesting part in these reactions is the fact that the hydrogen molecule, in spite of its distance from the furan molecule being rather high (~2.6 Å in 2,5-dihydrofuran and ~3.2 Å in 2,3-dihydrofuran) and not corresponding to normal bonds, is still part of the entire complex via Van der Waals interactions. This expresses itself by three phenomena: (1) the number of vibrational frequencies of the complex is the same as that of the original molecule, namely, 2,5-dihydrofuran and 2,3-dihydrofuran. (2) The vibrational frequency of the hydrogen molecule in all cases is around 4295 cm^{-1} that is lower than the 4368 cm^{-1} of the unbound molecule.

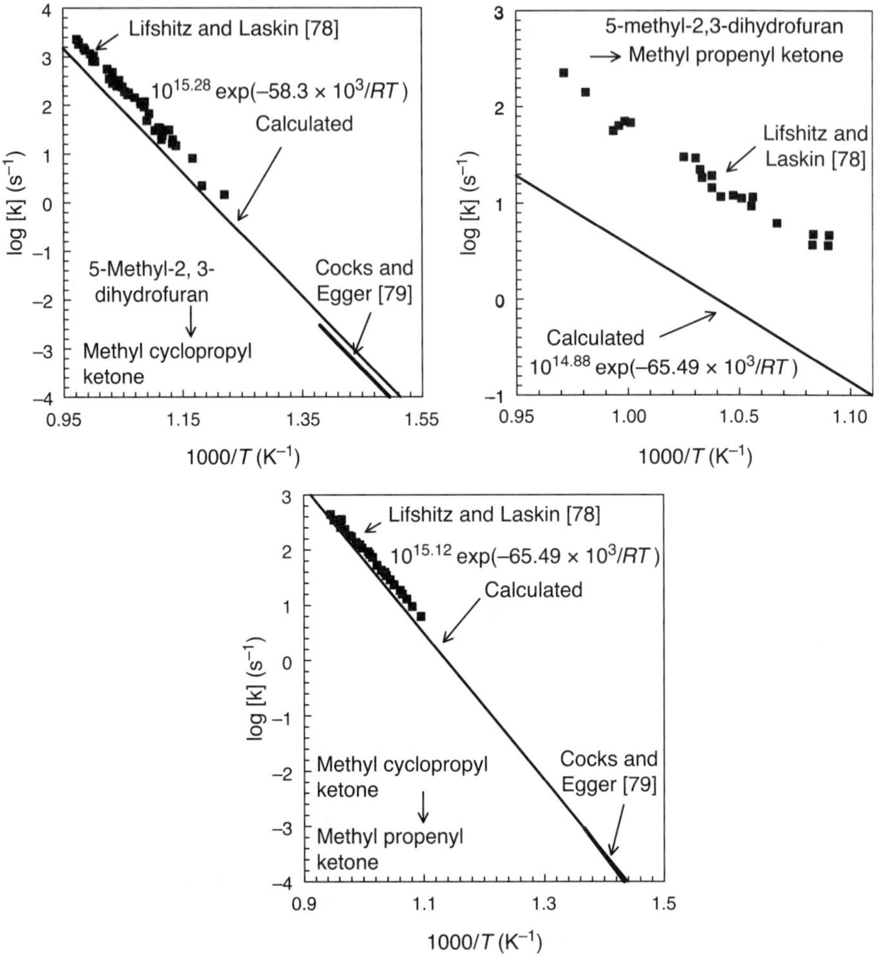

FIGURE 6.13 Comparison between the calculated [77] and experimental [78,79] rate constants of 5-methyl-2,3-dihyrofuran isomerizations and its products' interconversion.

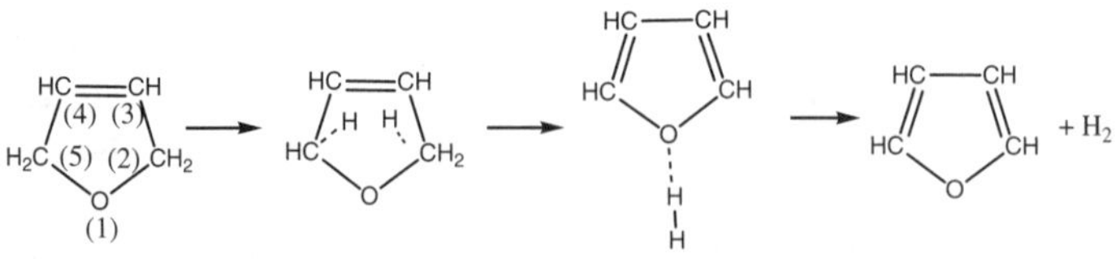

2,5-Dihydrofuran (**0.0**) TS (**49.2**) Furan·H_2(**5.1**) Furan + H_2(**4.9**)

FIGURE 6.14 Reaction scheme for H_2 elimination from 2,5-dihydrofuran. All the energy values are calculated relative to the reactant.

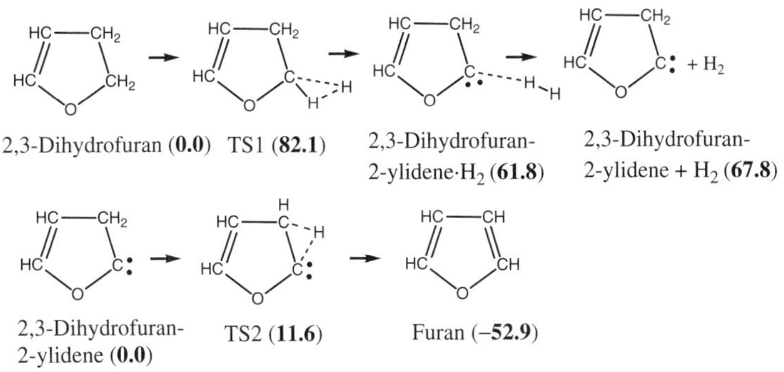

FIGURE 6.15 Reaction scheme for H_2 elimination from 2,3-dihydrofuran. All the energy values are calculated relative to the reactant.

(3) The most important phenomenon, however, is the fact that the entire entropy of the complex molecule, although higher than the original one, is considerably lower than the combined entropy of the product molecule + H_2. This means that the translational Sackur–Tetrode entropy has not yet been gained. The comparison between the entropy values of the original molecules, the complex molecules, and the product molecules + H_2 is shown in Table 6.1.

In 2,5-dihydrofuran the interaction is between the hydrogen molecule and the oxygen. In 2,3-dihydrofuran the bonds are between the hydrogen molecule and either the carbene structure or the C–C π-bond. This can be seen in Figure 6.16.

It should be mentioned that 2,3-dihydrofuran isomerizes via ring opening at a considerable lower barrier than its H_2 elimination process. Figure 6.17 shows a comparison between the calculated rate constant [84] of H_2 elimination from 2,5-dihydrofuran and the experimental results [81,82]. The agreement is excellent.

6.4.4 Dihydrobenzofuran and *iso*-Dihydrobenzofuran

Dihydrobenzofuran is a derivative of 2,3-dihydrofuran where a fused benzene ring replaces the C(4)=C(5) double bond in the ring. *iso*-Dihydrobenzofuran is similar to 2,5-dihydrofuran where the benzene ring replaces the double bond in the C(3)=C(4) position. The four molecules are shown in Figure 6.18.

TABLE 6.1 A Comparison Between the Entropy Values (in cal/(K mol)) of the Original Molecule (A), the Van der Waals Complex (B), and the Final Product + H_2 (C)

Original Molecule	Complex Type	A	B	C
2,5-Dihydrofuran	Furan$\cdots H_2$	69.3	82.9	96.5
2,3-Dihydrofuran	Furan$\cdots H_2$	67.6	84.6	98.0
	2,3-Dihydrofuran 2 yliden$\cdots H_2$	67.6	86.3	96.5

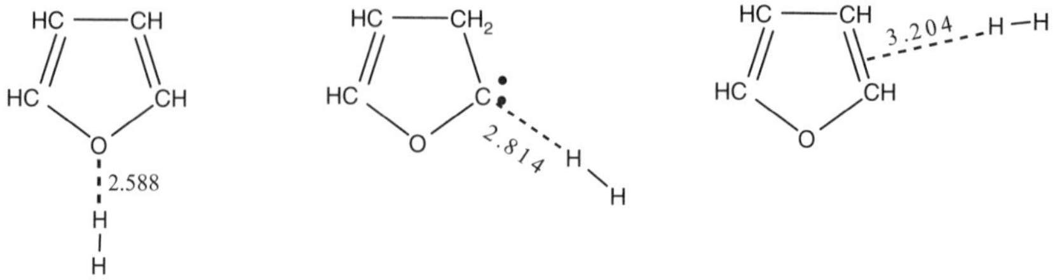

FIGURE 6.16 Molecular complexes 2,3- and 2,5-dihydrofurans with hydrogen molecule.

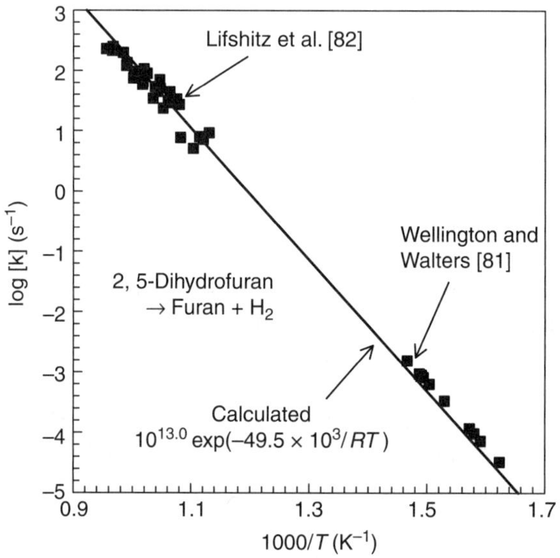

FIGURE 6.17 Comparison between the calculated rate constant [84] of H_2 elimination from 2,5-dihydrofuran and the experimental results [81,82].

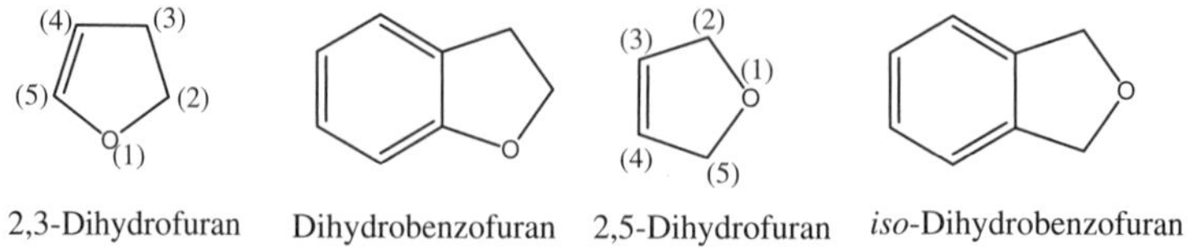

FIGURE 6.18 Dihydrofuran and dihydrobenzofuran structures.

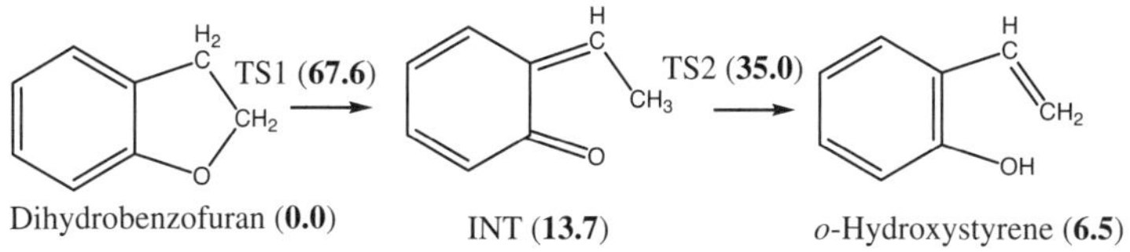

FIGURE 6.19 Reaction scheme for dihydrobenzofuran → *o*-hydroxystyrene isomerization. All the energy values are calculated relative to the reactant.

Both dihydrobenzofuran and *iso*-dihydrobenzofuran undergo a single isomerization reaction [85–87]. Dihydrobenzofuran isomerizes to *o*-hydroxystyrene [85], and *iso*-dihydrobenzofuran to *o*-tolualdehyde [86]. It is of interest to compare the isomerization pathways of dihydrobenzofuran to that of 2,3-dihydrofuran (see Figure 6.12, reaction (1)). The pathway of the 2,3-dihydrofuran is a single-step process with one transition state in which the C(2)–O(1) bond opens with an H-atom shift with no intermediates. The pathway of dihydrobenzofuran, on the other hand, is a two-step process, although its first transition state is very similar to the one in the 2,3-dihydrofuran isomerization. The compound that follows this transition state results in a loss of resonance energy and is thus unstable. It further isomerizes with a rather low barrier (35.0–13.7 kcal/mol) to form the final product, *o*-hydroxystyrene, regaining the lost resonance energy (Figure 6.19).

In *iso*-dihydrobenzofuran, the formation of *o*-tolualdehyde, which is the final isomerization product, requires a C–O bond cleavage and an H-atom shift that cannot take place simultaneously owing to steric hindrance. Therefore, it has three transition states and two intermediates. In the first step, the C–O bond opens; in the second, a rotation of the $-CH_2-O^{\bullet}$ group occurs so that the hydrogen atoms in the $-CH_2$ group that is connected to C(8) can come closer to the methylene group that is attached to C(9) in order to facilitate the H-atom shift to form the methyl group and then the *o*-tolualdehyde. The pathway is shown in Figure 6.20.

Rate constants for the two isomerization processes were calculated using $\Delta S^{\#}$ and $\Delta E^{\#}$ as described in Section 6.1.3. The comparison with the experimental values in a log k versus $1/T$ plot is shown in Figure 6.21. As can be seen, the agreement is very good.

The comparison between the two rate constants in Figure 6.21 demonstrates an interesting feature. The closed-shell pathway of dihydrobenzofuran should have proceeded much faster (via much lower barriers) than in the biradical one of *iso*-dihydrobenzofuran. In reality, they are quite similar. The reason is that the loss of

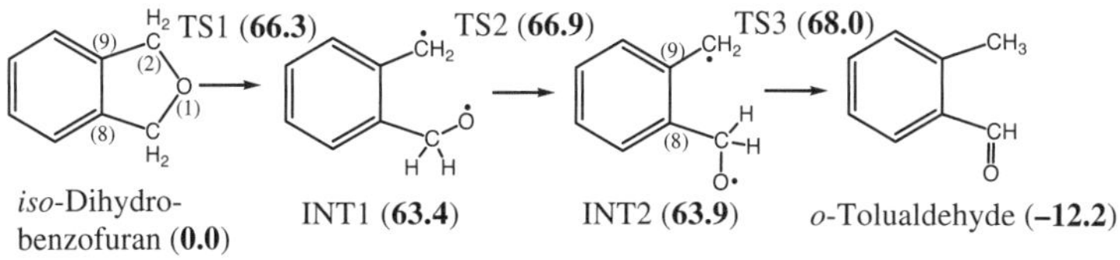

FIGURE 6.20 Reaction scheme for *iso*-dihydrobenzofuran → *o*-tolualdehyde isomerization. All the energy values are calculated relative to the reactant.

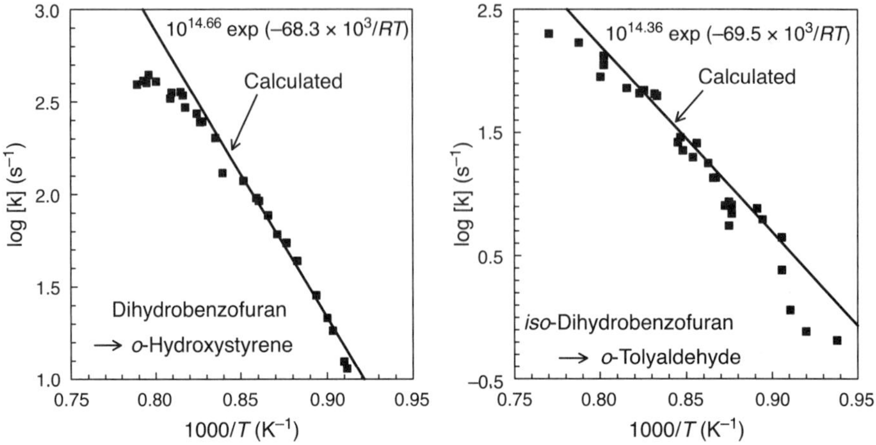

FIGURE 6.21 Comparison between the calculated [87] and experimental rate constants of dihydrobenzofuran [85] and *iso*-dihydrobenzofuran [86] isomerizations.

aromaticity (in TS1, INT, Figure 6.19) has approximately the same effect as the high-energy level of the intermediate biradical species (TS1, INT1, TS2, INT2, TS3 in Figure 6.20).

6.5 NAPHTHYL ACETYLENE–NAPHTHYL ETHYLENE

In this section, studies on the reactions of naphthyl radicals with unsaturated aliphatic hydrocarbons toward the production of three fused rings and of aliphatic residues attached to naphthalene are described. This is an effort to examine the process of polycyclic aromatic hydrocarbon (PAH) growth [88–99]. The source of naphthyl radicals for the experimental study was naphthyl iodide, in view of its low C–I bond dissociation energy. It dissociates very fast following the reflected shock heating. The unsaturated aliphatic hydrocarbons that have been studied were ethylene [28] and acetylene [29].

The reactions between naphthyl radical and both ethylene and acetylene are attachment reactions. The reaction with ethylene produces naphthyl ethylenyl radical and with acetylene the product is naphthyl acetylenyl, both with almost no barrier [28,29,88,96]. The attachment products are shown in Figure 6.22.

The potential energy surfaces of the two processes are shown in Figure 6.23. As can be seen, the surfaces are very similar. Both radicals react via two parallel channels. In one channel (via TS1), both eject an H-atom to produce naphthyl ethylene and naphthyl acetylene with energy barriers of 35 and 38 kcal/mol, respectively. In the second channel (via TS2) with energy barriers of 17.3 and 19.3 kcal/mol, they isomerize to acenaphthalenyl (INTeth) in surface (II) and acenaphthylenyl (INTac) in surface (I) toward the formation of acenaphthene and acenaphthylene. Since the

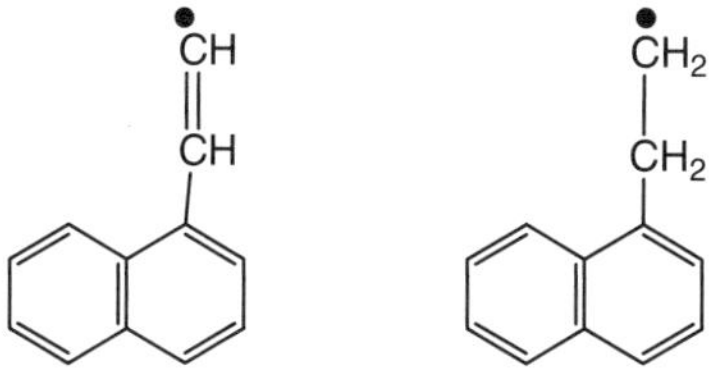

1-Naphthyl acetylenyl 1-Naphthyl ethylenyl

FIGURE 6.22 1-Naphthyl acetylenyl and 1-naphthyl ethylenyl structures.

energy barrier of the first channel on the surface is much higher than that of the second channel (~38 versus 19 kcal/mol), the second channel is the preferable route. Nevertheless, a small concentration of naphthyl ethylene is still obtained in the experiment, whereas naphthyl acetylene is not obtained at all [29].

The question that arises is why naphthyl ethylene is formed in the reaction of naphthyl radicals with ethylene, whereas naphthyl acetylene is not formed at all in the reaction with acetylene, although the barriers from 1-naphthyl ethylenyl and 1-naphthyl acetylenyl for their production are very close (~17 and 19 kcal/mol in the first step, and ~30 (9.8 + 19.7) and 28 (18.8 + 9.1) kcal/mol in the second step) (Figure 6.23).

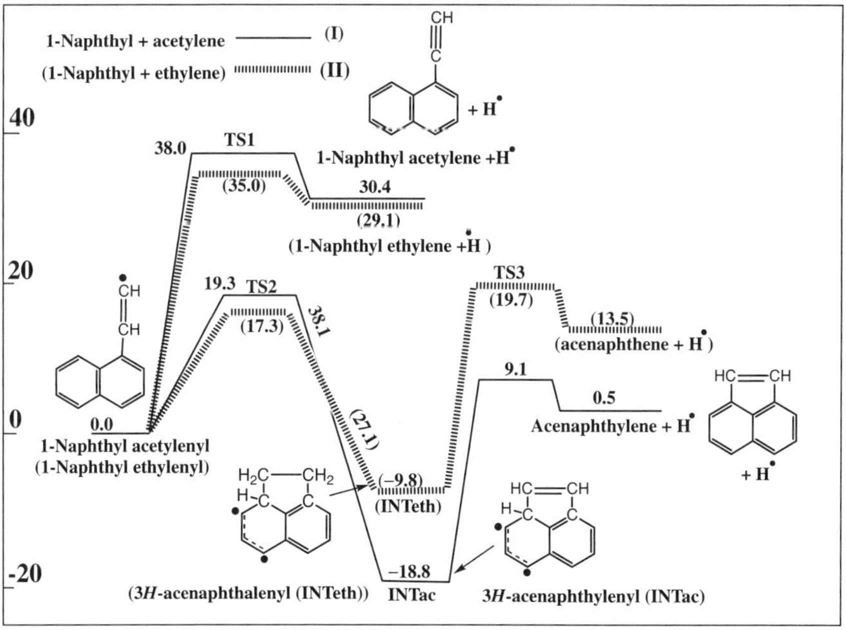

FIGURE 6.23 Potential energy surfaces for 1-naphthyl acetylenyl(I) and 1-naphthyl ethylenyl(II) reactions.

The only significant difference between the ethylene and acetylene pathways is the barriers of the reverse reactions back to 1-naphthyl ethylenyl and 1-naphthyl acetylenyl via TS2. Whereas for ethylene it is 27 kcal/mol (9.8 + 17.3), it is 38 kcal/mol for acetylene (18.8 + 19.3) (Figure 6.23). This means that the concentration level of 1-naphthyl ethylenyl is enough to produce naphthyl ethylene with the barrier of 35 kcal/mol following H-atom ejection [29]. This significant difference is the result of the difference between the thermochemical stabilities of the two intermediates INTeth and INTac relative to the reactant radical, which are ~10 and 19 kcal/mol, respectively. Whereas the formed five-member ring in acenaphthalenyl (INTeth) does not participate in the resonance of the naphthalene part of the molecule, in INTac, it is a part of the resonance of the entire molecule owing to the existence of the C=C double bond [29] and the electron delocalization between the three rings of the intermediate.

The last step for both intermediates, INTeth and INTac, is the ejection of an H-atom forming acenaphthene and acenaphthylene.

6.6 RING EXPANSION PROCESSES

Another interesting group of reactions is the ring expansion processes. This is an important channel in transferring aliphatic to aromatic compounds at high temperatures that prevail in combustion environment [100–103]. It can be considered as the beginning of PAH production [104–115]. In this chapter, the ring expansion processes in the four compounds shown in Figure 6.24 will be described.

The ring expansion processes do not normally occur in the molecular form. The process takes place from a radical species following a loss of a hydrogen atom. In most cases, the loss is from the methyl group to form a methylene group by either an ejection or abstraction. At the end, a very fast loss of an additional hydrogen atom by ejection takes place to form the expanded ring. There are several isomers for both the radical species and the molecule itself. The lowest energy pathways are in the isomers where the methylene group is connected to an sp^3 carbon or to the nitrogen atom in the ring [115–119].

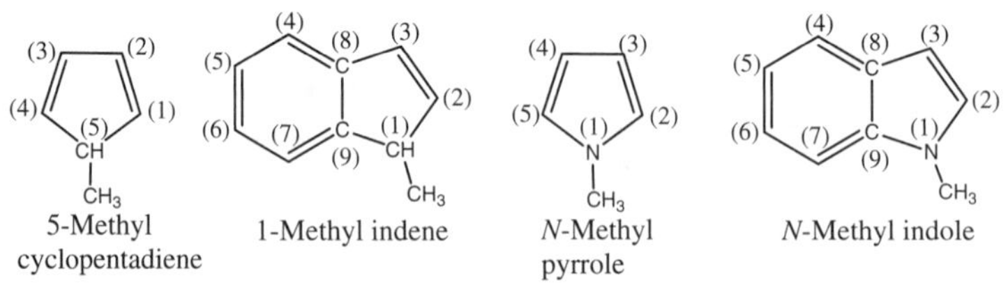

FIGURE 6.24 Molecules for which ring expansion processes are reported and their atom numbering.

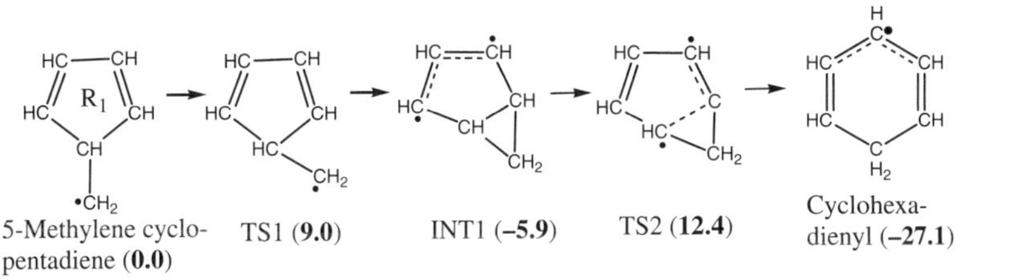

FIGURE 6.25 Reaction scheme of the lowest energy pathway for ring expansion of methylene cyclopentadiene. All the energy values are calculated relative to the reactant.

6.6.1 Methylcyclopentadiene

Methylcyclopentadiene has three isomers. In one isomer, hydrogen atom is replaced by a methyl group in position 5 (sp^3 carbon) (Figure 6.24) and in the other two isomers the methyl group replaces a hydrogen atom in positions 1 and 2 (sp^2 carbon). The radical species, from which ring expansion will occur, are obtained by a loss of a hydrogen atom from the methyl group.

The lowest barrier for ring expansion exists in the reaction from the isomer where the $-CH_2$ group is connected to the sp^3 carbon in the ring, namely, 5-methylene cyclopentadiene (R_1) (Figure 6.25) [102,116]. Those radicals in which the hydrogen atom is removed either from the ring or from a methylene group that is connected to an sp^2 carbon in the ring, namely, 1- and 2-methylene cyclopentadiene, have much higher barriers, and isomerization to R_1 is a favorable step toward the ring expansion. In view of the high symmetry of the methylcyclopentadiene ring, all the isomerizations take place just by H-atom shift without the need to move either methylene or methyl groups.

Figure 6.25 describes the lowest energy pathway as obtained by the quantum chemical calculations. As can be seen, it is a two-step mechanism with one intermediate and two transition states. In the first step, two fused rings, three- and five-member, are formed and in the second step, the C–C bond common to both rings is broken toward the formation of the six-member ring, which after an ejection of a hydrogen atom forms benzene. A similar mechanism was proposed recently for *tert*-butyl-1,3-cyclopentadiene ring expansion [120].

Since the thermal reactions of methylcyclopentadiene include also considerable extent of decomposition at the same temperature range where the ring expansion takes place, the calculated rate constants had to be incorporated into a large kinetics scheme. Whereas their influence on the decomposition channels was rather small, the sensitivity toward the production of benzene (and toluene) was very high. The comparison with the experimental yields of benzene and toluene where the calculated rate constants are part of the kinetics scheme is shown in Figure 6.26. The agreement is quite good [102].

It should be mentioned that fulvene can also yield benzene by ring expansion [116,121]. However, in contrast to methylcyclopentadiene, two hydrogen atoms are not ejected from the molecule as fulvene is an isomer of benzene. The barrier for

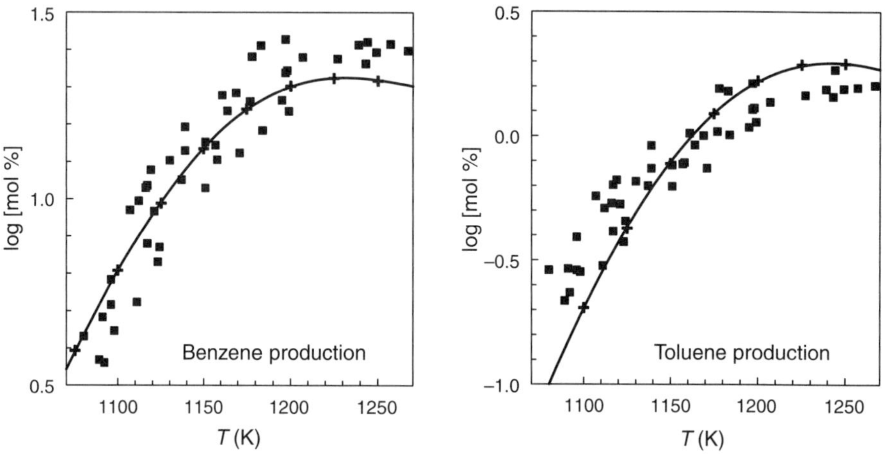

FIGURE 6.26 Comparison between the kinetic model calculations and experimental yields of two methylene cyclopentadiene ring expansion products [102].

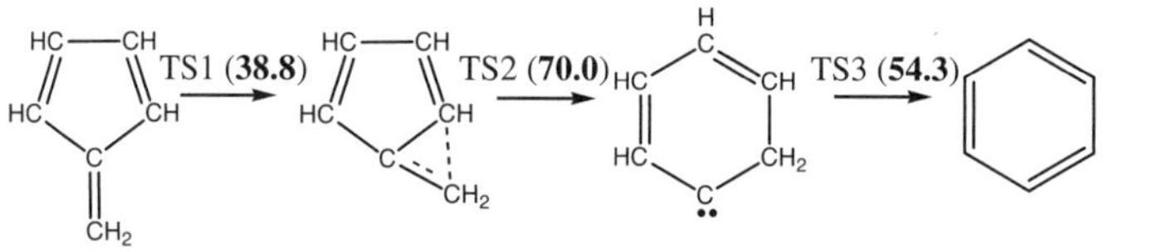

FIGURE 6.27 Reaction scheme of the ring expansion pathway starting from fulvene. All the energy values are calculated relative to the reactant.

the isomerization, however, is rather high since the six-member ring intermediate has a carbene structure. The scheme for this isomerization is shown in Figure 6.27.

6.6.2 Methyl Pyrrole

Methylpyrrole is a five-member ring compound similar to methylcyclopentadiene except that the sp^3 carbon in the ring is replaced by a nitrogen atom. Similar to methylcyclopentadiene, the molecule has three isomers. In one, the methyl group replaces the hydrogen atom that is bound to the nitrogen, forming *N*-methylpyrrole (Figure 6.24). In the other two, hydrogen atoms that are bound to sp^2 carbons in the ring, in positions 2 and 3, are replaced by a methyl group. (Note that the atom numbering in methylpyrrole is different than that of methylcyclopentadiene.) In the radical species, there are more isomers, depending upon the position from where the H-atom was removed. Here too, ring expansion will take place only from the radical isomers [122] having a CH_2 group that is formed by the loss of hydrogen from the methyl group [118,123]. The isomer where the CH_2 group is attached to the nitrogen atom has the lowest barrier and isomerization to this structure is a preferable step

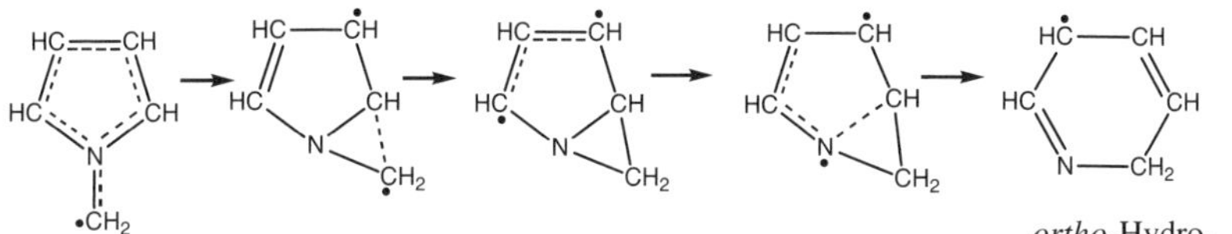

FIGURE 6.28 Reaction scheme of the lowest energy pathway of methylene pyrrole ring expansion. All the energy values are calculated relative to the reactant.

toward the ring expansion [118]. In all cases the final product is hydropyridyl that following an ejection of a hydrogen atom yields pyridine. The scheme can be seen in Figure 6.28.

The ring expansion pathways in methylene cyclopentadiene and methylene pyrrole radical systems are very much the same (Figures 6.25 and 6.28). However, the energy levels of both the transition states and the intermediates are very much different. The reason for this behavior is that there is a huge loss of resonance in both the transition states and the intermediate of *N*-methylene pyrrole pathway, whereas this resonance does not exist in 5-methylene cyclopentadiene. The resonance in *N*-methylene pyrrole results from the electron delocalization in its structure. Therefore, the energy level of TS1 (Figure 6.25), for example, in 5-methylene cyclopentadiene is considerably lower than the one in *N*-methylene pyrrole (Figure 6.28), a fact that influences the entire energetics of the pathway. The isomerizations of 2- and 3-methylene pyrrole to *N*-methylene pyrrole are rather fast despite the fact that they require a movement of the CH_2 group in the ring [124]. The best channel for ring expansion from these isomers is thus isomerization followed by expansion [118,123].

6.6.3 Methylindene and Methylindole

Other interesting ring expansion reactions are those where the initial radicals consist of five-member rings fused to benzene. Examples are methylene indene and methylene indole, shown in Figure 6.24. Here too, the ring expansion will take place when the radical is obtained by the removal of the hydrogen atom from the methyl group connected to an sp^3 carbon or the nitrogen [117,119,125,126]. Following ring expansion and ejection of an additional hydrogen atom, naphthalene is formed from methylene indene, and quinoline and/or isoquinoline are obtained from methylene indole.

It has been found that the attached benzene ring to both methylene cyclopentadiene and methylene pyrrole does not alter the expansion pathways when compared to that of methylene cyclopentadiene and methylene pyrrole.

The ring expansion process must be preceded by the formation of a new three-member ring fused to the five-member ring of the indene molecule in both 1- and 2-methylene indene (INT1 in Figure 6.29). The lowest energy pathway for ring expansion, however, occurs when the methylene group is in position 1. When the methylene group is in position 2, isomerization to INT1 takes place via one intermediate (INT2 in Figure 6.29) and two transition states. It is accompanied by

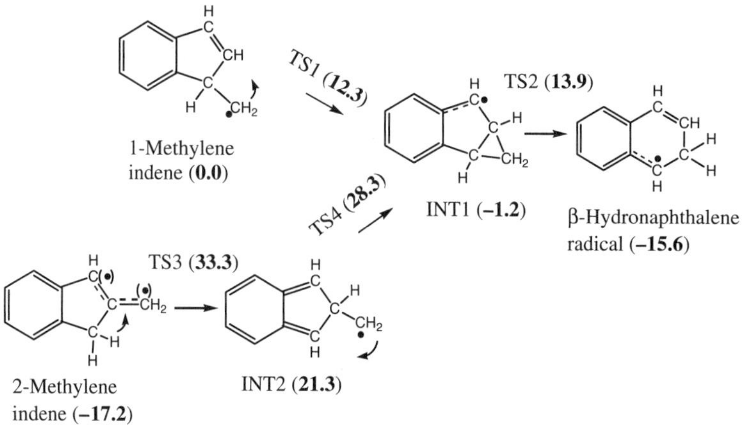

FIGURE 6.29 Reaction scheme of the lowest energy pathway of 1- and 2-methylene indole ring expansion and their interisomerization. All the energy values are calculated relative to the 1-methylene indene.

a considerably higher barrier (17.2 + 33.3 = 50.5 kcal/mol) owing to the complete loss of the resonance of the benzene ring. Interisomerization between the two isomers, 1- and 2-methylene indene, can also take place via the intermediate INT1.

In methylene indene owing to the fact that positions 1 and 3 (Figure 6.24) are identical with respect to the proximity to the benzene ring, there is no need to move a methylene group for ring expansion. Hydrogen shift is enough in all the three isomers similar to the situation in 5-methylene cyclopentadiene.

In methylene indole, however, owing to the presence of the nitrogen atom, positions 1 and 3 are not identical, so that the isomerization and ring expansion from position 3 require a methylene group movement [119]. The energetics on the methylene indole surface toward ring expansion does not differ very much from the one on the methylene pyrrole surface. The situation is different, however, when the energetics on the methylene indene surface is compared with the one on the methylcyclopentadiene surface in the final step. The reason for this behavior is the inability to create an electron delocalization (resonance) when a benzene ring is fused to the methylene cyclopentadiene ring. It should be mentioned that there is evidence that isomerization takes place also from position 7 to position 1 in methylene indole (Figure 6.24), when the methylene group is on the benzene ring. Details of the lowest energy pathways of the four compounds are shown in Table 6.2.

Experimental studies of the thermal reactions of both methylene indene and methylene indole were performed using the single pulse shock tube, where isomerization, decomposition, and ring expansion processes were observed [115–117,119]. In 2-methylene indene a complete kinetics scheme containing 71 elementary steps was constructed to take into account all the mentioned processes to which the rate constants that were evaluated on the basis of the quantum chemical calculations were inserted [115,117]. The production rate of naphthalene was highly sensitive to the isomerization and ring expansion rate constants. In 2-, 3-, and 7-methylindole,

TABLE 6.2 Energetics for Four Ring Expansion Lowest Energy Pathways in kcal/mol

Reactant	TS1	INT1	TS2	Product
5-Methylene cyclopentadiene	9.0	−5.9	12.4	−27.1
1-Methylene indene	12.3	−1.2	13.9	−15.6
N-Methylene pyrrole	33.9	19.7	32.2	−4.6
1-Methylene indole	32.8	18.2	29.7	−1.5

The values here are relative to those of the reactants.

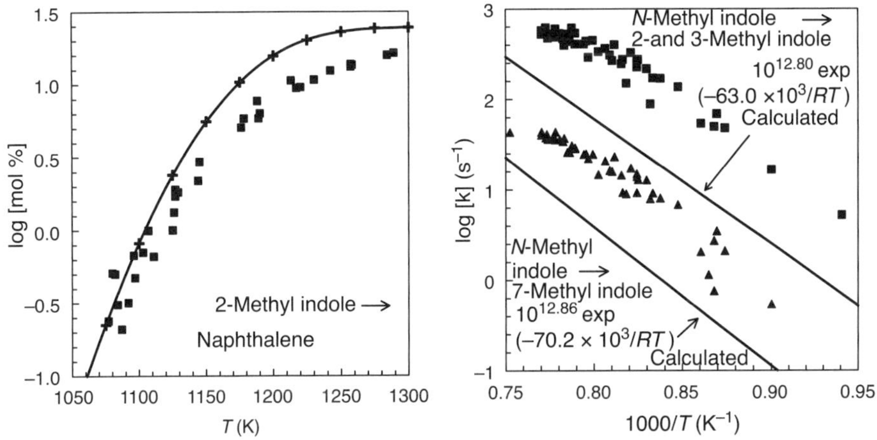

FIGURE 6.30 Comparison between the kinetic model calculation and the experimental yields of 2-methylindene ring expansion (a) [115] and between the calculated and the experimental rate constants of methylindole interisomerizations (b) [119].

a limited reaction scheme based only on the calculated rate constants was constructed and the experimental results were compared with the calculations. As can be seen in 2-methylene indene as the reactant, the agreement was good in the case of 2-, 3-, and 7-methylindole corresponding to an error equivalent to ~3.5 kcal/mol. This is shown in Figure 6.30.

6.7 BENZOXAZOLE–BENZISOXAZOLES

The benzoxazole–benzisoxazoles family of molecules is a group of molecules where five-member rings containing both nitrogen and oxygen are fused to benzene. This family of molecules has three isomers shown in Figure 6.31.

The thermochemical stability of the three isomers is quite different from one another [127]. The relatively weak N–O bond in 1,2-benzisoxazole introduces some instability and its heat of formation is 33.3 kcal/mol at 298 K [128]. 2,1-Benzisoxazole that has also weak N–O bond is much less stable with a heat of

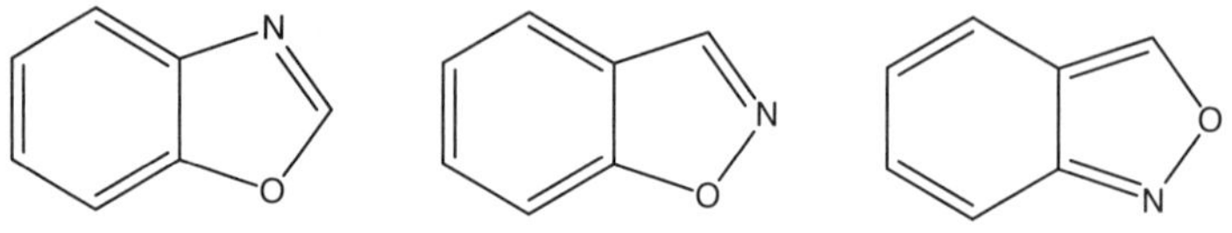

Benzoxazole 1,2-Benzisoxazole 2,1-Benzisoxazole (anthranil)

FIGURE 6.31 Benzoxazole and two benzisoxazole structures.

formation of 41.2 kcal/mol at the same temperature [128], owing to the loss of the benzene ring resonance energy. Benzoxazole, on the other hand, is the most stable of the three isomers as there is no loss of resonance energy in the molecule on one hand and the weak N–O bond does not exist. Its heat of formation is 10.8 kcal/mol at 298 K [129].

6.7.1 Benzoxazole

In benzoxazole, in view of the fact that a weak N–O bond does not exist, a stronger C–O bond has to be broken. This, together with an H-atom shift from the carbon atom to the oxygen, initiates an isomerization process. With one transition state it forms *o*-hydroxy-*iso*-benzonitrile that further isomerizes to *o*-hydroxybenzonitrile [130]. This is shown in Figure 6.32.

As the temperature is increased, the *o*-hydroxybenzonitrile undergoes ring contraction via CO elimination and the formation of cyclopentadiene carbonitrile. The pathway of this process is shown in Figure 6.33.

Figure 6.34 shows a comparison between the experimental and the calculated rate constants of the two final products of benzoxazole isomerization and decomposition [130]. The agreement is good.

6.7.2 1,2-Benzisoxazole

Over the temperature range where the reactions of 1,2-benzisoxazole have been studied with the single pulse shock tube (900–1040 K), a single isomerization product, *o*-hydroxybenzonitrile, was experimentally observed [131]. At higher temperatures, fragmentation begins to take place.

There are two channels that lead to the isomerization. One channel is rather simple [131]. It is composed of two transition states and one intermediate. It has,

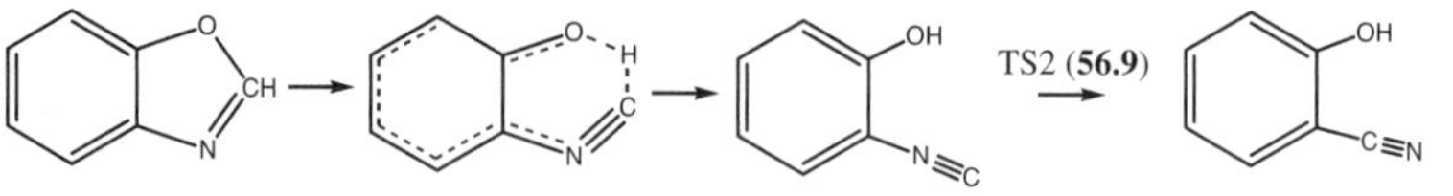

Benzoxazole (**0.0**) TS1 (**62.0**) *o*-Hydroxy-isobenzonitrile (**21.9**) *o*-Hydroxy-benzonitrile (**–0.4**)

FIGURE 6.32 Reaction scheme for benzoxazole → *o*-hydroxybenzonitrile isomerization. All the energy values are calculated relative to the reactant.

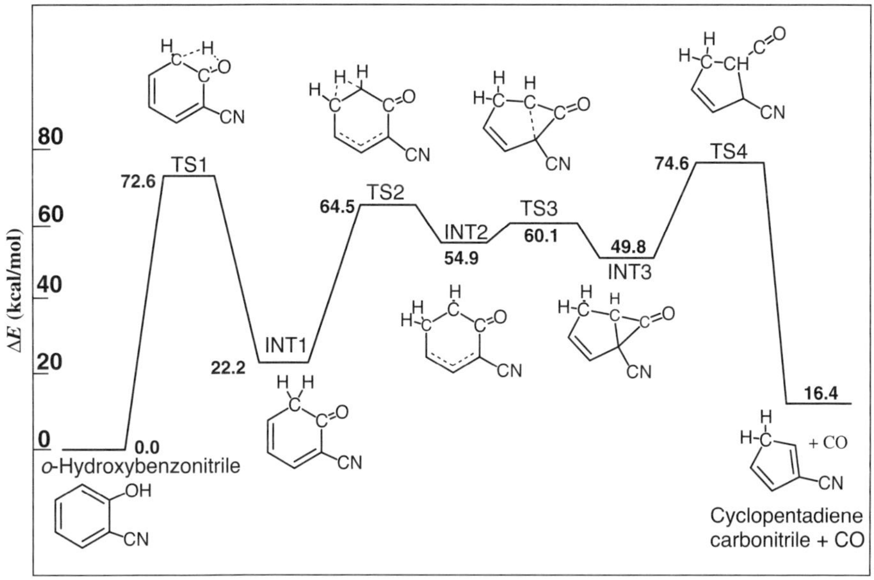

FIGURE 6.33 Potential energy surface of *o*-hydroxybenzonitrile decomposition.

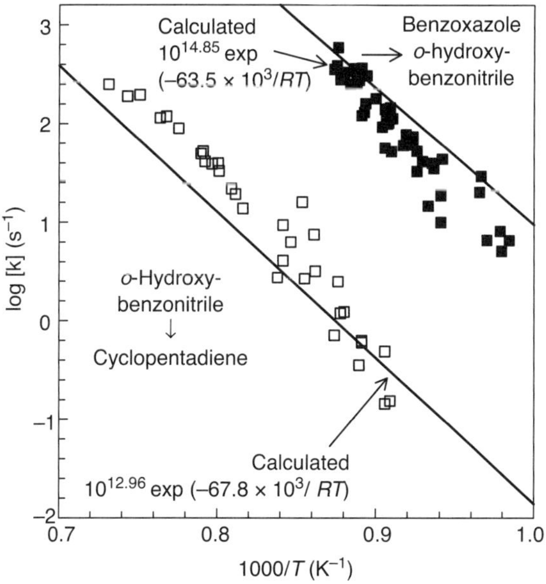

FIGURE 6.34 Comparison between the experimental and the calculated rate constants of benzoxazole isomerization and decomposition [130].

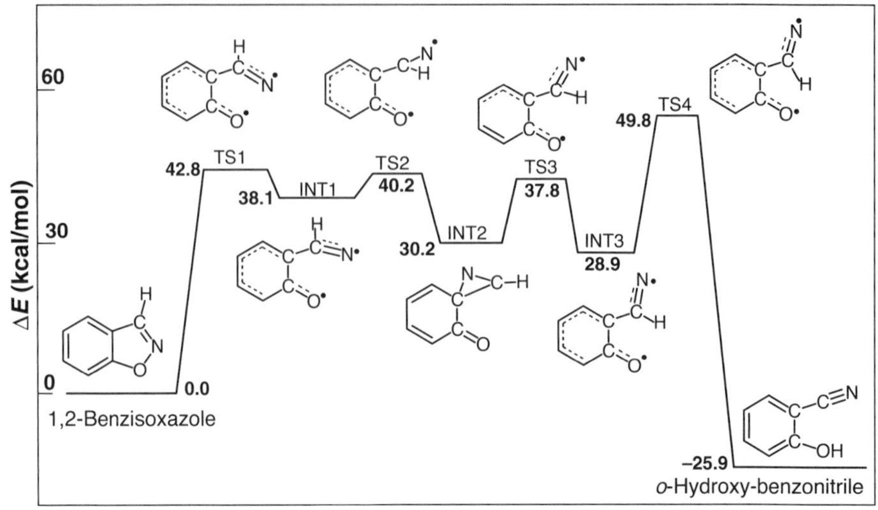

FIGURE 6.35 Potential energy surface of 1,2-benzisoxazole isomerization.

however, a rather high barrier so that it does not contribute to the isomerization process and will not be discussed here. The highest barrier on the second surface is lower by approximately 12 kcal/mol than the highest one on the first surface. This surface, however, is quite complicated as it is composed of four transition states and three intermediates. The surface is shown in Figure 6.35.

In order to evaluate the isomerization rate constant (via the second channel), a kinetics scheme containing all the four elementary steps, forward and backward, has been constructed. Their rate constants were calculated from the $\Delta S^{\#}$ and $\Delta E^{\#}$ for each elementary step. Together with the thermochemistry of the reactant, product, and all the intermediates on the surface, the scheme could be computer modeled to evaluate the product yields, and then the final rate constant for the reactant → product reaction. Figure 6.36 shows a comparison between the results of the experiment and the calculations. The difference is about a factor of 3 that corresponds to a disagreement of approximately 2.5 kcal/mol.

6.7.3 2,1-Benzisoxazole—Intersystem Crossing

2,1-Benzisoxazole (anthranil) is the less stable isomer of all the three and reacts at a considerably lower temperature range (~800–1000 K) at ~1 ms reaction time. The quantum chemical calculations show that it has two isomerization channels. One leads to the production of 1,3-cyclopentadiene-5-carbonitrile,1-aldehyde and the second channel to *o*-cyclo hexadienylidene ketene imine (CHKI). The two isomerization products are shown in Figure 6.37.

1,3-Cyclopentadiene-5-carbonitrile,1-al (channel 1) [133] was not detected experimentally since its production barrier of ~62 kcal/mol is too high for the temperature range covered. Also, when formed at low yields at the higher end of the temperature range, it will immediately decompose to cyclopentadiene carbonitrile

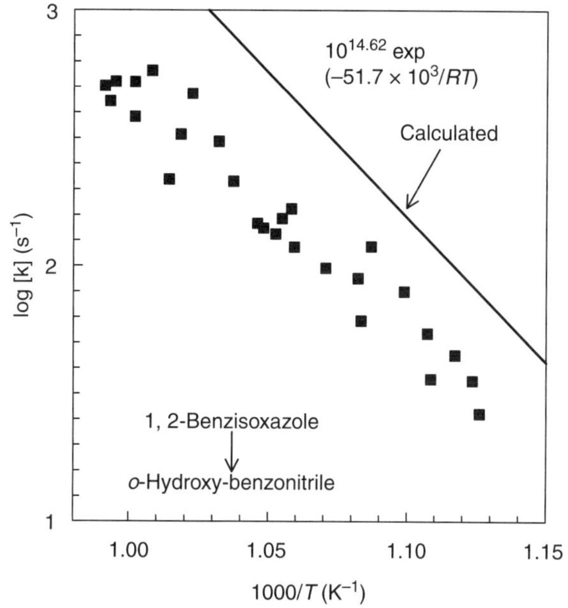

FIGURE 6.36 Comparison between the experimental and the calculated rate constants of 1,2-benzisoxazole isomerization [131].

together with CO elimination with a rather low barrier of ~21 kcal/mol. This channel, therefore, does not contribute to the 2,1-benzisoxazole isomerizations and will not be discussed here. It is not clear, however, why the second isomer, CHKI (channel 2), was not detected, probably owing to some analytical problem. The potential energy surface showing among other intermediates the CHKI production is shown in Figure 6.38.

The main final products of the isomerization–decomposition reactions of 2,1-benzisoxazole (anthranil), which must proceed via CHKI, are cyclopentadiene carbonitrile and phenylnitrene. Both are accompanied by CO elimination [132,133]. Phenylnitrene, being an unstable biradical [134–136] absorbs hydrogen on its way to the G.C. and turns into aniline. The scheme is shown in Figure 6.39.

To account for the production of both aniline (phenylnitrene) and cyclopentadiene carbonitrile via the intermediate CHKI (Figure 6.39), it needs to undergo a CO elimination process that comes together with the production of *o*-phenylimine biradical (Figure 6.38).

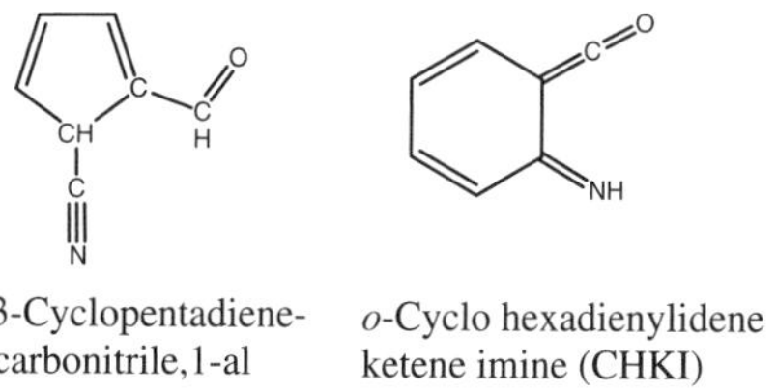

FIGURE 6.37 Two isomerization products of 2,1-benzisoxazoxazole.

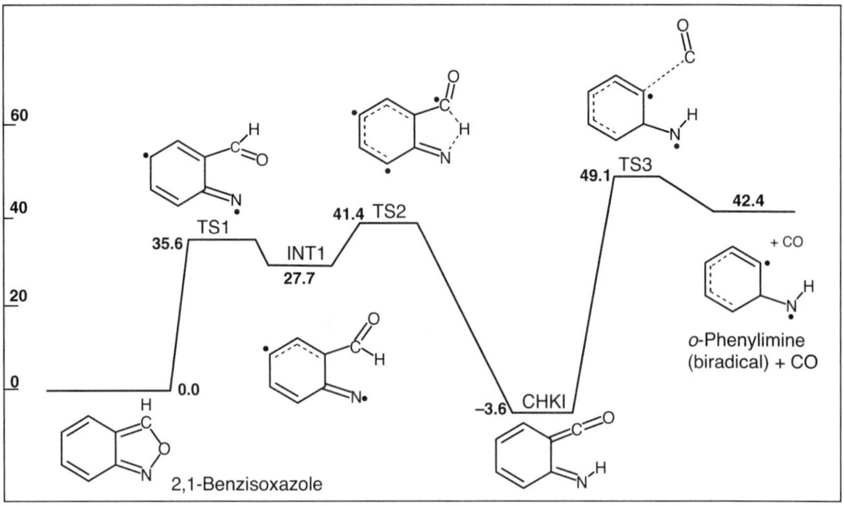

FIGURE 6.38 Potential energy surfaces of 2,1-benzisoxazole decomposition. The biradical phenylimine is produced when CO is eliminated.

o-Phenylimine has now two reaction channels. One leads to the formation of 6-iminofulvene with a low barrier of 22.8 kcal/mol (TS6, Figure 6.40). The second channel produces phenylnitrene by H-atom shift with a much higher barrier of approximately 36 kcal/mol. This large difference in the barriers with roughly the same pre-exponential factors practically prevents the formation of phenylnitrene (particularly in the low temperature range) that is required to precede the formation of aniline. The isomerization pathway from 6-iminofulvene to cyclopentadiene carbonitrile has one transition state, where H-atom is transferred from the nitrogen atom to the cyclopentadiene ring. It has however a high barrier of $38.2 + 17.6 = 56$ kcal/mol (TS7, Figure 6.40), which again cannot contribute much to the process.

Despite the high barriers involved, a reaction scheme was constructed based on the evaluated rate constants. They were calculated at various points over the temperature range covered in the experiment and were transferred to Arrhenius-type equations for

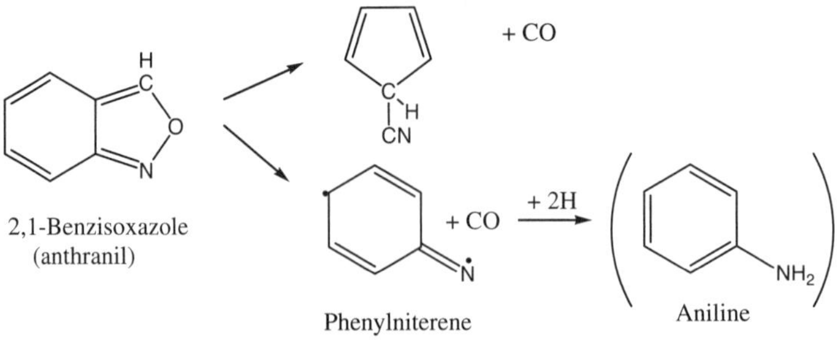

FIGURE 6.39 Final products of 2,1-benzisoxazole decomposition.

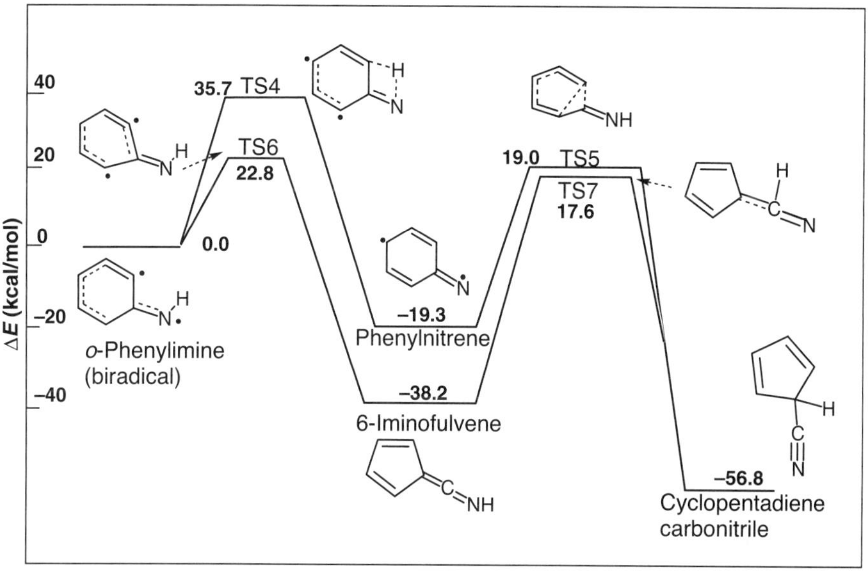

FIGURE 6.40 Two channels on singlet potential energy surfaces of ring contraction in *o*-phenylimine.

comparison with the experimental results [133]. This is shown in Figure 6.41. As can be seen, the agreement is very poor, particularly at the low end of the temperature range. The deviation is about a factor of 50 at ~900 K.

In view of the failure of the singlet surfaces to account for the production of the final products, and since all the species, except for CHKI, are open-shell biradicals, it was decided to calculate the reaction pathways on a triplet surface as the energy level of the triplets in many cases, when they contain nitrogen atoms, is lower than that of the

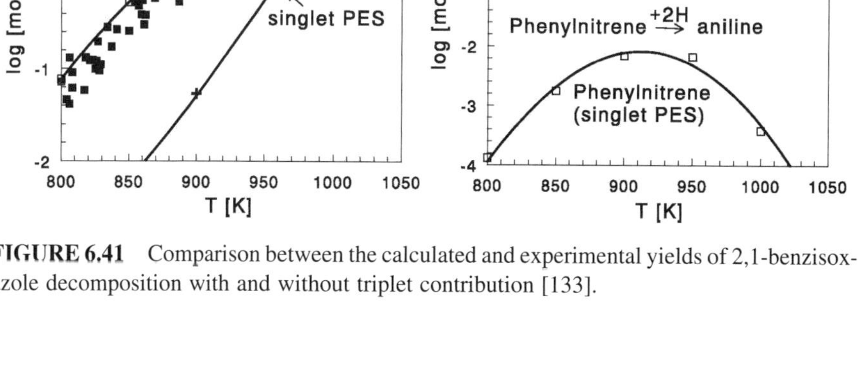

FIGURE 6.41 Comparison between the calculated and experimental yields of 2,1-benzisoxazole decomposition with and without triplet contribution [133].

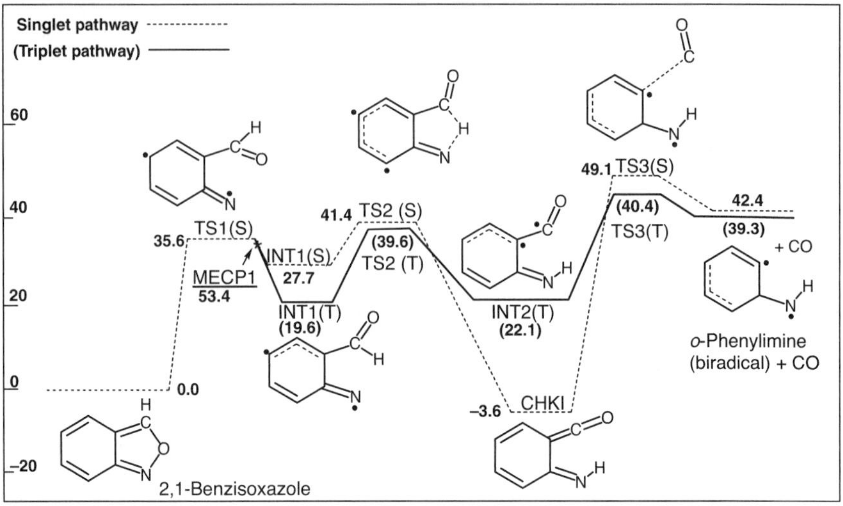

FIGURE 6.42 Triplet (——) and singlet (. . . .) potential energy surfaces of 2,1-benzisoxazole decomposition.

singlet open-shell biradicals [134–146]. We have thus calculated all the reaction pathways leading to the final products on triplet potential energy surfaces as can be seen in Figures 6.42 and 6.43. Indeed, all the barriers were much lower, never higher than ∼40 kcal/mol.

The next step was the calculation of the MECP from a singlet, open-shell biradical intermediate INT1 (Figure 6.42) to the triplet state. It was then followed by the calculation of the singlet → triplet crossing probability. The exact procedure has been described in Section 6.1.1. The final situation is such that both pathways—the singlet and the triplet—participate in the process since a considerable part of the species still remains in the singlet state and proceeds on the singlet surface.

The rate constants of all the steps on the triplet potential energy surface were added to the kinetics scheme that was based on the singlet surface only including all the singlet → triplet and the triplet → singlet crossing rate constants. The kinetics scheme is shown in Table 6.3.

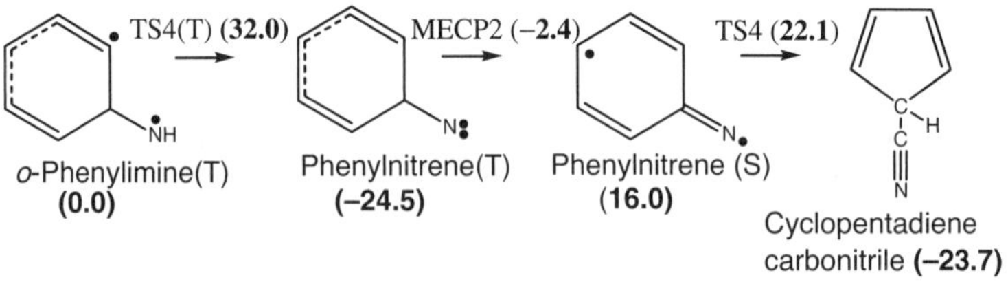

FIGURE 6.43 Reaction scheme of phenylimine ring contraction toward cyclopentadiene carbonitrile formation. Triplet → singlet crossing back of phenylnitrene is part of the process. All the energy values are calculated relative to the *o*-phenylimine(T).

TABLE 6.3 Arrhenius Parameters of the Elementary Steps on the Singlet and Triplet PES Including the S → T and T → S Intersystem Crossing Rate Constants

	A^a	$E_a^{\ b}$
Pathways on the singlet PES		
Anthranil → INT1	6.11×10^{14}	37.9
INT1 → INT2 (CHKI)	9.90×10^{12}	14.5
INT2 (CHKI) → phenylimine + CO	3.14×10^{15}	55.1
Phenylimine → 6-iminofulvene	4.68×10^{13}	24.1
6-Iminofulvene → cyclopentadiene carbonitrile	1.29×10^{14}	57.6
Phenylimine → phenylnitrene	3.17×10^{13}	36.8
Phenylnitrene → cyclopentadiene carbonitrile	9.73×10^{13}	39.9
Pathways on the triplet PES including the S → T and T → S crossings		
Anthranil → INT1(T)[c]	5.88×10^{14}	38.5
INT1(T) → INT2(T)	1.19×10^{13}	21.1
INT2(T) → phenylimine(T) + CO	3.46×10^{15}	20.5
Phenylimine(T) → phenylnitrene(T)	1.81×10^{13}	32.9
Phenylnitrene(T) → phenylnitrene(S)[c]	1.26×10^{11}	22.9

[a] Pre-exponential factor in units of s^{-1}.
[b] Activation energy in kcal/mol.
[c] Crossing steps.

A computer simulation was then performed and the comparison with the experimental values is shown in Figure 6.41. As can be seen, the addition of the triplet pathway has a significant contribution to the production rates of the final products and the agreement is very good. Note that the concentration of the phenylnitrene is much higher than that of aniline as not all of it is being transferred to the latter. It is high enough, however, to account for the concentration of the aniline.

The final conclusion is that triplet states can contribute a great deal to isomerization and decomposition reactions when biradical mechanism exists, and thus cannot be ignored.

6.8 CONCLUSION

Reactions of a large variety of organic ring compounds including heterocyclics, containing nitrogen, oxygen, and both nitrogen and oxygen, are described in this chapter. The reactions include isomerizations, decompositions, and ring expansions. The main effort, in addition to the description of reaction pathways, was the evaluation of rate constants for final reactant → product transitions but also for rate constants of reactions of significant species that play the role of active intermediates along the reaction coordinates.

The various methods of the quantum chemical calculations and the evaluation of rate parameters of the various elementary steps on the potential energy surface are discussed in detail. The evaluation required the knowledge of the energetics of the transition states and the intermediates, and their entropy, the degeneracy of their reaction coordinates, and other parameters depending on the particular case. The single pulse shock tube experiments are also briefly described. Finally, the experimental results and the results obtained by the quantum chemical calculations are compared and discussed.

In several cases the reactant to product transition involved a complicated pathway that included a number of transition states and intermediates on the potential energy surface. Examples are given for a number of such reaction systems. In order to evaluate the final rate parameters for the reactant to product transition in such cases, the overall pathway on the potential energy surface had to be treated as an independent kinetics scheme to be computer modeled by normal numerical integrations. This has to take into account all the forward and backward rate constants. The latter are determined by equilibrium constants between the various steps on the surface. All of these methods of calculations and the experimental methods used are described in Section 6.1.

Many of the reactions of the ring compounds that are described in this chapter, both theoretically and experimentally, simulate in one way or another the processes that prevail in combustion. One of the major mechanisms of aliphatic to aromatic transition is the expansion of a five-member ring with a small aliphatic tail to the six-member benzene structure. Section 6.6, in addition to the basic chemical kinetics, has a significant value to applied research.

APPENDIX: Table of Calculated Unimolecular Rate Constants

	Reaction	Temperature Range (K)	Rate Constant (s^{-1}) Energy (kcal/mol)	M^a	Section
1	Cyclopropane → propylene	700–1250	$10^{15.91} \exp(-66.7 \times 10^3/RT)$	A	6.2.1
2	Cyclopropane carbonitrile → *cis*-crotonitrile	900–1050	$10^{14.93} \exp(-60.3 \times 10^3/RT)$	B	6.2.2
3	Cyclopropane carbonitrile → *trans*-crotonitrile	900–1050	$10^{15.06} \exp(-61.3 \times 10^3/RT)$	B	6.2.2
4	Cyclopropane carbonitrile → vinyl acetonitrile	900–1050	$10^{15.04} \exp(-61.9 \times 10^3/RT)$	B	6.2.2
5	Cyclopropane carbonitrile → methacrylonitrile	900–1050	$10^{14.81} \exp(-72.4 \times 10^3/RT)$	B	6.2.2
6	Propylene oxide → acetone	800–1250	$10^{13.54} \exp(-54.2 \times 10^3/RT)$	C	6.2.3
7	Propylene oxide → propanal	800–1250	$10^{14.04} \exp(-54.5 \times 10^3/RT)$	C	6.2.3
8	Propylene oxide → allyl alcohol	800–1250	$10^{13.34} \exp(-60.2 \times 10^3/RT)$	C	6.2.3
9	Propylene oxide → methyl vinyl ether	800–1250	$10^{14.11} \exp(-59.3 \times 10^3/RT)$	C	6.2.3
10	Pyrrole → vinyl crotonitrile	1050–1350	$10^{15.04} \exp(-80.4 \times 10^3/RT)$	B	6.3.1
11	Pyrrole → *cis*-crotonitrile	1050–1350	$10^{13.60} \exp(-69.3 \times 10^3/RT)$	B	6.3.1
12	*cis*-Crotonitrile → *trans*-crotonitrile	1050–1350	$10^{13.92} \exp(-61.1 \times 10^3/RT)$	B	6.3.1
13	*cis*-Crotonitrile → vinyl acetonitrile	1050–1350	$10^{14.00} \exp(-70.8 \times 10^3/RT)$	B	6.3.1

(*Continued*)

	Reaction	Temperature Range (K)	Rate Constant (s^{-1}) Energy (kcal/mol)	M^a	Section
14	*trans*-Crotonitrile → vinyl acetonitrile	1050–1350	$10^{14.04}\exp(-71.8 \times 10^3/RT)$	B	6.3.1
15	Pyrrole → pyrrolenine	1050–1350	$10^{13.45}\exp(-43.1 \times 10^3/RT)$	B	6.3.1
16	Pyrrolenine → *cis*-crotonitrile	1050–1350	$10^{14.26}\exp(-64.2 \times 10^3/RT)$	B	6.3.1
17	Pyrrolenine → vinyl acetonitrile	1050–1350	$10^{14.71}\exp(-71.6 \times 10^3/RT)$	B	6.3.1
18	Indole → 2*H*-indole	1100–1550	$10^{13.11}\exp(-48.8 \times 10^3/RT)$	C	6.3.2
19	2*H*-indole → 3*H*-indole	1100–1550	$10^{13.28}\exp(-18.8 \times 10^3/RT)$	C	6.3.2
20	3*H*-indole → *o*-toluisocyanide	1100–1550	$10^{14.08}\exp(-77.4 \times 10^3/RT)$	C	6.3.2
21	*o*-Toluisocyanide → *o*-tolunitrile	1100–1550	$10^{12.89}\exp(-35.2 \times 10^3/RT)$	C	6.3.2
22	Indole → phenyl acetonitrile	1100–1550	$10^{14.22}\exp(-76.7 \times 10^3/RT)$	C	6.3.2
23[b]	Indole → *o*-tolunitrile	1100–1550	$10^{14.48}\exp(-84.8 \times 10^3/RT)$	C	6.3.2
24	2,3-Dihydrofuran → cyclopropane carboxaldehyde	850–1050	$10^{14.55}\exp(-53.6 \times 10^3/RT)$	B	6.4.1
25	2,3-Dihydrofuran → propenyl aldehyde	850–1050	$10^{14.47}\exp(-61.3 \times 10^3/RT)$	B	6.4.1
26	5-Methyl-2,3-dihydrofuran → methyl cyclopropyl ketone	850–1050	$10^{15.40}\exp(-56.8 \times 10^3/RT)$	B	6.4.2
27	5-Methyl-2,3-dihydrofuran → methyl propenyl ketone	850–1050	$10^{15.70}\exp(-63.6 \times 10^3/RT)$	B	6.4.2
28	Methyl cyclopropyl ketone → methyl propenyl ketone	850–1050	$10^{15.12}\exp(-60.8 \times 10^3/RT)$	B	6.4.2
29	2,5-Dihydrofuran → furan + H_2	600–1100	$10^{13.04}\exp(-49.5 \times 10^3/RT)$	B	6.4.3
30	2-Methyl-2,5-dihydrofuran → 2-methyl-furan + H_2	600–1100	$10^{12.81}\exp(-48.4 \times 10^3/RT)$	B	6.4.3
31	2,3-Dihydrofuran → furan + H_2	600–1100	$10^{13.76}\exp(-82.6 \times 10^3/RT)$	B	6.4.3
32	Dihydrobenzofuran → *o*-hydroxystyrene	1050–1350	$10^{14.65}\exp(-68.3 \times 10^3/RT)$	C	6.4.4
33	*iso*-Dihydrobenzofuran → *o*-tolualdehyde	1050–1350	$10^{14.36}\exp(-69.5 \times 10^3/RT)$	C	6.4.4
34	1-Naphthylacetylene → acenaphthene	1000–1250	$10^{12.55}\exp(-55.9 \times 10^3/RT)$	C	6.5
35	5-Methyl cyclopentadienyl → 5-methylene cyclopentadiene	600–1400	$10^{12.48}\exp(-50.4 \times 10^3/RT)$	B	6.6.1
36	5-Methylene cyclopentadiene → cyclohexadienyl	600–1400	$10^{13.15}\exp(-17.4 \times 10^3/RT)$	B	6.6.1
37	1-Methylene cyclopentadiene → cyclohexadienyl	600–1400	$10^{13.40}\exp(-73.4 \times 10^3/RT)$	B	6.6.1
38	2-Methylene cyclopentadiene → cyclohexadienyl	600–1400	$10^{12.76}\exp(-71.2 \times 10^3/RT)$	B	6.6.1

(*Continued*)

(*Continued*)

	Reaction	Temperature Range (K)	Rate Constant (s^{-1}) Energy (kcal/mol)	M^a	Section
39	2-Methylene cyclopentadiene → 1-methylene cyclopentadiene	600–1400	$10^{13.57}\exp(-31.5\times10^3/RT)$	B	6.6.1
40	1-Methylene cyclopentadiene → 5-methylene cyclopentadiene	600–1400	$10^{13.87}\exp(-42.0\times10^3/RT)$	B	6.6.1
41	1-Methylene cyclopentadiene → 5-methylene cyclopentadiene → cyclohexadienyl	600–1400	$10^{13.58}\exp(-40.8\times10^3/RT)$	B	6.6.1
42	*N*-Methylene pyrrole → *o*-hydropyridyl	1000–1400	$10^{13.04}\exp(-35.0\times10^3/RT)$	A	6.6.2
43	2-Methylene pyrrole → *N*-hydropyridyl	1000–1400	$10^{13.30}\exp(-72.5\times10^3/RT)$	A	6.6.2
44	3-Methylene pyrrole → *N*-hydropyridyl	1000–1400	$10^{13.20}\exp(-71.1\times10^3/RT)$	A	6.6.2
45	1-Methyl indenyl → 1-methylene indene	1000–1300	$10^{13.34}\exp(-52.5\times10^3/RT)$	B	6.6.3
46	2-Methylindene → 1-methylindene	1000–1300	$10^{13.46}\exp(-58.5\times10^3/RT)$	B	6.6.3
47	1-Methylindene → 3-methylindene	1000–1400	$10^{13.86}\exp(-38.6\times10^3/RT)$	B	6.6.3
48	*N*-Methylindole → 2-methylindole	1050–1350	$10^{13.06}\exp(-64.2\times10^3/RT)$	B	6.6.3
49[b]	*N*-Methylindole → 2- and 3-methylindole	1050–1350	$10^{12.80}\exp(-63.0\times10^3/RT)$	B	6.6.3
50[b]	*N*-Methylindole → 7-methylindole	1050–1350	$10^{12.86}\exp(-70.2\times10^3/RT)$	B	6.6.3
51	*N*-Methylindole → hydro-quinoline	1050–1350	$10^{12.21}\exp(-31.9\times10^3/RT)$	B	6.6.3
52	*N*-Methylindole → hydro-isoquinoline	1050–1350	$10^{12.11}\exp(-38.2\times10^3/RT)$	B	6.6.3
53	Benzoxazole → *o*-hydroxy benzonitrile	1000–1400	$10^{14.85}\exp(-63.5\times10^3/RT)$	B	6.7.1
54	Benzoxazole → *o*-hydroxy-*iso*-benzonitrile	1000–1400	$10^{15.24}\exp(-65.5\times10^3/RT)$	B	6.7.1
55	*o*-Hydroxy-*iso*-benzonitrile → *o*-hydroxybenzonitrile	1000–1400	$10^{13.33}\exp(-32.5\times10^3/RT)$	B	6.7.1
56	*o*-Hydroxybenzonitrile → cyclopentadiene carbonitrile	1000–1400	$10^{12.96}\exp(-67.8\times10^3/RT)$	B	6.7.1
57[b]	1,2-Benzisoxazole → *o*-hydroxybenzonitrile	850–1050	$10^{14.62}\exp(-51.7\times10^3/RT)$	B	6.7.2
58	CHKI → phenylimine + CO	800–1050	$10^{15.50}\exp(-55.1\times10^3/RT)$	B	6.7.3
59	Phenylimine → 6-iminofulvene	800–1050	$10^{13.67}\exp(-24.1\times10^3/RT)$	B	6.7.3
60	6-Iminofulvene → cyclopentadiene carbonitrile	800–1050	$10^{14.11}\exp(-57.6\times10^3/RT)$	B	6.7.3

[a] Method used in calculations: A, B3LYP/cc-pVDZ; B, QCISD(T)//B3LYP/cc-pVDZ; C, CCSD(T)//B3LYP/cc-pVDZ.

[b] Calculated and experimental rate constants have some kcal/mol difference.

ACKNOWLEDGMENT

The authors wish to thank Mrs. Carmen Tamburu for technical assistance.

REFERENCES

1. M. J. Frisch, G. W. Trucks, H. B. Schlegel, G. E. Scuseria, M. A. Robb, J. R. Cheeseman, J. A. Montgomery, Jr., T. Vreven, K. N. Kudin, J. C. Burant, J. M. Millam, S. S. Iyengar, J. Tomasi, V. Barone, B. Mennucci, M. Cossi, G. Scalmani, N. Rega, G. A. Petersson, H. Nakatsuji, M. Hada, M. Ehara, K. Toyota, R. Fukuda, J. Hasegawa, M. Ishida, T. Nakajima, Y. Honda, O. Kitao, H. Nakai, M. Klene, X. Li, J. E. Knox, H. P. Hratchian, J. B. Cross, V. Bakken, C. Adamo, J. Jaramillo, R. Gomperts, R. E. Stratmann, O. Yazyev, A. J. Austin, R. Cammi, C. Pomelli, J. W. Ochterski, P. Y. Ayala, K. Morokuma, G. A. Voth, P. Salvador, J. J. Dannenberg, V. G. Zakrzewski, S. Dapprich, A. D. Daniels, M. C. Strain, O. Farkas, D. K. Malick, A. D. Rabuck, K. Raghavachari, J. B. Foresman, J. V. Ortiz, Q. Cui, A. G. Baboul, S. Clifford, J. Cioslowski, B. B. Stefanov, G. Liu, A. Liashenko, P. Piskorz, I. Komaromi, R. L. Martin, D. J. Fox, T. Keith, M. A. Al-Laham, C. Y. Peng, A. Nanayakkara, M. Challacombe, P. M. W. Gill, B. Johnson, W. Chen, M. W. Wong, C. Gonzalez, and J. A. Pople, *Gaussian 03*, Gaussian, Inc. , Wallingford, CT (2004).
2. A. D. Becke, *J. Chem. Phys.* **98**, 5648 (1993).
3. C. Lee, W. Yang, and R. G. Parr, *Phys. Rev.* **B37**, 785 (1988).
4. T. H. Dunning, Jr., *J. Chem. Phys.* **90**, 107 (1989).
5. R. A. Kendall, T. H. Dunning, Jr., and R. J. Harrison, *J. Chem. Phys.* **96**, 6796 (1992).
6. T. H. Dunning, Jr., *J. Chem. Phys.* **90**, 1007 (1989).
7. D. Cremer, E. Kraka, and P. G. Szalay, *Chem. Phys. Lett.* **292**, 97 (1998).
8. G. E. Scuseria and H. F. Schaefer, III, *J. Chem. Phys.* **90**, 3700 (1989).
9. J. A. Pople, M. Head-Gordon, K. Raghavarchari, and G. W. Trucks, *Chem. Phys. Lett.* **87**, 5968 (1987).
10. J. A. Pople, M. Head-Gordon, and K. Raghavachari, *J. Chem. Phys.* **90**, 4635 (1989).
11. M. W. Schmidt, K. K. Baldridge, J. A. Boatz, S. T. Elbert, M. S. Gordon, J. H. Jensen, S. Koseki, N. Matsunaga, K. A. Nguyen, S. Su, T. L. Windus, M. Dupuis, and J. A. Montgomery, *J. Comput. Chem.* **14**, 1347 (1993).
12. H. B. Schlegel, *J. Comput. Chem.* **3**, 214 (1982).
13. C. Peng and H. B. Schlegel, *Isr. J. Chem.* **33**, 449 (1993).
14. S. Shaik, H. B. Schlegel, and S. Walfe, *Theoretical Aspects of Physical Organic Chemistry: The S_N2 Mechanism*, John Wiley & Sons, Inc. , New York (1992).
15. H. Eyring, *J. Chem. Phys.* **3**, 107 (1935).
16. M. G. Evans and M. Polanyi, *Trans. Faraday Soc.* **31**, 875 (1935).
17. E. Wigner, *Z. Phys. Chem.* **B19**, 203 (1932).
18. F. Louis, C. A. Gonzales, R. Huie, and M. J. Kurylo, *J. Phys. Chem. A* **104**, 8773 (2000).
19. P. George, J. P. Glusker, and W. Bock, *J. Phys. Chem. A* **104**, 11347 (2000).
20. H. Nakamura and C. Zhu, *Comments Atom. Mol. Phys.* **32**, 249 (1996).

21. A. M. Kuznetsov, *Stochastic and Dynamic Views of Chemical Reaction Kinetics in Solutions*, Presses polytechniques et universitaires romandes, Lausanne (1999).
22. J. S. Hess and D. J. Doren, *J. Phys. Chem. B* **106**, 8206 (2002).
23. D. Fedorov, S. Koseki, M. W. Schmidt, and M. S. Gordon, *Int. Rev. Phys. Chem.* **22**, 551 (2003).
24. W. Tsang and A. Lifshitz, Single pulse shock tube, in G. Ben-Dor. , O. Igra, , and T. Elperin, (eds), *Handbook of Shock Waves III* Academic Press (2001), p. 107.
25. E. F. Green and J. P. Toennies, *Chemical Reactions in Shock Waves*, Edward Arnold Publishers Ltd. , London (1964).
26. W. Tsang, *J. Chem. Phys.* **40**, 1171 (1964).
27. W. Tsang and A. Lifshitz, *Int. J. Chem. Kinet.* **30**, 621 (1998).
28. A. Lifshitz, C. Tamburu, and F. Dubnikova, *J. Phys. Chem. A* **112**, 925 (2008).
29. A. Lifshitz, C. Tamburu, and F. Dubnikova, *J. Phys. Chem. A* **113**, 10446 (2009).
30. C. Doubleday, Jr., *J. Chem. Phys.* **100**, 3520 (1996).
31. C. Doubleday, Jr., K. Bolton, G. H. Peslherbe, and W. L. Hase, *J. Am. Chem. Soc.* **118**, 9922 (1996).
32. F. Dubnikova and A. Lifshitz, *J. Phys. Chem. A* **102**, 3299 (1998).
33. Y. Yamaguchi, H. F. Schaefer, III, and J. E. Baldwin, *Chem. Phys. Lett.* **185**, 143 (1991).
34. J. E. Baldwin, Y. Yamaguchi, and H. F. Schaefer, III, *J. Phys. Chem.* **98**, 7513 (1994).
35. E. V. Waage and B. S. Rabinovitch, *J. Phys. Chem.* **76**, 1695 (1972).
36. A. Goodrow, A. Bell, and M. Head-Gordon, *J. Chem. Phys.* **130**, 244108 (2009).
37. H. F. Bettinger, J. C. Rienta-Kiracofe, B. C. Hoffman, H. F. Schaefer, III, J. E. Baldwin, and P. v. R. Schleyer, *Chem. Commun.* 1515 (1999).
38. R. Hoffman, *J. Am. Chem. Soc.* **90**, 1475 (1968).
39. S. Kato and K. Morokuma, *Chem. Phys. Lett.* **65**, 19 (1979).
40. E. F. Hayes and A. K. Q. Siu, *J. Am. Chem. Soc.* **93**, 2090 (1971).
41. J. A. Horsley, Y. Jean, C. Moser, L. Salem, R. M. Stevens, and J. S. Wright, *J. Am. Chem. Soc.* **94**, 279 (1972).
42. S. W. Benson, *Thermochemical Kinetics: Methods for the Estimation of Thermochemical Data and Rate Parameters*, Wiley-Interscience Publications, New York (1976).
43. C. Doubleday, Jr., J. McIver, Jr., and M. Page, *J. Am. Chem. Soc.* **104**, 6533 (1982).
44. P. J. Hay, W. J. Hunt, and W. A. Goddard, *J. Am. Chem. Soc.* **94**, 638 (1972).
45. S. S. Getty, E. R. Davidson, and W. T. Borden, *J. Am. Chem. Soc.* **114**, 2085 (1992).
46. A. Skancke, L. J. Schaad, and B. A. Hess, Jr., *J. Am. Chem. Soc.* **110**, 5315 (1988).
47. F. Dubnikova and A. Lifshitz, *J. Phys. Chem. A* **102**, 5876 (1998).
48. D. A. Luckraft and P. J. Robinson, *Int. J. Chem. Kinet.* **5**, 137 (1973).
49. A. Lifshitz, J. H. Shweky, J. H. Kiefer, and S. S. Sidhu, Shock waves, in K. Takayama. (ed.), *Proceedings of the 18th International Symposium on Shock Waves, Sendia, Japan*, 1991, Springer, Berlin, p. 825.
50. M. C. Flowers, *J. Chem. Soc., Faraday Trans.* **73**, 1927 (1977).
51. A. Lifshitz and C. Tamburu, *J. Phys. Chem.* **98**, 1161 (1994).
52. F. Dubnikova and A. Lifshitz, *J. Phys. Chem. A* **104**, 4489 (2000).
53. X. S. Xu, M. X. Jin, and D. Ding, *J. Mol. Struct. (Theochem)* **638**, 215 (2003).

54. T. Thoemmes, S. Zuercher, A. Wix, A. Reitzmann, and B. Kraushaar-Czametszki, *Appl. Catal. A* **318**, 160 (2007).
55. H. Shinoda, Y. Mori, and M. Mizuguchi, *Int. J. Quant. Chem.* **106**, 952 (2006).
56. F. Liu, L. Sheng, F. Qi, H. Gao, C. Li, Y. Zhang, S. Yu, K.-C. Lau, and W.-K. Li, *J. Phys. Chem.* **103**, 8179 (1999).
57. J. M. Coxon, R. G. A. R. Maclagan, A. Rauk, A. J. Thorpe, and D. Whalen, *J. Am. Chem. Soc.* **119**, 4712 (1997).
58. H. Shinoda, M. Sayama, Y. Mori, and M. Mineyuki, *Internet Electron. J. Mol. Des.* **4**, 698 (2005).
59. A. Lifshitz, C. Tamburu, and A. Suslensky, *J. Phys. Chem.* **93**, 5802 (1989).
60. J. C. Mackie, M. B. Colket, III, P. F. Nelson, and M. Esler, *Int. J. Chem. Kinet.* **23**, 733 (1991).
61. F. Dubnikova and A. Lifshitz, *J. Phys. Chem. A* **102**, 10880 (1998).
62. L. Zhai, X. Zhou, and R. Liu, *J. Phys. Chem. A* **103**, 3917 (1999).
63. G. B. Bacskay, M. Martoprawiro, and J. C. Mackie, *Chem. Phys. Lett.* **300**, 321 (1999).
64. M. P. Sammes and A. R. Katritsky, *Adv. Heterocycl. Chem.* **32**, 233 (1982).
65. A. Laurent, P. Mison, A. Nafti, and N. Pellisier, *Tetrahedron Lett.* **23**, 655 (1982).
66. M. P. Sammes, M. W. L. Chung, and A. R. Katritzky, *J. Chem. Soc., Perkin Trans.* **1** 1773 (1985).
67. X. Hong, L. Zhang, T. Zhang, and F. Qi, *J. Phys. Chem. A* **113**, 5397 (2009).
68. G. da Silva and J. W. Bozzelli, *J. Org. Chem.* **73**, 1343 (2008).
69. R. Liu, T. T.-S. Huang, J. Tittle, and D. Xia, *J. Phys. Chem.* **104**, 8368 (2000).
70. A. Laskin and A. Lifshitz, *J. Phys. Chem. A* **101**, 7787 (1997).
71. F. Dubnikova and A. Lifshitz, *J. Phys. Chem. A* **105**, 3605 (2001).
72. C. Ruchardt, M. Meier, K. Haaf, J. Pakusch, E. K. A. Wolber, and B. Muller, *Angew. Chem.* **30**, 893 (1991).
73. D. Wang, X. Qian, and Q. Zhang, *Chem. Phys. Lett.* **266**, 560 (1997).
74. O. Yamauchi, M. Takani, K. Toyada, and H. Masuda, *Inorg. Chem.* **29**, 1856 (1990).
75. S. Guidotti, I. Camurati, F. Focante, L. Angellini, G. Moscardi, L. Resconi, R. Leardini, D. Nanni, P. Mercandelli, A. Sironi, T. Beringhelli, and D. Maggioni, *J. Org. Chem.* **68**, 5445 (2003).
76. A. Lifshitz, M. Bidani, and S. Bidani, *J. Phys. Chem.* **93**, 1139 (1989).
77. F. Dubnikova and A. Lifshitz, *J. Phys. Chem. A* **106**, 1026 (2002).
78. A. Lifshitz and A. Laskin, *J. Phys Chem.* **98**, 2341 (1994).
79. A. T. Cocks and K. W. Egger, *J. Chem. Soc., Perkin Trans.* **2**, 199 (1973).
80. J. A. Rubin and S. V. Filseth, *J. Chem. Educ.* **46**, 57 (1969).
81. C. A. Wellington and W. D. Walters, *J. Am. Chem. Soc.* **83**, 4888 (1961).
82. A. Lifshitz, M. Bidani, and S. Bidani, *J. Phys. Chem.* **90**, 6011 (1986).
83. T. L. James and C. A. Wellington, *J. Chem. Soc. A* 2398 (1968).
84. F. Dubnikova and A. Lifshitz, *Int. J. Chem. Kinet.* **33**, 685 (2001).
85. A. Lifshitz, A. Suslensky, and C. Tamburu, *Proc. Combust. Inst.* **28**, 1733 (2000).
86. A. Lifshitz, A. Suslensky, and C. Tamburu, *J. Phys. Chem. A* **105**, 3148 (2001).

87. F. Dubnikova and A. Lifshitz, *J. Phys. Chem. A* **106**, 9278 (2002).
88. H. Richter, O. A. Mazyar, R. Sumathi, W. H. Green, J. B. Howard, and J. W. Bozzelli, *J. Phys. Chem. A* **105**, 1561 (2001).
89. H. Wang and M. Frenklach, *J. Phys. Chem.* **98**, 11465 (1994).
90. A. Violi, A. F. Sarofim, and T.N. Truong, *Combust. Flame* **126**, 1506 (2001).
91. I. V. Tokmakov and M. C. Lin, *J. Am. Chem. Soc.* **125**, 11397 (2003).
92. M. Frenklach, D. W. Clary, W. C. Gardiner, and S. E. Stein, *Proc. Combust. Inst.* **20**, 887 (1984).
93. I. V. Tokmakov and M. C. Lin, *J. Phys. Chem. A* **108**, 9697 (2004).
94. A. Fahr and S. E. Stein, *Proc. Combust. Inst.* **22**, 1023 (1988).
95. T. Yu and M. C. Lin, *Combust. Flame* **100**, 169 (1995).
96. J. Cioslovski, M. Schimeczek, P. Piskorz, and D. Moncrieff, *J. Am. Chem. Soc.* **121**, 3773 (1999).
97. M. Kamphus, M. Braun-Unkhoff, and K. Kohse-Hoeinghaus, *Combust. Flame* **152**, 28 (2008).
98. A. Lifshitz, C. Tamburu, and F. Dubnikova, *Proc. Combust. Inst.* **31**, 241 (2007).
99. I. D. Mackie and R. P. Johson, *J. Org. Chem.* **74**, 499 (2009).
100. P. Frank, J. T. Herzler, and C. Wahl, *Proc. Combust. Inst.* **25**, 833 (1994).
101. S. Sharma and W. H. Green, *J. Phys. Chem. A* **113**, 8871 (2009).
102. A. Lifshitz, C. Tamburu, A. Suslensky, and F. Dubnikova, *Proc. Combust. Inst.* **30**, 1039 (2005).
103. J. A. Miller and S. J. Klippenstein, *J. Phys. Chem. A* **107**, 7783 (2003).
104. C. L. Rasmussen, M. S. Skjoth-Rasmossen, A. D. Jensen, and P. Glarborg, *Proc. Combust. Inst.* **30**, 1023 (2005).
105. E. R. Ritter, J. W. Bozzelli, and A. M. Dean, *J. Phys. Chem.* **94**, 2493 (1990).
106. M. Yahyaoui, M. H. Hakka, P. A. Glaude, and F. Battin-Leclerc, *Int. J. Chem. Kinet.* **40**, 25 (2008).
107. C. S. McEnally and L. D. Pfefferle, *Combust. Sci. Technol.* **131**, 323 (1998).
108. E. Ikeda, R. S. Tranter, J. H. Kiefer, R. D. Kern, H. J. Singh, and Q. Zhang, *Proc. Combust. Inst.* **28**, 1725 (2000).
109. A. Laskin and A. Lifshitz, *Proc. Combust. Inst.* **27**, 313 (1998).
110. M. Lu and J. A. Mulholland, *Chemosphere* **42**, 625 (2001).
111. L. V. Moskaleva, A. M. Mebel, and M. C. Lin, *Proc. Combust. Inst.* **26**, 521 (1996).
112. C. S. McEnally and L. D. Pfefferle, *Combust. Flame* **148**, 210 (2007).
113. F. Battin-Lecterc, *Prog. Energy Combust. Sci.* **34**, 440 (2008).
114. Q. Wang, Y. Zhang, W. J. Rogers, and M. S. Mannan, *J. Hazard. Mater.* **165**, 141 (2009).
115. A. Lifshitz, C. Tamburu, A. Suslensky, and F. Dubnikova, *J. Phys. Chem. A* **108**, 3430 (2004).
116. F. Dubnikova and A. Lifshitz, *J. Phys. Chem. A* **106**, 8173 (2002).
117. F. Dubnikova and A. Lifshitz, *Isr. J. Chem.* **43**, 325 (2003).
118. F. Dubnikova and A. Lifshitz, *J. Phys. Chem. A* **104**, 530 (2000).
119. F. Dubnikova and A. Lifshitz, *J. Phys. Chem. A* **108**, 5962 (2004).

120. W. S. McGivern, J. A. Manion, and W. Tsang, *J. Phys. Chem. A* **110**, 12822 (2006).
121. L. K. Madden, A. M. Mebel, M. C. Lin, and C. F. Melius, *J. Phys. Org. Chem.* **9**, 801 (1996).
122. R. Liu, T. T.-S. Huang, J. Tittle, and D. Xia, *J. Phys. Chem. A* **104**, 8368 (2000).
123. A. Lifshitz, I. Shweky, and C. Tamburu, *J. Phys. Chem.* **97**, 4442 (1993).
124. A. Doughty, J. C. Mackie, and G. B. Bacskay, *Chem. Phys. Lett.* **221**, 267 (1994).
125. Y. Kuninobu, H. Ueda, A. Kawata, and K. Takai, *J. Org. Chem.* **72**, 6749 (2007).
126. A. Brandi, F. Pisaneshi, and M. Gensini, *Org. React. Mech.* 493 (2008).
127. C. Domene, L. W. Jenneskens, and P. W. Fowler, *Tetrahedron Lett.* **46**, 4077 (2005).
128. M. A. R. Matos, M. S. Miranda, V. M. F. Morais, and J. F. Liebman, *Eur. J. Org. Chem.* 2004, **3340** (2004).
129. W. V. Steele, R. D. Chirico, S. E. Knipmeyer, and A. Nguyen, *J. Chem. Thermodyn.* **24**, 449 (1992).
130. A. Lifshitz, C. Tamburu, A. Suslensky, and F. Dubnikova, *J. Phys. Chem. A* **110**, 4617 (2006).
131. A. Lifshitz, C. Tamburu, A. Suslensky, and F. Dubnikova, *J. Phys. Chem. A* **110**, 11677 (2006).
132. Y. Z. He, J. P. Cui, W. G. Mallard, and W. Tsang, *J. Am. Chem. Soc.* **110**, 3754 (1988).
133. A. Lifshitz, C. Tamburu, A. Suslensky, and F. Dubnikova, *J. Phys. Chem. A* **110**, 8248 (2006).
134. T. Sakaizumi, H. Kikuchi, O. Ohashi, and I. Yamaguchi, *Bull. Chem. Soc. Jpn.* **60**, 3903 (1987).
135. C. Thetaz and W. D. Crow, *J. Am. Chem. Soc.* **98**, 1258 (1976).
136. C. Wentrup, C. Mayor, and R. Gleiter, *Helv. Chim. Acta* **55**, 2628 (1972).
137. N. P. Cristsan and M. S. Platz, *Adv. Phys. Org. Chem.* **36**, 255 (2001).
138. P. S. Drzaic and J. I. Brauman, *J. Am. Chem. Soc.* **106**, 3443 (1984).
139. P. S. Drzaic and J. I. Brauman, *J. Phys. Chem.* **88**, 5285 (1984).
140. K. Ozawa, T. Ishida, K. Fuke, and K. Kaya, *Chem. Phys. Lett.* **150**, 249 (1988).
141. J. M. Galbraith, P. P. Gaspar, and W. T. Borden, *J. Am. Chem. Soc.* **124**, 11669 (2002).
142. G. Smolinsky, E. Wasserman, and W. A. Yager, *J. Am. Chem. Soc.* **84**, 3220 (1962).
143. O. Castell, V. M. Garcia, C. Bo, and R. Caballol, *J. Comp. Chem.* **17**, 42 (1996).
144. S.-J. Kim, T. P. Hamilton, and H. F. Schaefer, III, *J. Am. Chem. Soc.* **114**, 5349 (1992).
145. D. A. Hrovat, E. E. Waali, and W. T. Borden, *J. Am. Chem. Soc.* **114**, 8698 (1992).
146. W. D. Crow, M. N. Paddon-Row, and D. S. Sutherland, *Tetrahedron Lett.* **13**, 2239 (1972).

CHAPTER 7

Challenges in the Computation of Rate Constants for Lignin Model Compounds

ARIANA BESTE

Computer Science and Mathematics Division, Oak Ridge National Laboratory, Oak RidgeTN, USA

A.C. BUCHANAN III

Chemical Sciences Division, Oak Ridge National Laboratory, Oak RidgeTN, USA

7.1 LIGNIN: A RENEWABLE SOURCE OF FUELS AND CHEMICALS

Increasing the amount of energy produced through clean and reliable sources has become one of the USA's highest priority research directions, as is evident from the Department of Energy FY 2010 Congressional Budget Request [1]. The political motivation stems from environmental awareness, economical benefits, and a national security incentive to minimize the dependence on foreign oil. Most of the world's energy needs are supplied by diminishing fossil fuel resources. While developed nations have shown a steady increase in energy consumption over the last decades, an accelerated growth rate is expected in Asia, particularly China and India. It is difficult to predict the exact time at which the worldwide oil production can no longer be increased, but some studies expect this to be as soon as 2020 [2]. We, therefore, need to focus our research efforts on renewable resources now to meet the energy demands of the future.

Combustion of fossil fuels for energy generation leads to greenhouse gas emission, which will have a devastating effect on the life cycle of our planet if further pollution is not significantly reduced. For instance, the melting of polar ice already causes the natural habitat of arctic wildlife to change and decrease, thereby jeopardizing animal populations [3]. Using alternative energy resources is imperative to protect life on

Rate Constant Calculation for Thermal Reactions: Methods and Applications, Edited by Herbert DaCosta and Maohong Fan.

earth. Potential solutions include nuclear energy generation, hydro, wind, and solar power stations, and the utilization of biomass. Fossil fuels are also used to produce chemicals and materials, which are an integral part of modern life. New, sustainable, and environmentally friendly ways to provide raw material for the chemical industry are of great importance as well.

Biomass as a resource for fuels and chemicals is a promising fossil fuel alternative: it is renewable and carbon dioxide release in the atmosphere from biomass combustion is reduced by the carbon dioxide consumption during plant growth. Other harmful gases emitted during fossil fuel processing are sulfur and nitro oxides. Biomass contains insignificant amounts of sulfur and bio-oil fuels generate more than 50% less nitrogen oxides than diesel oil in a gas turbine [4]. To ease transport and to obtain a precursor for further refining, biomass is converted to bio-oil. Raw materials range from wood chips, grasses, and agricultural waste to algae. Most advantageous, economically and environmentally, is the use of waste or industrial by-products. Lignin, for instance, is accessible in large quantities as a by-product of the pulping industry, bark is removed from trees to manufacture lumber, sawdust is available from saw mills, and agricultural residues contain olive husks, corncobs, sugar cane bagasse, nut shells, and more. The enormous number of possibilities makes the optimization of bio-oil production and processing difficult because the nature of the feed influences the final product.

If bio-oil is generated from virgin biomass that is not harvested for an alternative main purpose, potentially negative aspects of biomass use have to be considered as well. Wood is an excellent source of biomass but deforestation has an adverse impact on the habitat, biodiversity, and aridity. Corn has been used extensively to make bioethanol, which is in competition with food cultivation [5]. If a significant percentage of the world's energy and material need is to be satisfied by biomass, large-scale supply and facilities are necessary. Fast-growing grasses have potentially higher annual energy yields than wood and are less demanding regarding soil and climate conditions. Algae are a promising pathway to biofuels because of their rapid growth and their potential ease to culture.

Essential for viable alternative energy and chemicals supply is a developed infrastructure including availability, waste management, transportation, and storage. Over the last century, fossil fuel power plants have flourished. Transportation and storage problems have been solved and the petrochemical industry has developed numerous processes to transform hydrocarbons with little functionalization into valuable materials on which modern life depends. Most of this success can be attributed to the deep chemical understanding of underlying reaction mechanisms, development of appropriate catalysts, and clever engineering.

In contrast to petroleum, the chemistry of biomass is based on oxygen-rich functional groups [2]. So-called first-generation biofuels (bioethanol, biodiesel) are utilized mainly as a 5–30% admixture in gasoline and diesel for common engines [6]. Alternative applications include the production of biodegradable, slow-release nitrogen fertilizer, meat-browning agents, and food flavor. Second-generation biofuels are obtained in an integrated lignocellulosic biorefinery essentially paralleling the machinery of a crude oil refinery. In a first step, biomass is separated into its primary components, which are then purified and further reacted applying chemical or biochemical methods to yield high-value products. Such refinery combines technologies for gasification, pyrolysis, syngas generation, biofuel production, Fischer–Tropsch synthesis, catalytic transformation to hydrogen, hydroprocessing, bio-oil separation yielding phenolic resins, and more, making efficient use of precious resources. Biorefineries are, however, at this stage merely concepts. Efficiently engineered production plants, refineries, and end-user facilities will have to be reinvented. Basic and applied research will be the driving force for biomass applications and their launch into a sustainable future. For further reading on biomass, we refer to recent reviews [2,4–9].

7.1.1 Origin and Chemical Structure

Plant biomass is a composite of oxygen-containing organic polymers, whose major components are cellulose, hemicelluloses, and lignin. The specific amount of each group differs for the various plant species. Cellulose is a high-molecular-weight linear polymer containing 5000–10,000 glucose units. Hemicellulose has a lower molecular weight ($\sim$150 U) and is a mixture of various polymerized monosaccharides. The third major component of biomass is lignin, which accounts for up to 33% of the mass of softwoods and up to 25% of the mass of hardwoods [4]. Lignin is an amorphous, very complicated, cross-linked, three-dimensional polyphenolic substance. It is an integral part of the cell walls linking fibrous cellulosic components during the last stages of cell differentiation. It reinforces cell walls, prevents water penetration, and sterilizes wounds by deposition. The exact structure of lignin is extremely hard to resolve. Figure 7.1 shows a representative partial structure of a lignin macromolecule that highlights key compositional units.

Lignin is formed through nonselective condensation of three phenolic alcohol units (monolignols) [11]: *p*-coumaryl alcohol, coniferyl alcohol, and sinapyl alcohol, depicted in Figure 7.2. (5-Hydroxyconiferyl units have also been detected in lignocellulosic materials in low amounts [12].) The ratio of the three major units changes with cell and plant type. Whereas lignin from grasses is made up of all three alcohols, lignin in softwood is predominantly built from coniferyl alcohol, and hardwood lignin from coniferyl and sinapyl alcohol. How exactly polymerization takes place *in vivo* is still subject of discussions. However, a completely random radical mechanism seems unlikely [11]. Enzyme-controlled polymerization can occur at the double bond or the phenolic oxygen leading to a variety of linkages. The three most dominant linkages are given in Figure 7.3: arylglycerol-β-aryl ether (β-O-4), phenylcoumaran (β-5), and biphenyl (5-5), which can be recognized as substructures in Figure 7.1. Figure 7.3 also indicates their distribution in softwood and hardwood. A variety of natural functional groups are present in lignin, most commonly methoxy, hydroxy, and carbonyl groups. The methoxy functional group is dominant and its abundance depends on the origin, for example, hardwood has a higher methoxy content than softwood.

As a component of whole biomass, lignin is obtained through an isolation process. Various lignins are available commercially. For instance, kraft lignin is a by-product

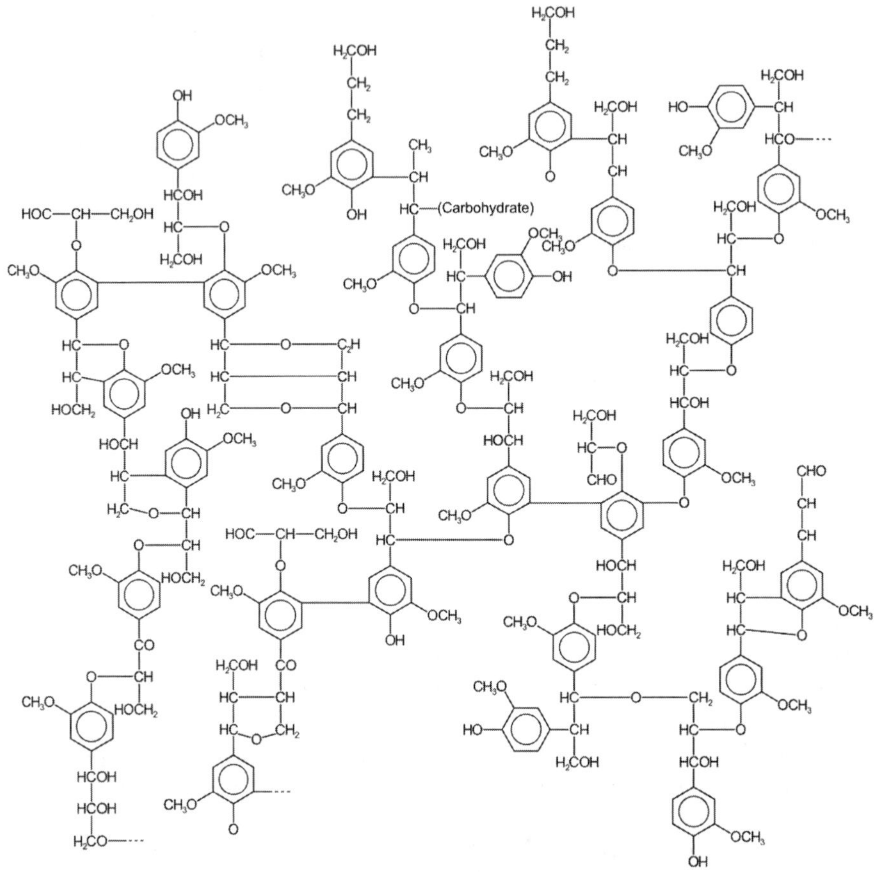

FIGURE 7.1 An example of a possible lignin substructure taken from Ref. [10].

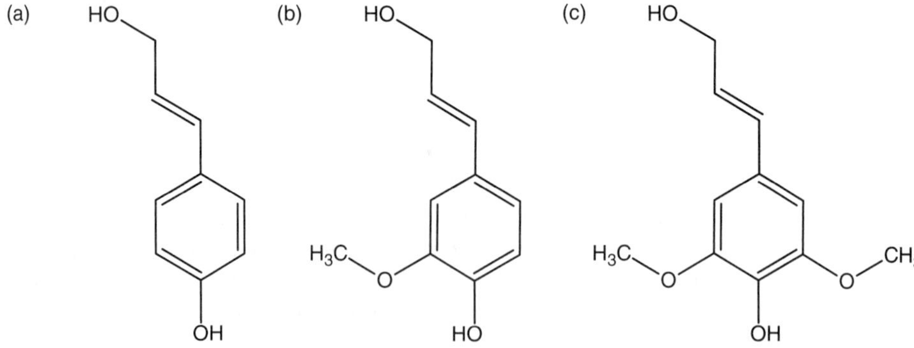

FIGURE 7.2 Monolignols: (a) *p*-coumaryl alcohol, (b) coniferyl alcohol, and (c) sinapyl alcohol.

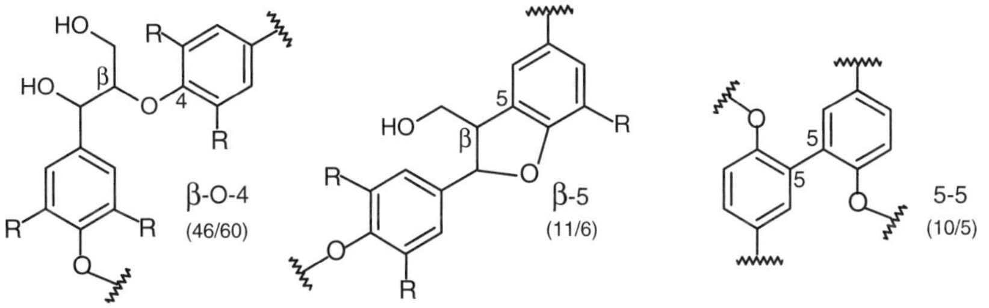

FIGURE 7.3 Most dominant linkages in lignin: arylglycerol-β-aryl ether (β-O-4), phenylcoumaran (β-5), and biphenyl (5-5). Their distribution in percentage is given in parentheses (softwood/hardwood) [11].

of the pulping industry and is separated from cellulose by dissolving wood chips in a sodium hydroxide solution forming black liquor. In the kraft process sodium sulfide is an additional component, whereas soda is used to obtain soda lignin. Milled wood lignin is extracted from wood meal by dry grinding or grinding in an organic solvent followed by treatment with a dioxane–water mixture. Application of sulfuric acid leaving solid lignin produces klason lignin. Other isolation techniques utilize enzymes, acid-catalyzed organic solvent, and periodate solutions [11]. Composition and reactivity of lignin obtained with different separation procedures vary.

7.1.2 Processing Techniques and Challenges

Most industrially accessible lignin is produced by the pulping industry and consumed as fuel. With only less than 5% alternative industrial use [13], other large-scale applications are currently not available. However, lignin has shown promising features as a natural resin in wood-based composites, as a natural absorbent for metal ions, dyes, pesticides, and more, and as a precursor for activated carbon [14]. Tanning agents were one of the first products fabricated from lignin and synthetic vanillin has been industrially synthesized from lignin since the 1930s. Efficient lignin conversion to hydrogen was demonstrated [15,16] and expanded polyurethane foam is an example of green plastic from lignin [17].

A prerequisite for the contribution of lignin to the innovations for greener energy and materials is a high-yield and selective processing technique. Lignin is difficult to structurally modify because of its chemical stability, and often harsh reaction conditions are needed for its depolymerization. A variety of processes have been applied to generate lower molecular weight compounds from lignin: pyrolysis, steam gasification, hydrogenation, oxidation, and biotechnological routes have been discussed [18]. Whereas gasification is the thermal transformation of lignin in reactive atmosphere, pyrolysis is the thermal degradation of lignin under inert conditions and has been extensively studied [15,19–21]. Oxidation of lignin in alkaline media results in aromatic aldehydes: vanillin from softwood and vanillin and syringaldehyde from hardwood [22]. Transition metal hydroxides and oxides can be used to catalytically oxidize lignin [23]. A reductive hydrogen atmosphere in the presence of metal catalysts allows the production of phenols by hydrogenation in organic solvents [24]. Another route to depolymerize lignin is the pyrolysis in acids with organic cosolvents [13,25]. Steam gasification can promote the formation of desired, low-molecular-weight gases because of steam reforming reactions of undesired higher weight species [16].

Pyrolysis is probably the most important processing technique, in part because pyrolysis reactions occur in other thermal lignin degradation as well, such as hydrogenation and gasification. The pyrolysis of lignin is highly complex due to the intricate structure of lignin and the dependence on origin [20], isolation technique [15], and reaction conditions, such as temperature, heating rate, and reaction time. It is further complicated by the fact that volatiles formed during primary pyrolysis undergo secondary reactions, such as coking and cracking. Generally, pyrolysis products can be divided into a gas phase (hydrocarbons, carbon monoxide, carbon dioxide), a liquid phase (water, alcohols, aldehydes, low-molecular-weight phenols), and a solid phase (high-boiling, phenolic compounds, char). At higher temperatures the yield of low-molecular-weight products increases but the decomposition chemistry becomes more complicated. Typically, fast pyrolysis conditions with rapid heating, short residence times, and moderate temperature (400–650°C) are advantageous, forming high amounts of liquid and less char.

Because of its complexity, the simulation of lignin pyrolysis uses simplified methods. While first-order and single reaction models were applied in early work, the distributed activation energy model assumes a set of irreversible first-order reactions with different activation energies [19,26]. Recently, a semidetailed model of the chemistry in the thermal degradation of lignin has been developed [21]. The initial lignin, intermediate, and product structures are characterized by copolymers formed from three model compounds representing the β-O-4 linkage. Devolatilization and depolymerization were expressed generically, leading to 500 elementary and lumped reactions. Crucial to the predictive value of the model is the use of accurate kinetic and thermodynamic parameters for the main reaction classes. For instance, the kinetic parameters of the initiation reactions were derived from bond energies, and kinetic parameters of propagation reactions were obtained using the assumption that forces between atoms are short ranged. An example where molecular-level detail was used to successfully simulate lignin pyrolysis is based on the statistical interpretation of the lignin structure and results from kinetic studies of model compounds [27,28]. Exceptions were computed water and methanol yields pointing to a neglect of important reaction pathways. Obviously, the less severe the assumptions in the simulations, the more precise they are. This is where an increased understanding of the detailed chemistry of model compounds and the determination of reliable kinetic parameters could be of great benefit.

7.2 MECHANISTIC STUDY OF LIGNIN MODEL COMPOUNDS

On one hand, the complexity of the lignin polymer, the structural differences by cell tissue and plant species, the changes undergone during the isolation process, and the harsh liquefaction conditions make the detailed mechanistic study of lignin depolymerization extremely difficult. On the other hand, to determine optimal processing techniques including the design of new selective catalysts and to push the limits of kinetic modeling, knowledge of the dynamics of the competing pathways in the thermal degradation of lignin is clearly necessary. To that end, conversion of model compounds representing the various linkages and substituent distributions occurring in lignin is studied. Insights obtained from model compounds can then be integrated into simulations of lignin degradation [27,28]. Generalizations from model compounds to lignin have to be done carefully, including increasingly complex systems. Whereas the value of the study of model systems is indisputable, challenges include higher reactivity of some model compounds compared with lignin [29], the possibility of lignin self-condensation, and potential cross-reactions between structural units represented by the model compounds.

7.2.1 Experimental Work

A variety of lignin model compounds have been studied experimentally, most of them representing the linkages shown in Figure 7.3, where truncated bonds are saturated with hydrogen or methyl groups [29–42]. Other important linkages include the α-O-4 and the β-1 linkage [32]. Investigations of even simpler models can also be useful because their chemical reactions tend to be less complex [29,43]. The β-O-4 linkage is by far the most common linkage and much research has been dedicated to it [33–42]. Although pyrolysis has obtained considerable attention [30,32–37,39,41,42], other techniques are investigated and developed in laboratories. Examples are the radiative degradation of β-O-4 model compounds [38], the thermochemolysis of β-5 models in tetramethylammonium hydroxide [31], acid-catalyzed thermal liquefaction [43], and the use of ionic liquids as reaction medium [29]. Confinement and restricted mass transport effects on the pyrolysis product selectivity were considered as well [35,36]. Often, the naturally occurring substituents incorporated in the model systems were found to have a strong effect on product distributions and mechanisms [31,34,37,41].

The simplest model compound for the β-O-4 linkage is phenethyl phenyl ether ($PhCH_2CH_2OPh$, PPE), shown in Figure 7.4. Several mechanisms were proposed for the thermolysis of PPE. Gilbert and Gajewski [44] proposed a free radical mechanism, which we will discuss in detail below. Klein and Virk [45] considered the six-centered retro-ene fragmentation to be the most plausible based on kinetic analysis. If a hydroxy substituent is located at the α-carbon, an oxirane mechanism becomes possible [46], and with the presence of a *para*-phenol and a hydroxymethyl group on the β-carbon, an additional pathway with a quinone methide as an intermediate opens [47]. At low pressure, a 1,2-elimination pathway can contribute [34], and at low pH, an acid-catalyzed mechanism with a cation as intermediate might be competitive. Scheme 7.1 demonstrates the various possibilities. Depending on the reaction conditions (temperature, solvent, concentration, heating rate, pH) and the location and type of the substituents, multiple mechanisms might be relevant simultaneously. However, at moderate temperatures and under fast pyrolysis conditions, neither the oxirane [48] nor the retro-ene [33] mechanism contributed significantly.

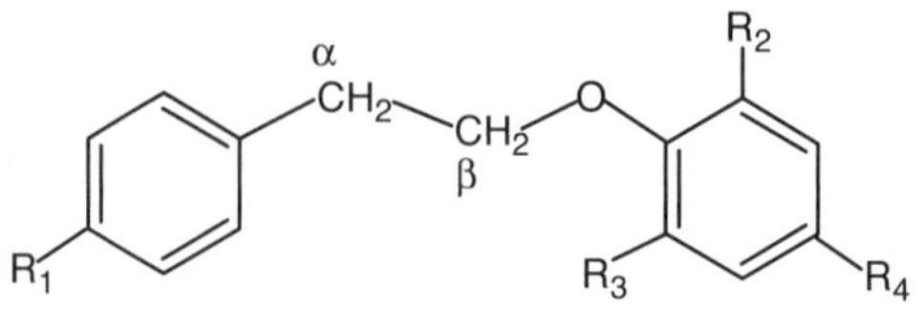

1 $R_1, R_2, R_3, R_4 = H$
2 $R_1 = OH, R_2, R_3, R_4 = H$
3 $R_1 = OCH_3, R_2, R_3, R_4 = H$
4 $R_1, R_2, R_3 = H, R_4 = OCH_3$
5 $R_1 = H, R_2 = OCH_3, R_3, R_4 = H$
6 $R_1 = H, R_2, R_3 = OCH_3, R_4 = H$
7 $R_1 = OH, R_2 = OCH_3, R_3, R_4 = H$

FIGURE 7.4 Phenethyl phenyl ether (PPE) with naturally occurring hydroxy and methoxy substituents.

Britt et al. [33] studied the thermolysis of PPE in the neat liquid, in solution with a nonreactive or with a hydrogen donor solvent, and in the gas phase from 330 to 425°C. Independent of reaction medium, the data were consistent with a free radical chain mechanism cycling through radicals formed at both the α- and β-carbons. They found that the cracking of PPE proceeds through two competitive pathways, defined as α/β-selectivity, with $PhOH + PhCH{=}CH_2$ (α) and $PhCHO + PhCH_3$ (β) formed as primary products. Minor amounts of $Ph(CH_2)_2Ph$, $Ph(CH_2)_3Ph$, PhH, and $PhCH_2CH_3$ were also detected. The average α/β-selectivity for the neat liquid, in solution, and the gas phase was measured to be 3.1 ± 0.3 at 375°C. If a hydrogen bonding or hydrogen donor solvent was present, the α/β-selectivity increased. Considering that the kinetic order was determined to be fractional, that the conversion rate increased by adding a free radical initiator, and that a significant isotope effect on product selectivities was observed for aliphatic deuterium-labeled species, Britt et al. proposed a detailed free radical chain mechanism, given in Scheme 7.2. The first reaction step is the homolytic cleavage of the weakest bond in the system, the carbon–oxygen bond, forming the phenethyl radical and one of the chain-carrying radicals, the phenoxy radical. The phenoxy radical then abstracts a hydrogen whether on the α- or β-position of PPE (reactions 2a and 2b), resulting in the α- or β-PPE radicals and PhOH, a primary product. The α-PPE radical undergoes fast β-scission (reaction 3) contributing to the radical chain with a phenoxy radical and yielding $PhCH{=}CH_2$, another primary product. The β-scission of the β-PPE radical is slow and it can potentially convert to the α-product through intermolecular hydrogen abstraction. However, because of the concentration independence of the α/β-selectivity, it was concluded that intermolecular hydrogen abstraction is not competitive with 1,2-phenyl migration (reaction 4). The β-scission of the migration product (reaction 5) is again fast forming the second chain-carrying radical, the benzyl radical, and PhCHO, a major reaction product. The benzyl radical abstracts a hydrogen on the α- or β-position of PPE (reactions 6a and 6b) giving the last primary product $PhCH_3$ and the PPE radicals, which can further react through reactions 3–5. Since $PhCH_2CH_3$ was observed to be a minor product, hydrogen abstraction by the phenethyl radical (reactions 7a and 7b) does not

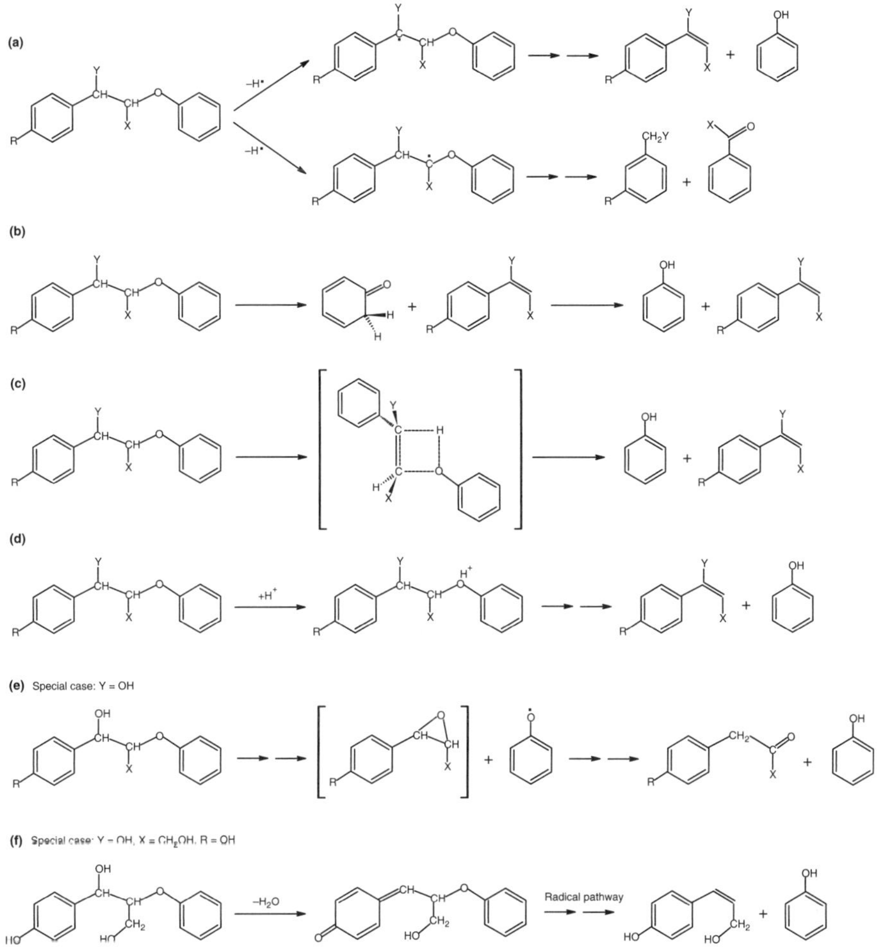

SCHEME 7.1 Potential mechanisms for the thermal decomposition of PPE (including substituents). (a) Radical chain, (b) retro-ene, (c) 1,2-elimination, (d) acid catalysis, (e) oxirane, and (f) quinone methide mechanism.

significantly contribute. At 375°C, the kinetic chain length is approximately 30–35 cycles long, where termination is predominantly caused by benzyl radical coupling. The measured α/β-selectivity is determined by the relative rates of hydrogen abstraction by phenoxy and benzyl radicals at the α- and β-carbons since β-scission and 1,2-phenyl migration from oxygen to carbon are rapid compared to hydrogen abstraction.

In subsequent work by Britt et al. [37], the influence of substituents on the pyrolysis mechanism of PPE in biphenyl solvent at moderate temperature (345°C) was considered. Naturally occurring methoxy and hydroxy substituents were located at either phenyl ring as depicted in Figure 7.4. It was found that the radical mechanism

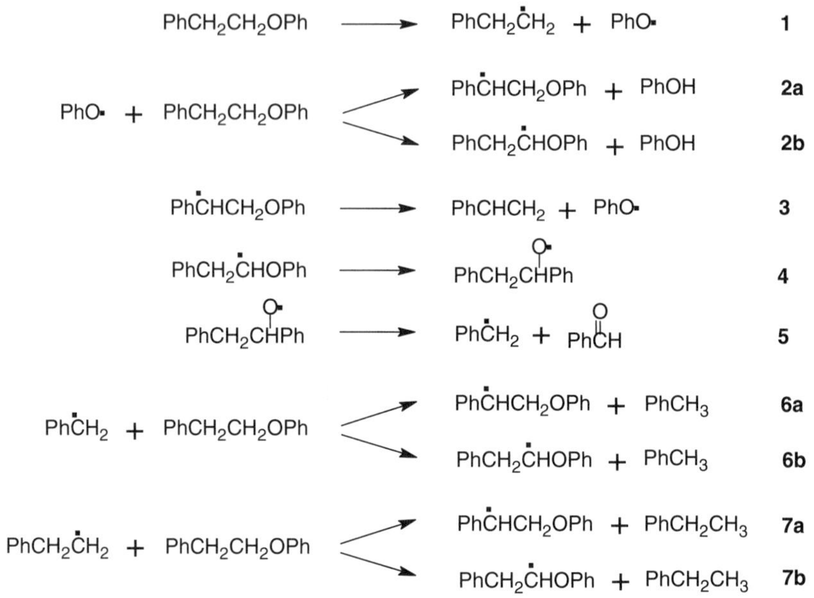

SCHEME 7.2 Radical chain mechanism for the pyrolysis of PPE [33].

still holds and that the effect of the substituents is an increase in the pyrolysis rate and to a smaller extent in the α/β-selectivity. Table 7.1 gives the measured values for conversion rates and α/β-selectivities for the PPE derivatives shown in Figure 7.4.

Substituent effects are complicated and can perturb all the key steps in the mechanism. Steric effects might enhance or reduce electronic effects. A substantial shortening of the kinetic chain length (as observed for **6** in Figure 7.4) makes the anticipation of substituent effects on product selectivities even more difficult because they are no longer determined by the radical chain propagation steps only. The strength of computational studies is the ability to probe individual reactions,

TABLE 7.1 Pyrolysis Rates and Selectivities in Biphenyl Solution at 345°C for Substituted PPE Shown in Figure 7.4 [37]

Compound	Rate (%/h)	α/β-Selectivity
1	1.7 ± 0.2	3.8 ± 0.3
2	13.1 ± 2.8	5.1 ± 0.1
3	3.2 ± 0.1	7.4 ± 0.3
4	10.7 ± 1.0	3.0 ± 0.03
5	8.0 ± 0.2	4.6 ± 0.1
6	42.1 ± 3.1	7.3 ± 1.9
7	41.9 ± 8.2	6.6 ± 0.6

which makes them extremely valuable in situations like this where collective effects on multiple reactions cannot be readily separated by experiment.

7.2.2 Computational Work

Ab initio computational methods provide sufficient accuracy to explore chemical reaction pathways. However, not many studies have tackled lignin model compounds, in part because they are of considerable size and reaction mechanisms are involved (see Schemes 7.1 and 7.2). Density functional theory (DFT) is commonly applied to investigate certain aspects of lignin model chemistry. For instance, the β-scission of the α-PPE radical with a hydroxy substituent on the α-carbon (reaction 3, Scheme 7.2) has been studied [38] and it was found that intramolecular hydrogen bonding stabilizes ground and transition states. Transition states and rate constants for the 1,2-migration of an aryl group (neophyl-like rearrangement) in ring-substituted 1,1-diarylalkoxyl radicals, which is similar to the phenyl-shift reaction of the β-PPE radical (reaction 4, Scheme 7.2), have been calculated [49,50] and it was concluded that the migration proceeds through an intermediate and that substituents strongly influence the reaction barriers. Another study, again using DFT, reports the response of a β-O-4 model to mechanical force [51]. Perturbation theory was applied to survey the conformational space of a small model compound (guaiacyl unit) and to obtain rotational barriers for the hydroxy and methoxy substituents; strong intramolecular hydrogen bonds were detected [52]. Fragmentation products of lignin models after bombardment with heavy metal ions were analyzed using Hartree–Fock molecular dynamics [53] and classical force fields were parameterized on the basis of second-order perturbation theory calculations [54].

Lignin formation through addition of monolignols is also an interesting subject because of the large variety of linkages and, yet, not random distribution. Contributing factors are thermodynamic control [55], elevated unpaired spin density at the O4 position of the monolignol radical [55,56] (which explains the dominance of the β-O-4 linkage), and solvent effects [56,57]. The principle of hard and soft acids has been applied to explain the intrinsic reactivity of monolignols [57,58]. The polymerization of monolignols is perceived to proceed through quinone methide intermediates, which are also proposed to be intermediates in lignin pyrolysis. A reaction mechanism for the unimolecular conversion of *o*-quinone methide was presented and transition state rate constants were calculated [59].

7.3 COMPUTATIONAL INVESTIGATION OF THE PYROLYSIS OF β-O-4 MODEL COMPOUNDS

A great amount of insight can be obtained by the computational study of the thermal decomposition of β-O-4 model compounds, representing the most common linkage in lignin. While experimental work determines overall product distributions and total rates of reaction [33,37], kinetic parameters of individual reactions and details of substituent effects on equilibrium and transition state structures are difficult to obtain experimentally. Computational methods allow the location of individual transition states, the calculation of activation barriers and entropic prefactors for targeted reaction steps, and electronic structure analysis of reactants, intermediates, products, and transition states to explain the effect of naturally occurring substituents.

7.3.1 Methodology

7.3.1.1 Overview Over the last few decades, quantum chemistry has evolved into a predictive tool for the calculation of energy differences. Of equal importance as a driving force for chemical reactions is the entropy that has received less attention from the computational community. The predictive value of computed rate constants depends not only on the accuracy of the electronic structure calculations but also on the correct description of the molecular entropy, including anharmonic effects in the vibrational modes, which is subject of Section 7.3.1.3.

The electronic structure method is chosen by accuracy and computational demand. High-level *ab initio* methods [60], predicting energy differences within a few tenths of kcal/mol, have become a standard tool for small molecules. With increasing system size, the computational cost of such methods can rapidly become prohibitive (polynomial scaling with the number of basis functions). For molecular systems of the size of a typical lignin model, DFT [61] is a cost-effective alternative. The accuracy of the DFT functional depends to some extent on its parameterization but errors, typically, do not exceed a few kcal/mol. It is advisable to calibrate the DFT calculations to high-level *ab initio* methods or to experimental results. The precision of the electronic structure method is very important when calculating absolute rate constants because a deviation in the activation energy of less than 1.5 kcal/mol at room temperature can alter the rate constant by a factor of 10. Errors of this order of magnitude are common in the computation of absolute rate constants [59,62,63]. Often, however, we are interested in relative rate constants, for example, when monitoring the progress of a reaction, the product formation is determined by the competing rates of intermediate formation and depletion, reverse reactions, alternative pathways, and so on. Most of the errors in electronic structure methods are systematic and the precision of relative rate constants can be expected to be much higher.

Another challenge when computing rate constants for reactions of lignin models is that with the introduction of substituents, the number of open channels in a narrow energy range increases because conformers differing by the orientation of a substituent can be very close in energy. Energetically close reaction pathways contribute to the rate constant with similar weight and have to be included. Depending on the number of substituents and how flexible the molecule is, the comprehensive search for transition states can be a tremendous task.

Computed kinetic parameters are incorporated into simulations to model the chemical mechanism or part of it. In some instances, simplifications can be introduced, such as the steady-state approximation, allowing the formulation of an analytical kinetic model, involving only few kinetic parameters (Section 7.3.2). In the general case, the rate constants of all participating reactions need to be included.

Numerical integration methods are then applied to follow reactant depletion and intermediate and product formation as a function of time (Section 7.3.3).

7.3.1.2 Transition State Theory Reaction dynamics can be treated quantum mechanically to high accuracy, but this is limited to small molecular systems (on the order of five atoms) [64]. For medium-sized systems (50–100 atoms), transition state theory is an appropriate alternative. Within transition state theory, a coordinate is defined describing the progress of the reaction, connecting reactants with products along a minimum energy path. This is demonstrated in Figure 7.5 for a reaction $A + BC \rightarrow AB + C$. Reactants and products are local minima for which normal mode analysis yields only real frequencies. The transition state ABC is located at the maximum along the reaction coordinate and is characterized as a saddle point, for which all other coordinates are at a minimum on the potential energy surface. Normal mode analysis at the transition state results in exactly one imaginary frequency, while all other modes correspond to real frequencies. Once the energetically highest point is passed, the reaction proceeds toward the products. Consequently, recrossing of the reaction barrier is not allowed. However, the dividing surface between reactants and products does not have to be at the saddle point, but can also be determined variationally to minimize recrossing effects. This is realized in variational transition state theory [65,66]. Alternatively, trajectory analysis can be used to obtain a correction for recrossing effects [67]. Within transition state theory, the motion across the reaction barrier is treated classically. Since tunneling is a quantum phenomenon, it is not accounted for. Multiplicative tunneling corrections for transition state rate constants, derived for different barrier shapes, are available [68–70]. Implicit in transition state theory is the assumption of separability of the reaction coordinate from all other degrees of freedom. Miller formulated a semiclassical transition state theory where this assumption is lifted [71,72]. Semiclassical approaches [73,74] can, in principle, incorporate interference and tunneling processes.

A rigorous derivation of transition state theory can be found in Ref. [75]; here, a more intuitive approach is adopted, similar to the outline in Ref. [76]. A central

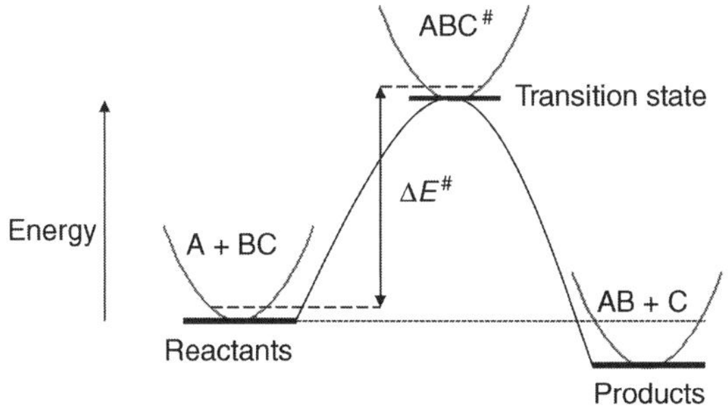

FIGURE 7.5 Schematic of the reaction progress along the reaction coordinate for the reaction $A + BC \rightarrow AB + C$; $\Delta E^{\#}$ is the zero-point-corrected activation barrier.

assumption in transition state theory is that the reactants A and B are in thermal equilibrium with the transition state ABC, described by the Boltzmann distribution:

$$\mathrm{A} + \mathrm{BC} \rightleftharpoons \mathrm{ABC} \rightarrow \mathrm{AB} + \mathrm{C} \tag{7.1}$$

The rate of reaction r is expressed as the rate constant k multiplied by the concentration of the reactants [A] and [BC]:

$$r = k[\mathrm{A}][\mathrm{BC}] \tag{7.2}$$

The concentration of the reactants is determined by the quasi-equilibrium between reactants and transition state, for which the equilibrium constant is given by the ratio of the concentration of the transition state [ABC] and the reactants. Employing statistical considerations, the equilibrium constant can also be expressed in terms of the molecular partition functions [77]:

$$K = \frac{[\mathrm{ABC}]}{[\mathrm{A}][\mathrm{BC}]} = \frac{Q_{\mathrm{ABC}}/V}{(Q_{\mathrm{A}}/V)(Q_{\mathrm{BC}}/V)} e^{-\beta\,\Delta E^{\#}} \tag{7.3}$$

where Q_{ABC}, Q_{A}, and Q_{BC} are the molecular partition functions of the transition state, reactant A, and reactant BC; V is the volume; β is given by $1/k_bT$ with k_b the Boltzmann constant and T is the temperature. $\Delta E^{\#}$ is the zero-point-corrected activation energy. Since every transition state once formed will lead to products, the rate of reaction is given by the concentration of the transition state [ABC] divided by its lifetime τ:

$$r = \frac{[\mathrm{ABC}]}{\tau} = k[\mathrm{A}][\mathrm{BC}] \tag{7.4}$$

where Eq. (7.2) is used. Solving for k and substituting into Eq. (7.3) leads to

$$k = \frac{[\mathrm{ABC}]}{[\mathrm{A}][\mathrm{BC}]\tau} = \frac{K}{\tau} = \frac{V}{\tau}\frac{Q_{\mathrm{ABC}}}{Q_{\mathrm{A}}Q_{\mathrm{BC}}} e^{-\beta\Delta E^{\#}} \tag{7.5}$$

The transition state exists only in the vicinity of the saddle point in a small distance d of the minimal energy path between reactants and products. If $\bar{v}_x$ is the average velocity of the transition state toward the products, defined as direction x, τ is given by

$$\tau = \frac{d}{\bar{v}_x} \tag{7.6}$$

The average velocity is determined by the Maxwell–Boltzmann velocity distribution in x direction $f(v_x)$:

$$f(v_x) = 2\sqrt{\frac{\mu_{\mathrm{ABC}}\beta}{2\pi}} e^{-[(\mu_{\mathrm{ABC}}v_x^2\beta)/2]} \tag{7.7}$$

where μ_{ABC} is the effective mass of the transition state. The additional factor of 2 in Eq. (7.7) results from considering forward velocities only. Taking the statistical average of the velocity leads to

$$\bar{v}_x = \int_0^\infty v_x f(v_x) dv_x = \sqrt{\frac{2}{\pi \mu_{\mathrm{ABC}} \beta}} \tag{7.8}$$

The gradient of the potential energy is very small in the saddle point region. Therefore, the force acting on the transition state within d is small and the particle in the box approximation can be applied for a box length d. The translational partition function of a particle in a one-dimensional box is written as

$$q_{\mathrm{ABC}}^{\mathrm{tr}} = \frac{1}{2}\sqrt{\frac{2\pi\mu_{\mathrm{ABC}}}{h^2\beta}}d \tag{7.9}$$

where h is Planck's constant. The factor 1/2 excludes contributions to the partition function from product to reactant motion. The molecular partition function of the transition state is a product of the translational partition function of Eq. (7.9) and the partition function of the remaining degrees of freedom $Q_{\mathrm{ABC}}^{\#}$:

$$Q_{\mathrm{ABC}} = q_{\mathrm{ABC}}^{\mathrm{tr}} Q_{\mathrm{ABC}}^{\#} \tag{7.10}$$

Equations (7.9) and (7.10) are substituted into Eq. (7.5):

$$k = \frac{V}{2\tau}\sqrt{\frac{2\pi\mu_{\mathrm{ABC}}}{h^2\beta}}d\frac{Q_{\mathrm{ABC}}^{\#}}{Q_{\mathrm{A}}Q_{\mathrm{BC}}}e^{-\beta\Delta E^{\#}} \tag{7.11}$$

τ is expressed through Eqs (7.6) and (7.8) leading to the final equation for the transition state rate constant:

$$k = \sigma\frac{V}{h\beta}\frac{Q_{\mathrm{ABC}}^{\#}}{Q_{\mathrm{A}}Q_{\mathrm{BC}}}e^{-\beta\Delta E^{\#}} \tag{7.12}$$

The factor σ is included to account for all indistinguishable ways the reactants can approach each other. Notice that due to the assumption of no recrossing of the barrier once a transition state is formed, the rate constant does not depend on the products.

Because the molecular partition functions are a measure of how many states are accessible to the molecule, the prefactor in Eq. (7.12) embodies entropic contributions to the rate constant, while the exponential term accounts for the energetic contributions. A general formula for the partition function is

$$Q = \sum_{n=1}^{\infty} g_n e^{-\beta\varepsilon_n} \tag{7.13}$$

where the sum runs over all states n of the molecule, g_n is the degeneracy of state n, and ε_n is the energy of state n. Molecules can exist in ground and excited electronic, vibrational, rotational, and translational states. The energy spacing between the electronic states is typically large and when only the electronic ground state is considered, the electronic partition function is the degeneracy of the ground state. In contrast, the energy spacing of the translational states is very small and summation over distinct states can be approximated by continuous integration. Because the energy scale of electronic and translational states is very different, their contributions to the molecular partition function are independent of each other and can be written as a product. Rotational and vibrational states, however, are much closer in energy. Common practice is to evaluate the rotational and vibrational partition functions separately. Rotations are usually treated within the rigid rotor approximation and vibrations are often considered to be harmonic. Within these approximations, the molecular partition function and the individual terms are given by the following:

Total molecular partition function:

$$Q = Q^{\mathrm{el}} Q^{\mathrm{vib}} Q^{\mathrm{rot}} Q^{\mathrm{tr}} \tag{7.14}$$

Electronic partition function:

$$Q^{\mathrm{el}} = g_{\mathrm{gr}} \tag{7.15}$$

where g_{gr} is the degeneracy of the ground state.

Vibrational partition function:

$$Q^{\mathrm{vib}} = \prod_{j=1}^{3N-6} \frac{e^{-[(\beta h\nu_j)/2]}}{1 - e^{-\beta h\nu_j}} \tag{7.16}$$

where ν_j is the harmonic frequency of mode j, applicable to nonlinear polyatomic molecules, and N is the number of atoms.

Rotational partition function:

$$Q^{\mathrm{rot}} = \sqrt{\frac{\pi T^3}{\Theta_{\mathrm{A}}\Theta_{\mathrm{B}}\Theta_{\mathrm{C}}}}, \qquad \Theta_J = \frac{\hbar^2}{2I_J k_b} \tag{7.17}$$

where I_J is the moment of inertia of a three-dimensional rotor applicable to asymmetric top.

Translational partition function:

$$Q^{\mathrm{tr}} = \left(\frac{2\pi\mu}{h^2\beta}\right)^{3/2} V \tag{7.18}$$

where μ is the effective mass of the molecule.

The rate constant in Eq. (7.12) is the rate constant for the classical barrier crossing along a reaction coordinate passing once through the dividing surface between reactants and products at the maximum point of the minimal energy path. To account for quantum mechanical effects, a correction factor κ can be introduced:

$$k = \kappa\sigma \frac{V}{h\beta} \frac{Q^{\#}_{\mathrm{ABC}}}{Q_{\mathrm{A}} Q_{\mathrm{BC}}} e^{-\beta\Delta E^{\#}} \tag{7.19}$$

If information on the reaction path is available, as, for instance, in variational transition state theory, this can be used to calculate κ [69,70]. In transition state theory, only the knowledge of the energy and its first and second derivatives at the reactant and transition state locations is needed and the barrier is typically approximated by a simple functional form. One possibility is to describe the reaction barrier by an Eckart potential [75] (also called a sech^2 potential, depending on the literature source). κ in Eq. (7.19) is defined as the ratio of transmitted quantum particles to classical particles and the resulting integral for the Eckart potential can be solved numerically. An approximate solution is the Wigner tunneling correction:

$$\kappa^{\mathrm{W}} = 1 + \frac{1}{24} (h\nu^{\#}\beta)^2 \tag{7.20}$$

where $\nu^{\#}$ is the absolute value of the imaginary frequency corresponding to barrier crossing.

To calculate rate constants for bimolecular systems, Eqs (7.19) and (7.20) are used. An equivalent theory can be derived for unimolecular reactions. In a unimolecular reaction only one reactant forms a transition state through radiation or in an apparent unimolecular reaction through collision, which results in product generation. Applying similar assumptions as above (classical motion when crossing the barrier, no recrossing events, and thermal equilibrium) leads to the canonical rate constant, given by

$$k = \kappa \frac{\sigma}{h\beta} \frac{Q^{\#}_{\mathrm{A}}}{Q_{\mathrm{A}}} e^{-\beta\Delta E^{\#}} \tag{7.21}$$

where $Q^{\#}_{\mathrm{A}}$ and Q_{A} are the molecular partition functions of the unimolecular transition state and the reactant, respectively. $\Delta E^{\#}$ is the reaction barrier, and κ, h, and β are defined as above. Since the translational partition functions cancel in Eq. (7.21), $Q^{\#}_{A}$ and Q_{A} contain only rotational, vibrational, and electronic contributions.

7.3.1.3 Anharmonic Corrections When evaluating the vibrational partition function, the harmonic approximation (Eq. (7.16)) is often used to describe vibrational motion. This, however, can lead to large errors, particularly in the treatment of low-lying frequencies [78]. Transition states for hydrogen abstraction on the simplest lignin model for the β-O-4 linkage, **1** in Figure 7.4, possess about 10 frequencies below 100 cm^{-1} and that number increases when substituents are added. The quantum vibrational partition function is given by a sum over states formula, with exponential dependence on the vibrational energy levels (Eq. (7.13)). This implies that the errors made in low-frequency modes have the largest impact on the partition function and, therefore, the rate constant. We do not expect the errors due to anharmonicity in the low vibrational modes to be systematic and the error size will depend on the particular reaction studied. Consequently, even when calculating relative rate constants, anharmonic effects should be incorporated.

The simplest way to take anharmonicity into account is within the independent mode approximation [79–81]. However, vibrational motion is independent only for the harmonic oscillator in normal mode coordinates, that is, anharmonicity in the vibration causes the modes to couple. The quantum anharmonic vibrational partition function can rigorously be calculated (including mode–mode coupling) using the sum over states formula (Eq. (7.13)) and vibrational configuration interaction, as has been demonstrated for ethane [82]. A more cost-effective approach is to retain the form of the harmonic vibrational partition function (Eq. (7.16)), but use anharmonic ground-state energies and fundamental frequencies [83]. Recently, a virtual vibrational self-consistent field method has been developed [84], which includes mode–mode coupling in an average fashion and which is capable of treating larger molecules. Alternative approaches utilize path integral methods (see Ref. [85], and references therein).

If the independent mode approximation is used to partially include anharmonic effects, one-dimensional anharmonic motion replaces specific normal modes in the harmonic vibrational partition function. Anharmonic motion can be modeled using a Morse potential [80], an oscillatory potential of a hindered rotation [81], or a potential of polynomial form [79]. (Partition functions per mode will be denoted by q.) Typically, only the lowest few modes are corrected because of the decreasing significance of the error contribution of high-energy states to the error in the partition function. A quantum expression of the one-dimensional anharmonic partition function can be obtained by calculating the energies of the Hamiltonian corresponding to the anharmonic potential and applying the sum over states formula (Eq. (7.13)).

An example of a one-dimensional, anharmonic potential for the lowest frequency mode of a transition state for hydrogen abstraction on lignin model **5** in Figure 7.4 is given in Figure 7.6. The potential was obtained by displacement along the normal mode j and by fitting to a fourth-order polynomial:

$$V(x_j) = a_0 + a_1 x_j + a_2 x_j^2 + a_3 x_j^3 + a_4 x_j^4 \qquad (7.22)$$

The harmonic potential is flatter than the anharmonic potential and the eigenvalue spacing is smaller. The harmonic approximation, therefore, overestimates the vibrational partition function compared to the quantum anharmonic partition function (at 400K: harmonic $q^{\text{vib}} = 40.2$; anharmonic $q^{\text{vib}} = 7.8$).

In the sum over states formula, excited states for the vibrational modes need to be included up to convergence. A more convenient integral expression is provided by classical or semiclassical theories. At high temperatures and low frequencies, the vibrational motion behaves increasingly classically and the semiclassical Wigner–Kirkwood expression is an excellent approximation to the quantum partition function [78] for low-frequency vibrations at pyrolysis temperatures. The semiclassical

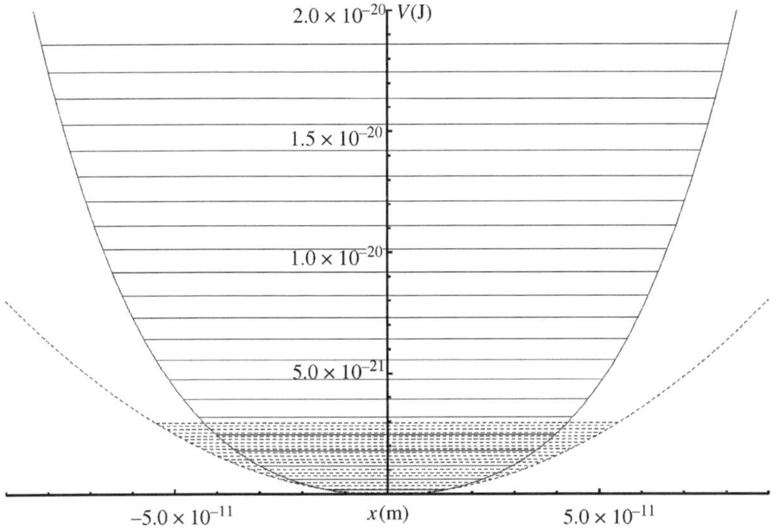

FIGURE 7.6 Anharmonic potential ($a_0 = -2.63 \times 10^{-25}$ J, $a_1 = 3.80 \times 10^{-14}$ J/m, $a_2 = 2.45 \times 10^{-2}$ J/m^2, $a_3 = 5.48 \times 10^6$ J/m^3, $a_4 = 5.29 \times 10^{18}$ J/m^4) with first 22 eigenvalues (solid line) and harmonic potential ($\nu = 6.9$ cm^{-1}) with first 22 eigenvalues (dashed line) of a transition state for hydrogen abstraction on lignin model **5** in Figure 7.4.

partition function for the one-dimensional anharmonic potential along normal mode coordinate x_j is given as an integral over the particle density $\rho_{\mathrm{WK}_j}(x_j)$:

$$q_{\mathrm{WK}_j} = \int \rho_{WK_j}(x_j)dx_j \tag{7.23}$$

The superscript "vib" is omitted with the understanding that the partition functions and particle densities correspond to vibrations. The particle density is expressed as a Wigner–Kirkwood perturbation expansion [86,87] in terms of a small parameter λ:

$$\lambda = \frac{h^2\beta^2}{8\mu_j\pi^2} \tag{7.24}$$

where μ_j is the effective mass for mode j $\mu_j = \left(\sum_i^{3N} l_{ij}^2\right)^{-1}$ with l being the vector of Cartesian displacement for mode j and N is the number of atoms. The second-order expansion in λ and fourth-order expansion in h of the Wigner–Kirkwood particle density is written as

$$\rho_{\mathrm{WK4}_j}(x_j) = \rho_{\mathrm{Cl}_j}(x_j)(1 + \lambda A_1 + \lambda^2 A_2) \tag{7.25}$$

The classical particle density is given by

$$\rho_{\mathrm{Cl}_j}(x_j) = h^{-1}\sqrt{\frac{2\pi\mu_j}{\beta}}e^{-\beta V(x_j)} \tag{7.26}$$

The expansion coefficients in Eq. (7.25) depend on the various derivatives of the potential:

$$A_1 = \frac{\beta}{12}(V')^2 - \frac{1}{6}V''$$
$$A_2 = \frac{\beta^2}{288}(V')^4 - \frac{11\beta}{360}(V')^2V'' + \frac{1}{40}(V'')^2 + \frac{1}{30}V'V''' - \frac{1}{60\beta}V^{\mathrm{IV}} \tag{7.27}$$

The total vibrational partition function is a product of the Wigner–Kirkwood partition functions for each corrected mode and the harmonic partition function of the remaining vibrations:

$$Q^{\mathrm{vib}} = \prod_{j=1}^{\eta} q_{\mathrm{WK}_j} \prod_{j=\eta}^{3N-6} \frac{e^{-[(\beta h\nu_j)/2]}}{1 - e^{-\beta h\nu_j}} \tag{7.28}$$

where η is the number of anharmonically corrected vibrations, assuming that the lowest η vibrations are corrected. Diagonal anharmonic effects are accounted for in Eq. (7.28); coupling between modes is included at the harmonic level only.

7.3.2 Analytical Kinetic Models

The pyrolysis mechanism of PPE and its derivatives given in Scheme 7.2 consist of bimolecular and unimolecular reactions. Applying transition state theory, we calculate the rate constants for the hydrogen abstraction reactions using Eq. (7.19) and the rate constants for reactions 1 and 3–5 using Eq. (7.21). The Wigner correction (Eq. (7.20)) is utilized to approximate quantum effects and the molecular partition functions are defined through Eqs (7.14), (7.15), (7.17), (7.18), and (7.28).

7.3.2.1 Parallel Reactions For medium-sized organic systems, there are often multiple conformers within an energy of 1 or 2 kcal/mol, which contribute to the reaction. If multiple reactants have to be considered, a good assumption at pyrolysis temperatures is thermal equilibrium between them. For instance, if conformers I and II are separated by an energy ΔE, the mole fractions f_{I} and f_{II} of the conformers are determined by the equilibrium constant:

$$K = \frac{f_{\mathrm{I}}}{f_{\mathrm{II}}} = \frac{Q_{\mathrm{I}}}{Q_{\mathrm{II}}} e^{-\Delta E\beta} \tag{7.29}$$

where Q_{I} is the molecular partition function of conformer I and Q_{II} is the molecular partition function of conformer II. Since $1 = f_{\mathrm{I}} + f_{\mathrm{II}}$, the mole fraction of conformer I is $f_{\mathrm{I}} = \frac{1}{(1+K)}$ and the mole fraction of conformer II is $f_{\mathrm{II}} = \frac{K}{1+K}$. The total rate constant is the sum of rate constants for each conformer weighted by their mole fractions, $k = f_{\mathrm{I}}k_{\mathrm{I}} + f_{\mathrm{II}}k_{\mathrm{II}}$, which is easily generalized to more than two reactants.

If multiple reaction channels are available for a reaction, which are characterized by conformationally distinct transition states, a rate constant is calculated for each reaction path and the total rate constant is the sum of the rate constants for the parallel occurring reactions. For a bimolecular reaction including p transition states, the total rate constant is given by

$$k = \frac{\kappa\sigma}{h\beta}\frac{V}{Q_{\mathrm{RA}}Q_{\mathrm{RB}}}\sum_{r=1}^{p} Q^{\#}_{\mathrm{TSr}}\, e^{-\beta\Delta E^{\#}_{\mathrm{TSr}}} \tag{7.30}$$

with an equivalent equation for unimolecular reactions derived from Eq. (7.21).

7.3.2.2 Series of First-Order Reactions

The phenyl rearrangement (reaction 4, Scheme 7.2) proceeds through an intermediate rather than a transition state. In this case, the unimolecular reaction becomes a series of first-order reactions:

$$\mathrm{R} \underset{k_{-1}}{\overset{k_1}{\rightleftharpoons}} \mathrm{I} \overset{k_2}{\longrightarrow} \mathrm{P} \tag{7.31}$$

where R is the reactant, P is the product, and I is the intermediate. Each transformation, from R to I and I to P, advances through a transition state with corresponding rate constants k_1 and k_2. Since the energy barrier for the reaction from the intermediate to the reactant is small, the reverse reaction with corresponding rate constant k_{-1} is also considered. In the steady-state approximation, the rate of change of a reaction intermediate is set to zero. If the concentration of an intermediate is low, the relative variation in its concentration is small and the steady-state approximation is expected to be valid when the rate of formation of the intermediate is much slower than the rate of consumption. For the intermediate of reaction 4 in Scheme 7.2, this is expected to hold and we set $\frac{d[\mathrm{I}]}{dt} = 0$. The rate of product formation is a function of k_2 and the concentration of the intermediate:

$$\frac{d[\mathrm{P}]}{dt} = k_2[\mathrm{I}] \tag{7.32}$$

The intermediate is formed from reactant R and is consumed through the reverse reaction and through the reaction yielding the product

$$\frac{d[\mathrm{I}]}{dt} = 0 = k_1[\mathrm{R}] - k_{-1}[\mathrm{I}] - k_2[\mathrm{I}] \tag{7.33}$$

Solving for [I] and substituting into Eq. (7.32) leads to

$$\frac{d[\mathrm{P}]}{dt} = \frac{k_1 k_2}{k_{-1} + k_2}[\mathrm{R}] \tag{7.34}$$

which gives for the total rate constant of Eq. (7.31), $k = \frac{k_1 k_2}{(k_{-1}+k_2)}$.

7.3.2.3 Product Selectivity for the Pyrolysis of PPE

For the pyrolysis mechanism of PPE and its derivatives, rates of conversion and total α/β-selectivities can be measured by experiment [33,37]. The corresponding computational quantities are obtained by solving the differential equations given by the rate expressions for the reactions shown in Scheme 7.2. Even if all rate constants were known, the resulting equations are too complicated to be solved analytically. However, we can again use the steady-state approximation for intermediates and obtain a simplified kinetic model.

In Scheme 7.2, we observe that all α-abstraction yields phenoxy radical, which abstracts hydrogen from PPE producing phenol. All β-abstraction yields benzyl radical, which abstracts hydrogen from PPE producing toluene. The overall α/β-selectivity is, therefore, given as the rate of formation of phenol divided by the rate of formation of toluene. We assume steady-state conditions for benzyl, the β-PPE radical, and the migration product of reaction 4 (PPA). We also presume that the contribution of the hydrogen abstraction by the phenethyl radical (reaction 7) is insignificant and that the interconversion between the α-PPE and the β-PPE radicals is negligible, both of which are supported by experiment [33]. Furthermore, reverse reactions are assumed not to participate.

The formation rates of the steady-state intermediates are given by

$$0 = \frac{d[\mathrm{PhCH_2}^\bullet]}{dt} = k_5[\mathrm{PPA}] - k_{6a}[\mathrm{PhCH_2}^\bullet][\mathrm{PPE}] - k_{6b}[\mathrm{PhCH_2}^\bullet][\mathrm{PPE}] \tag{7.35}$$

$$0 = \frac{d[\beta\mathrm{PPE}]}{dt} = k_{2b}[\mathrm{PhO}^\bullet][\mathrm{PPE}] + k_{6b}[\mathrm{PhCH_2}^\bullet][\mathrm{PPE}] - k_4[\beta\mathrm{PPE}] \tag{7.36}$$

$$0 = \frac{d[\mathrm{PPA}]}{dt} = k_4[\beta\mathrm{PPE}] - k_5[\mathrm{PPA}] \tag{7.37}$$

All rate constants are labeled according to Scheme 7.2. [PPA] can be eliminated by adding Eqs (7.35) and (7.37):

$$k_4[\beta\mathrm{PPE}] = k_{6a}[\mathrm{PhCH_2}^\bullet][\mathrm{PPE}] + k_{6b}[\mathrm{PhCH_2}^\bullet][\mathrm{PPE}] \tag{7.38}$$

which is substituted into Eq. (7.36) leading to

$$\frac{k_{6a}}{k_{2b}} = \frac{[\mathrm{PhO}^\bullet]}{[\mathrm{PhCH_2}^\bullet]} \tag{7.39}$$

The total α/β-selectivity also depends on the phenoxy/benzyl ratio:

$$\frac{d[\mathrm{PhOH}]/dt}{d[\mathrm{PhCH_3}]/dt} = \frac{k_2[\mathrm{PPE}][\mathrm{PhO}^\bullet]}{k_6[\mathrm{PPE}][\mathrm{PhCH_2}^\bullet]} = \frac{k_2[\mathrm{PhO}^\bullet]}{k_6[\mathrm{PhCH_2}^\bullet]} \tag{7.40}$$

where k_2 is the combined rate constant for α- and β-abstraction by phenoxy and k_6 is the combined rate constant for α- and β-abstraction by benzyl. We define x to be the fraction of conversion through the α-pathway of phenoxy abstraction and y the fraction of conversion through the β-pathway; y and x are defined correspondingly for reaction 6. k_{2b} and k_{6a} can then be expressed through k_2 and k_6, $k_{2b} = (1 - x)k_2$ and $k_{6a} = yk_6$, which is substituted together with Eq. (7.39) into Eq. (7.40):

$$\frac{d[\mathrm{PhOH}]/dt}{d[\mathrm{PhCH_3}]/dt} = \frac{k_2 k_{6a}}{k_6 k_{2b}} = \frac{k_2 y k_6}{k_6(1 - x)k_2} = \frac{y}{1 - x} \quad (7.41)$$

x and y are obtained from the individual α/β-selectivities for the phenoxy and benzyl abstractions:

$$\frac{x}{1 - x} = \frac{k_{2a}}{k_{2b}} \quad \text{and} \quad \frac{y}{1 - y} = \frac{k_{6a}}{k_{6b}} \quad (7.42)$$

The overall α/β-selectivity of the thermal decomposition of PPE is determined by the relative constants for the hydrogen abstraction reactions by phenoxy and benzyl only. This is an important result because rate constants are extremely sensitive to errors in the reaction barriers and significant errors can occur due to the anharmonicity in the vibrations. Generally, errors in reaction barriers calculated with *ab initio* or density functional methods are systematic and the anharmonic correction introduced in Section 7.3.1.3 reduces the nonsystematic errors in the prefactors. When calculating relative rate constants, errors cancel, particularly for the ratios of rate constants in Eq. (7.42), because k_{2a} and k_{2b} are rate constants for reactions with the same reactants and very similar transition states (analogously, for k_{6a} and k_{6b}).

7.3.3 Numerical Integration

The kinetic model of Section 7.3.2.3 was derived assuming steady states for radical intermediates, that the interconversion between α- and β-PPE radicals is insignificant, that the radical chain length is sufficiently long, and that reverse reactions can be neglected. For PPE, these assumptions are experimentally supported, but the introduction of substituents can modify the chemical behavior. If these conditions are not met, the complete reaction scheme for the pyrolysis of PPE and its derivatives, as shown in Scheme 7.2, has to be modeled involving the calculations of rate constants for a large number of reactions. Depending on the nature and location of substituents, for each reaction multiple reactants (each following multiple reaction paths) need to be considered. In addition, reverse reactions and termination reactions might have to be included as well. This can be become a computationally very demanding task.

Given the rate constants of all participating reactions, the combination of the corresponding rate equations yields a system of coupled differential equations. These equations can be solved using numerical integration schemes, for example, the Runge–Kutta method [88]. Changes of reactant, intermediate, or product concentration can be followed in time. As an example, we modeled the radical chain propagation for the pyrolysis of PPE including reactions 2–6 of Scheme 7.2. This implies that reverse reactions, radical conversion, abstraction by the phenethyl radical, and termination reactions are assumed to be negligible during the time the reactions are followed, but the steady-state condition for intermediates is not imposed. We used the rate constants calculated in Section 7.4.3.2 for the hydrogen abstraction reactions by phenoxy (reactions 2a and 2b) and by benzyl (reactions 6a and 6b) and in

TABLE 7.2 Initial Conditions and Kinetic Parameters Used to Model the Chain Propagation of PPE (Reactions 2–6 of Scheme 7.2)

C_0 (mol/L) PPE	1.0
C_0 (mol/L) $PhO^•$	1.0×10^{-3}
k_{2a}	1.0×10^3
k_{2b}	9.1×10^2
k_3	1.0×10^8
k_4	3.2×10^6
k_5	1.0×10^8
k_{6a}	8.9×10^2
k_{6b}	2.4×10^2

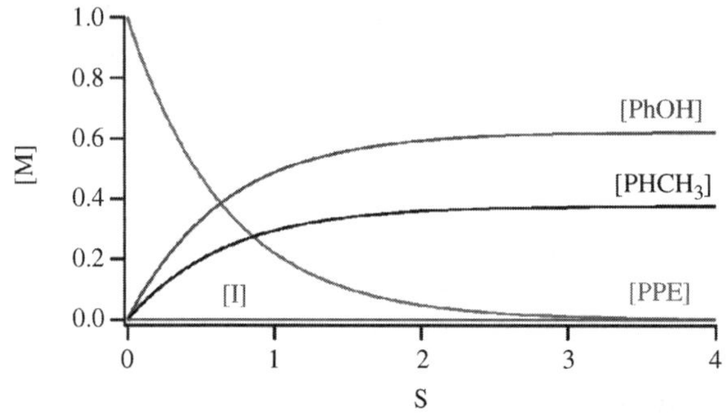

FIGURE 7.7 Numerical integration of rate equations for reactions 2–6 of Scheme 7.2, performed with the program Qinetic [89]. The concentration of the intermediates I (benzyl, β-PPE radical, migration product of reaction 4) coincides with the x-axis.

Section 7.4.4 for the phenyl migration (reaction 4). β-Scission is assumed to be two orders of magnitude faster than phenyl migration. The input parameters for the numerical integration are listed in Table 7.2. Figure 7.7 shows the concentration of PPE, phenol, and toluene as a function of time. The concentration of the reactant, PPE, decreases as the reaction progresses and the product concentrations of toluene and phenol increase until a steady state is reached. Also included in Figure 7.7 are the concentrations of the intermediates, for which steady states are assumed in the analytical kinetic model of Section 7.3.2.3 (benzyl, β-PPE radical, migration product of reaction 4). We observe that the intermediate concentrations are very small and the steady-state approximation is valid. The final product ratio of phenol to toluene is 1.7, which is consistent with the α/β-selectivity calculated in Section 7.4.3.2, where we applied the kinetic model derived in Section 7.3.2.3.

7.4 CASE STUDIES: SUBSTITUENT EFFECTS ON REACTIONS OF PHENETHYL PHENYL ETHERS

Based on the mechanism proposed by Britt and coworkers (Scheme 7.2), we started a systematic study of the kinetic details of the pyrolysis of PPEs [79,90–92].

Experimental results were largely independent of the reaction medium [33] and gas-phase calculations are, therefore, sufficient. The thermal decomposition of PPE and several PPE derivatives has been investigated under a range of pyrolysis conditions [33–35]. Fast, high-temperature techniques highlight unimolecular transformations, such as bond scission and intramolecular rearrangements, while slow pyrolysis in the liquid phase allows additional bimolecular and radical chain processes to occur. Under both pyrolysis conditions, the pyrolysis rate and product distribution were found to be substantially influenced by naturally occurring substituents, such as hydroxy and methoxy substituents.

Tracing the origin of substituent effects can be difficult because multiple reaction steps (e.g., homolytic initiation, competitive hydrogen abstraction, radical rearrangement, radical scission) can be perturbed. In the following sections we probe the rates of individual reaction steps using the methods outlined above. Our main goal is the understanding of how naturally occurring substituents affect product selectivities.

7.4.1 Computational Details

All calculations are carried out with the NWChem program package [93]. In earlier work, summarized in Section 7.4.3.1, we utilized the B3LYP functional [94,95]. All other calculations are based on potential energy surfaces explored with the M06-2X, which is part of the family of kinetic functionals developed by Zhao and Truhlar [96]. A mixed basis set is used for geometry optimizations and frequency calculations. A 6-31G* basis is centered on each atom except on atoms where the unpaired electron is located (delocalization into π-systems and neighboring atoms in transition states are included), for which diffuse functions are added (6-31 ++ G**). Energies at stationary points are computed with the 6-311 ++ G** basis set and energy differences used in energy profiles are zero-point-corrected. To ensure that the conformational space is well sampled, we used a variety of starting structures for reactants and transition states.

Rate constants are calculated with our own Python code, which is interfaced to NWChem. Bimolecular rate constants are computed according to Eqs (7.19) and (7.20) except in our earlier work (Section 7.4.3.1) where we did not include tunneling corrections. (Simple tunneling corrections mostly cancel for the relative rate constants in Eq. (7.42).) Unimolecular rate constants are given by Eqs (7.20) and (7.21) and anharmonic effects are incorporated via Eq. (7.28) for low-frequency vibrations up to 110 cm^{-1}. This implies that on the order of 10 modes per transition state of hydrogen abstraction and about 5 modes per transition state of phenyl migration and reactants are anharmonically corrected. We employ the analytical kinetic model described in Section 7.3.2.3 to obtain α/β-selectivities in Section 7.4.3.

7.4.2 Initiation: Homolytic Cleavage

The initiation reaction (reaction 1, Scheme 7.2) of the radical chain mechanism for the pyrolysis of PPE is the homolytic cleavage of the oxygen–β-carbon bond. At higher temperatures the cleavage of the α-carbon–β-carbon bond is a possible alternative.

TABLE 7.3 Bond Dissociation Enthalpies (BDEs) at 298K in kcal/mol of the Oxygen–β-Carbon and the α-Carbon–β-Carbon Bond of PPE Derivatives 1–6 in Figure 7.4

	C–O Dissociation		C–C Dissociation	
	BDE (kcal/mol)	ΔBDE (kcal/mol)	BDE (kcal/mol)	ΔBDE (kcal/mol)
PPE (**1**)	69.54	0.00	77.06	0.00
*p*HO-PPE (**2**)	69.59	−0.05	76.56	0.50
pCH_3O-PPE (**3**)	69.60	−0.06	76.55	0.51
PPE-*p*OCH_3 (**4**)	64.07	5.47	76.68	0.38
PPE-*o*OCH_3 (**5**)	64.41	5.13	76.96	0.10
PPE-di-*o*OCH_3 (**6**)	60.92	8.62	78.01	−0.95

Also included are bond dissociation enthalpy differences (ΔBDEs) relative to PPE.

The breaking of a bond yielding radical products is expected to have a small reverse barrier. As a consequence, the reaction energy is similar to the reaction barrier and has been used as an input parameter in kinetic simulations of lignin pyrolysis [21]. For the homolytic cleavage, the reaction enthalpy is equal to the bond dissociation enthalpy (BDE).

We calculated the oxygen–β-carbon and α-carbon–β-carbon BDEs for the PPE derivatives **1–6** in Figure 7.4 at 298K [92]. The BDEs reported in Table 7.3 are based on the energetically lowest conformer. Also included in the table are the bond dissociation enthalpy differences (ΔBDE) relative to PPE. Relative BDEs can be calculated to higher accuracy than absolute BDEs, in particular when the calculations are based on isodesmic reactions, as is the case here. We observe that the α-carbon–β-carbon BDE in PPE is 7.52 kcal/mol larger than the oxygen–β-carbon BDE.

This can be used to estimate the ratio of oxygen–β-carbon and α-carbon–β-carbon cleavage contributing to the pyrolysis of PPE at different temperatures. We assume that the Arrhenius prefactors for the oxygen–β-carbon and α-carbon–β-carbon cleavage are the same and that the difference in activation energies is equivalent to the difference in BDEs since the energy barrier for recombination is expected to be small. The ratio of oxygen–β-carbon and α-carbon–β-carbon cleavage is then given as $k_{CO}/k_{CC} = e^{\Delta BDE/RT}$. At a typical slow pyrolysis temperature of 375°C, the ratio is estimated to be 365:1, indicating very little contribution from α-carbon–β-carbon cleavage. At higher temperatures, typical of fast pyrolysis, the α-carbon–β-carbon cleavage gains influence, for example, the ratio is reduced to 80:1 at 600°C. The effect of the substituents on the homolysis can be substantial, but the difference between the oxygen–β-carbon and α-carbon–β-carbon BDE is too large to allow for a significant contribution of the α-carbon–β-carbon homolysis.

The oxygen–β-carbon BDE is considerably lowered and, therefore, the initial homolysis step accelerated by substituents located at the phenyl ring adjacent to the ether oxygen. A methoxy substituent in *ortho* or *para* position reduces the oxygen–β-carbon BDE by about 5 kcal/mol. An additional methoxy substituent in di-*o*-methoxy PPE reduces the BDE further by 3.5 kcal/mol. Substituents on the opposite phenyl ring have an insignificant effect on the oxygen–β-carbon BDEs. The

α-carbon–β-carbon BDE is much less influenced by oxygen substituents situated at either phenyl ring. The seeming outliner in Table 7.3, the negative ΔBDE of about -0.95 kcal/mol for the α-carbon–β-carbon BDE of PPE derivative **6**, can be explained by steric hindrance in di-*o*-methoxy phenoxymethyl radical [92].

7.4.3 Hydrogen Abstraction Reactions and α/β-Selectivities

Competitive hydrogen abstraction reactions by phenoxy and benzyl radicals on the α- or β-carbon of PPE are the core reactions of the radical chain propagation (Scheme 7.2) in the thermal degradation of PPE and are the only reactions needed in the analytical kinetic model derived in Section 7.3.2.3 to calculate overall α/β-selectivities. Because hydrogen abstractions are the rate-determining steps, effects of substituents on hydrogen abstraction have the most impact on overall product distributions and pyrolysis rates.

7.4.3.1 PPE and PPE Derivatives with Substituents on Phenethyl Group

Our first set of β-O-4 lignin model compounds includes PPE and PPE derivatives **2** (*p*HO-PPE) and **3** (*p*CH_3O-PPE) in Figure 7.4, where a hydroxy and a methoxy substituent are located in *para* position of the phenyl ring in the phenethyl group of PPE [79,90]. The first step in the calculation of the rate constants for the hydrogen abstraction reactions by phenoxy and by substituted benzyl radicals is the determination of relevant transition states. PPE has two equivalent α-hydrogen atoms and two equivalent β-hydrogen atoms that can be abstracted by the radicals and the phenoxy and benzyl radicals have two equivalent sides of attack. These equivalencies are taken into account through symmetry numbers. However, when substituents are introduced, the number of distinct conformers increases. Not only can the substituent be orientated in multiple ways in the reactants, but also the hydrogen atoms in PPE are not equivalent anymore and the radical attack can occur from two distinct sides. Regardless of the substituent pattern, the reactants can approach each other from various directions. This makes the conformational space, which has to be sampled carefully to locate all contributing transition states, large. For instance, we searched for transition states of the α-pathway for the hydrogen abstraction by *p*-hydroxybenzyl on *p*HO-PPE (reaction 6a for derivative **2**) starting from 20 geometries, similarly for the β-pathway. If the conformers are very close in energy (within a few tenths of kcal/mol), it is not necessary to assign a reaction path to each transition state but a symmetry number can be reintroduced. However, this can only be determined after the conformers have been localized.

For each pathway and each PPE derivative, we located multiple transition states. The transition states of PPE, *p*HO-PPE, and *p*CH_3O-PPE are very similar with only slight variations in geometry. Even the energetic order of the transition states is different on only one occasion. As an example, the transition states for the hydrogen abstraction by phenoxy on *p*HO-PPE are shown in Figure 7.8, and the transition states for the hydrogen abstraction by *p*HO-benzyl on *p*HO-PPE are given in Figure 7.9. The hydrogen–radical and hydrogen–ether distances are similar for the α- and β-pathways of the hydrogen abstraction by the phenoxy radical but differ

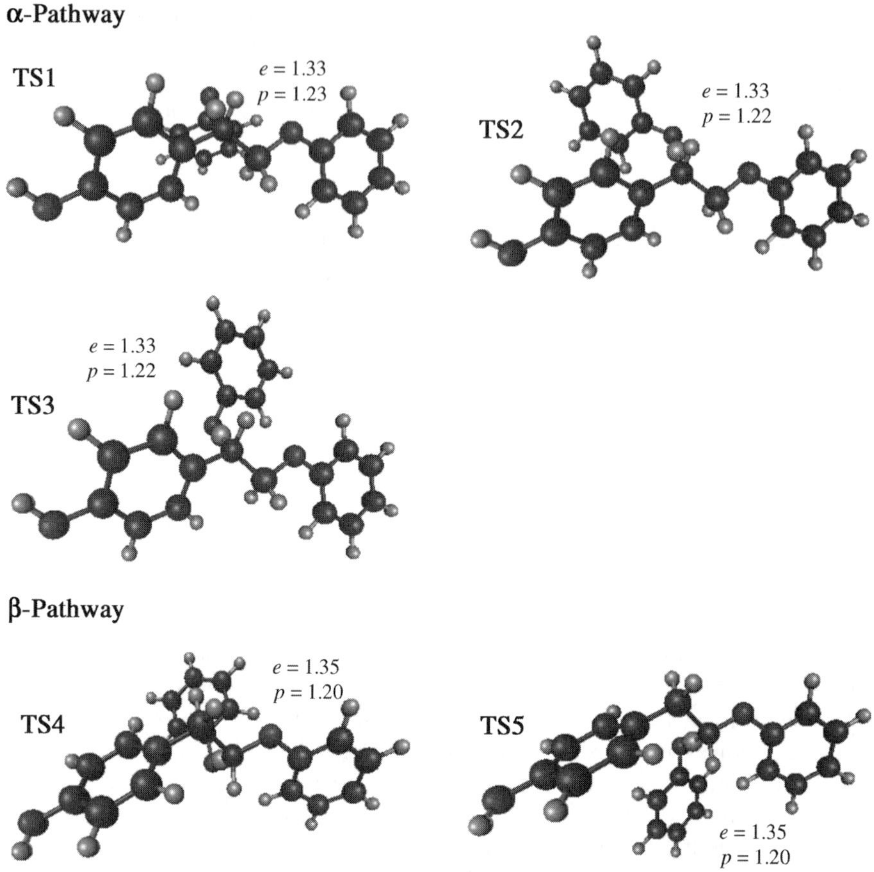

FIGURE 7.8 Transition state geometries for hydrogen abstraction by phenoxy radical on hydroxy PPE (reaction 2, Scheme 7.2); *e*: hydrogen–PPE bond lengths in angstrom and *p*: hydrogen–phenoxy bond lengths in angstrom.

for the α- and β-pathways of the hydrogen abstraction by the benzyl radical, where the α-transition states of the benzyl abstraction are symmetric with respect to the position of the transferred hydrogen atom. Figures 7.8 and 7.9 show transition states for the abstraction of one of the two distinct α- or β-hydrogens and one of the two distinct sides of radical attack. The corresponding structures for abstraction of the other hydrogen atom or for the opposite radical side are energetically very close. The largest energetic differences of up to 0.19 kcal/mol occurred for the hydrogen abstraction by pCH$_3$O-benzyl on pCH$_3$O-PPE for the reaction paths where the two aromatic rings are closest (TS9 in Figure 7.9) and steric effects are largest. Assuming equivalency between hydrogens and radical sides results in the same α/β-selectivity compared to including all conformers for the hydrogen abstraction by pCH$_3$O-benzyl on pCH$_3$O-PPE.

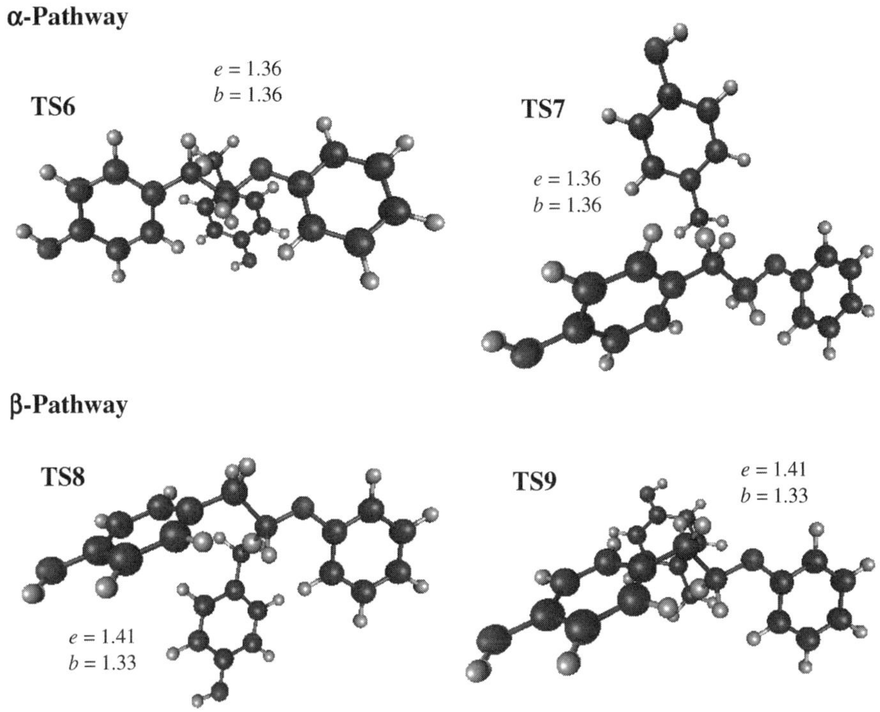

FIGURE 7.9 Transition state geometries for hydrogen abstraction by *p*-hydroxy benzyl radical on hydroxy PPE (reaction 6, Scheme 7.2); *e*: hydrogen–PPE bond lengths in angstrom and *b*: hydrogen–benzyl bond lengths in angstrom.

Figure 7.10 displays the reaction profile for the hydrogen abstraction by phenoxy and benzyl radicals on PPE and the PPE derivatives *p*HO-PPE and *p*CH_3O-PPE. The α-PPE radicals are stabilized relative to the β-PPE radicals by delocalization of the unpaired electron into the aromatic ring [79]. The substituent effect on the α-PPE radical can be deduced from the reaction profile of the phenoxy abstraction (left side of Figure 7.10); for the benzyl abstraction (right side of Figure 7.10), the benzyl radical also carries a substituent. The stability of the β-PPE radical is not influenced by the substituents, but the α-PPE radical is stabilized by the hydroxy and methoxy substituents.

The delocalization of the unpaired electron is less effective in the transition states and most of the spin density is found in the hydrogen bridge [79]. Opposing the delocalization effect is the polar effect, which occurs in the β-phenoxy transition states [79]. The electrophilic phenoxy radical polarizes the transition states by drawing charge into the phenoxy ring and the β-transition states are stabilized relative to the α-transition states by charge donation from the adjacent ether oxygen. The calculation of partial charges in the transition states suggests that the polar effect decreases in the order PPE > *p*HO-PPE ≥ *p*CH_3O-PPE [79], which corresponds to a small increase in the α/β-barrier difference ($\Delta E_{\alpha\beta}$): PPE, $\Delta E_{\alpha\beta} = 0.3$ kcal/mol; *p*HO-PPE, $\Delta E_{\alpha\beta} = 1.2$ kcal/mol; *p*CH_3O-PPE, $\Delta E_{\alpha\beta} = 1.3$ kcal/mol.

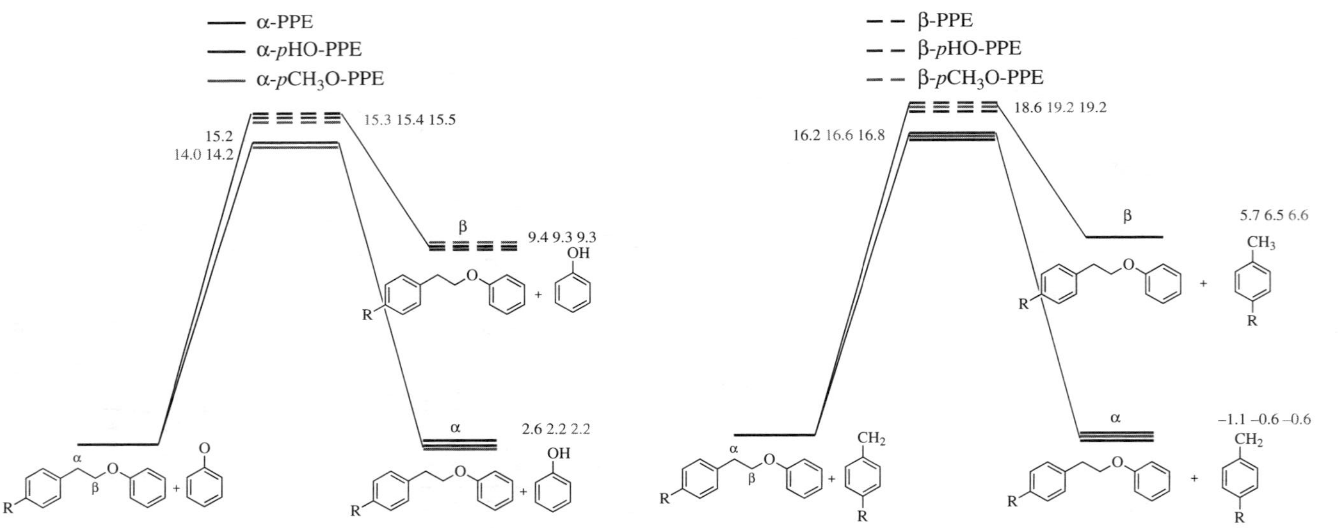

FIGURE 7.10 Reaction profile of the hydrogen abstraction by phenoxy (left) and by benzyl (right) on PPE, *p*HO-PPE, and *p*CH_3O-PPE; only the energetically lowest transition states are included.

The Arrhenius plots for the hydrogen abstraction in the temperature range of 580–660K can be found in Figure 7.11. Solid lines denote α-abstraction and dashed lines indicate β-abstraction. The calculated rate constants contain contributions from all transition states shown in Figures 7.8 and 7.9 for *p*HO-PPE and the corresponding transition states for PPE and *p*CH_3O-PPE. They are taken into account using Eq. (7.30). Prefactors and activation energies extracted from Figure 7.11 are given in Table 7.4. For the hydrogen abstraction by the electrophilic phenoxy radical, we observe a stronger influence of the substituent on the α-pathway than on the β-pathway. The abstraction on the α-carbon is accelerated by the substituents and is fastest for methoxy. The abstraction on the β-carbon, however, becomes slower. This results in an increase of the α/β-selectivity for the phenoxy abstraction when the oxygen substituents are introduced. For the benzyl abstraction, which proceeds through a less polarized transition state, the substituents decrease the reaction rates for the α- and β-pathways and the α/β-selectivity is less altered. Table 7.5 records the individual α/β-selectivities for the hydrogen abstraction reactions by phenoxy k_{2a}/k_{2b} and benzyl k_{6a}/k_{6b} together with total α/β-selectivities according to Eq. (7.41). Experimental values are also included. The computed total α/β-selectivities slightly underestimate the experimental values but the experimental trend of the α/β-selectivities in the pyrolysis of the substituted ethers is well reproduced; PPE < *p*HO-PPE < *p*CH_3O-PPE.

7.4.3.2 PPE and PPE Derivatives with Substituents on Phenyl Group Adjacent to Ether Oxygen The second set of β-O-4 lignin model compounds is composed of PPE and PPE derivatives **4** (PPE-*p*OCH_3), **5** (PPE-*o*OCH_3), and **6** (PPE-di-*o*OCH_3) in Figure 7.4, where methoxy substituents are located on the phenyl ring adjacent to the ether oxygen [91]. PPE is included again because the results of Section 7.4.3.1 were obtained using the B3LYP functional, while, here, we report calculations utilizing one of Zhao and Truhlar's kinetic functionals, the M06-2X functional [96].

The starting point of the transition state searches is the sampling of the equilibrium structures of PPE and its derivatives. Figure 7.12a–d shows their lowest energy conformers. Notice that PPE, PPE-*p*OCH_3, and PPE-*o*OCH_3 share the same lowest energy conformer. The steric hindrance introduced by the addition of another methoxy group in PPE-di-*o*OCH_3 makes a conformer favorable where the phenyl ring adjacent to the ether oxygen is rotated by 90°. The energy profiles for the rotation of the phenyl ring adjacent to the ether oxygen showed that for PPE, and similarly for PPE-*p*OCH_3, the only minima along the rotational coordinate are found at 0 and 180° [91], which are the conformers shown in Figure 7.12a, b, and f. PPE-*o*OCH_3 possesses additional minima at 105 and 255°, which are 0.6 kcal/mol higher in energy than the minima at 0 and 180° and separated by an energy barrier of 2.3 kcal/mol. The additional conformer is given in Figure 7.12e. The rotation of the phenyl ring by 180° in PPE-*o*OCH_3 leads to a maximum on the potential energy surface, that is, the conformer, where the methoxy group is on the bottom in Figure 7.12c is not stable. For PPE di *o*OCH_3 the only minimum is found at 90°, that is, the equivalent conformer to PPE and its *para*- and *ortho*-substituted derivatives do not exist. All PPE conformers

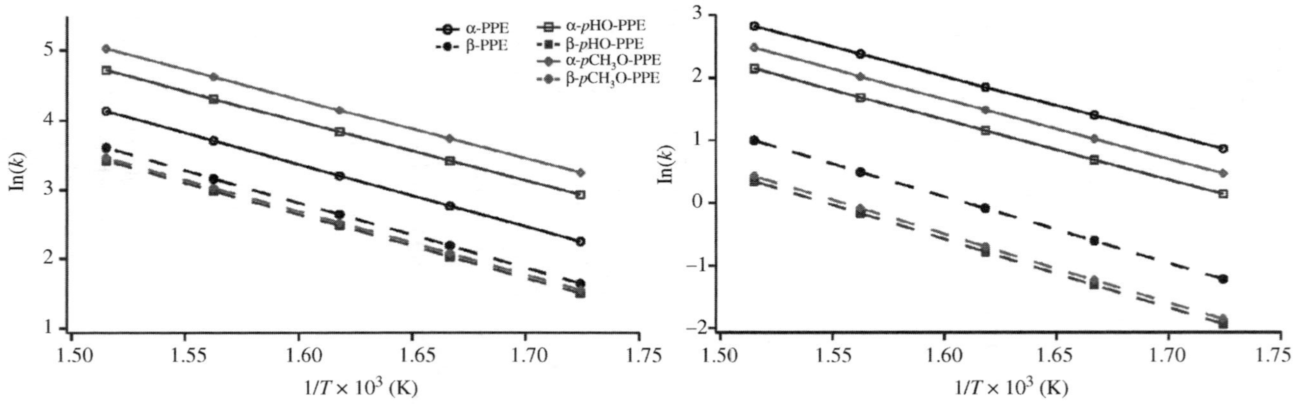

FIGURE 7.11 Arrhenius plots for the rate constants of the α- and β-pathways of the hydrogen abstraction on PPE, pHO-PPE, and pCH$_3$O-PPE by the phenoxy radical (left) and the benzyl radical (right) in the temperature range 580–660K.

TABLE 7.4 Prefactors and Activation Energies Extracted from the Arrhenius Plots in Figure 7.11

	PPE		*p*HO-PPE		pCH_3O-PPE	
	ln(*A*)	E_a (kcal/mol)	ln(*A*)	E_a (kcal/mol)	ln(*A*)	E_a (kcal/mol)
Phenoxy						
α	17.9	18.0	17.7	17.1	18.0	17.0
β	17.8	18.6	17.3	18.2	17.3	18.1
Benzyl						
α	17.1	18.7	16.8	19.2	17.1	19.2
β	17.1	21.1	16.9	21.7	17.0	21.8

Temperature range 580–660K.

given in Figure 7.12 show the phenyl ring opposite to the ether oxygen in the same orientation. The rotation of this phenyl ring in PPE does not lead to additional minima.

Another concern is the orientation of the methoxy groups in conformers (b)–(d) shown in Figure 7.12. A rotation of the methoxy group in PPE-$pOCH_3$ of 180° results in the minimum structure (f) shown in Figure 7.12. Conformer (f) is 0.3 kcal/mol higher in energy than conformer (b) with a rotational barrier of 2.3 kcal/mol. In contrast, the rotation of the methoxy group by 180° in PPE-$oOCH_3$ and either or both methoxy groups in PPE-di-$oOCH_3$ does not yield an equilibrium structure. Based on this analysis, we consider the six conformers displayed in Figure 7.12 to be reactants for the hydrogen abstraction reactions. The mole fractions of conformers (c) and (e) and conformers (b) and (f) are determined through Eq. (7.29).

The second reactants in the hydrogen abstraction reactions are the benzyl and the phenoxy radical; only the latter carries substituents. Initially, the conformer of the phenoxy radicals depends on the conformer of the PPE derivatives because they are formed through PPE cleavage, that is, reactions 1 and 3 of Scheme 7.2. The barriers for the rotation of the methoxy group are high, that is, 5.5 kcal/mol for the rotation of the methoxy group in PPE-$pOCH_3$ and 6.1 kcal/mol for the rotation of the methoxy group in PPE-$oOCH_3$. Because of the high-energy barriers for the methoxy rotation in the phenoxy radicals, we only consider phenoxy conformers corresponding to the dissociation products of the conformers shown in Figure 7.12.

TABLE 7.5 α/β-Selectivities for the Hydrogen Abstraction Reactions by Phenoxy and Benzyl Radicals on PPE, *p*HO-PPE, and pCH_3O-PPE at 618K

	PPE		*p*HO-PPE		pCH_3O-PPE	
	This Work	Experiment	This Work	Experiment	This Work	Experiment
Phenoxy	1.8		3.9		5.1	
Benzyl	7.0		6.9		8.9	
Total	2.4	3.8 ± 0.3	4.3	5.1 ± 0.1	5.5	7.4 ± 0.3

Experimental values from Ref. [37].

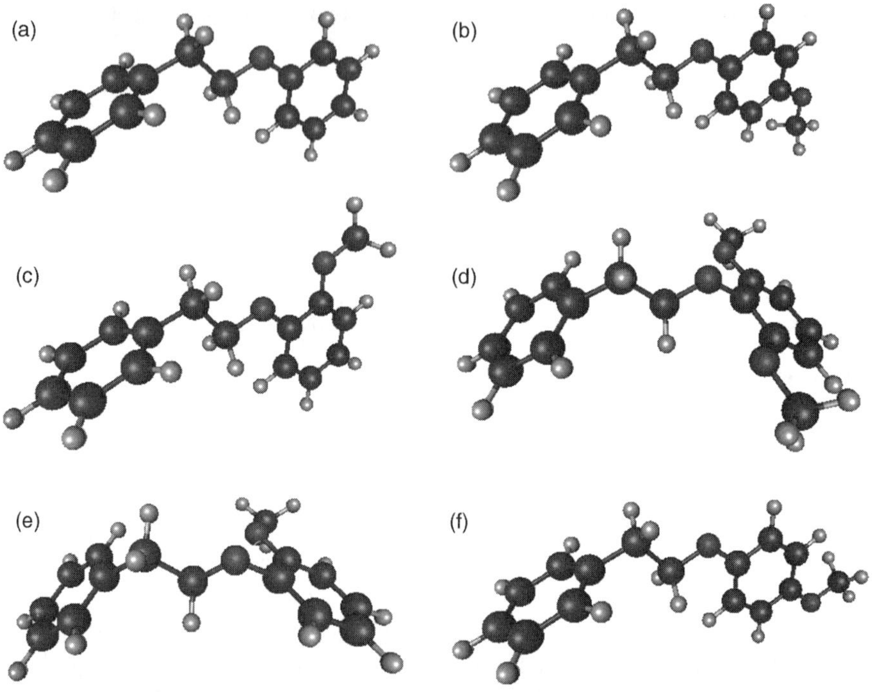

FIGURE 7.12 Equilibrium geometries for (a) PPE; (b and f) PPE-*p*OCH_3; (c and e) PPE-*o*OCH_3; and (d) PPE-di-*o*OCH_3.

In contrast to the PPE derivatives considered in Section 7.4.3.1, breaking of symmetry because of substituents has larger effects on the energies of the transition states. Various conformers resulting from the abstraction of two distinct hydrogens at the α-side and the β-side and the two sides of phenoxy attack are considered and included in the calculation of rate constants through Eq. (7.30). In total, we identified 117 transition states for hydrogen abstraction reactions on the conformers shown in Figure 7.12 [91]. We found very similar transition state structures for the different PPE derivatives, close in appearance to the transition states shown in Figures 7.8 and 7.9. Their energetic order is often influenced by the substituent. The distinct rotation of the phenyl ring adjacent to the ether oxygen in conformers (a)–(c) and (f) shown in Figure 7.12 on one hand and conformers (d) and (f) shown in Figure 7.12 on the other hand remains in the transition states. Generally, stronger steric effects can be observed when substituents are located in an *ortho* position.

Figure 7.13 displays the reaction profiles for the hydrogen abstraction reactions based on the lowest energy transition states. We notice a moderate substituent effect on the reaction energies and barriers of the benzyl abstraction. Whereas the *p*-methoxy substituent has little influence, the *o*-methoxy substituent lowers the α- and β-barriers. With an additional *o*-methoxy group in PPE-di-*o*OCH_3, the β-barrier decreases further while the α-barrier remains at the value for PPE-*o*OCH_3. The α/β-barrier difference varies within 0.6 kcal/mol. In contrast, the reaction energies and barriers for the α- and

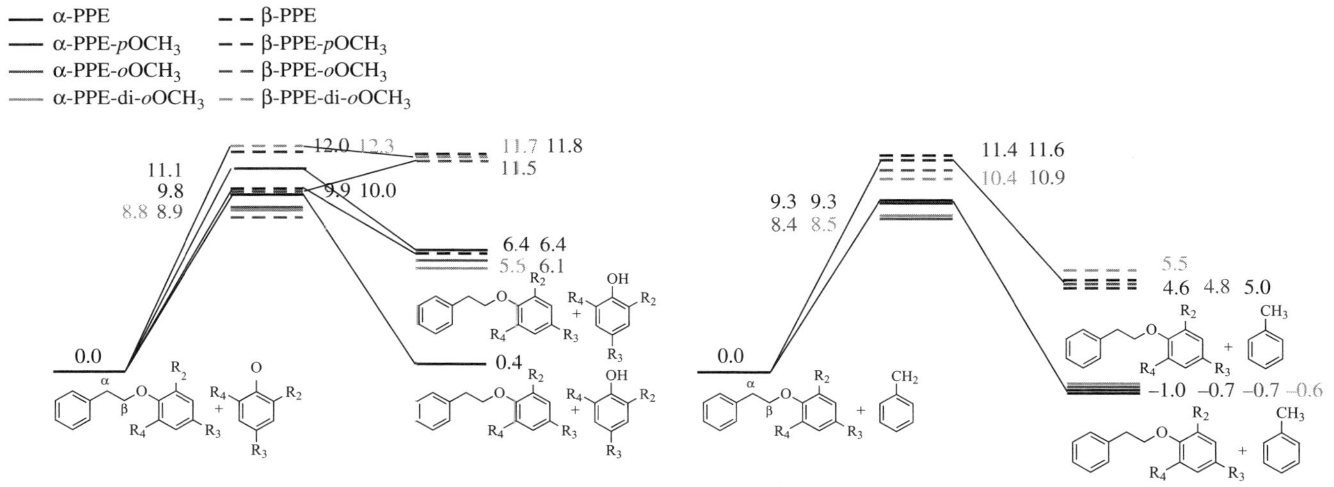

FIGURE 7.13 Reaction profile of the hydrogen abstraction by phenoxy (left) and by benzyl (right) on PPE, PPE-pOCH$_3$, PPE-oOCH$_3$, and PPE-di-oOCH$_3$; only the energetically lowest transition states are included.

β-channels of the phenoxy abstraction are strongly influenced by the substituents. The *p*-methoxy substituent increases the reaction barriers for the α- and β-channels. The *o*-methoxy group decreases the barrier for the α-abstraction, whereas the β-barrier stays the same for PPE-*o*OCH_3 and increases for PPE-di-*o*OCH_3. The latter might be caused by an increased steric effect in the β-transition state. The α/β-barrier difference varies over a wide range of 0.2–3.5 kcal/mol, indicating a potentially wide spread of the α/β-selectivity within the PPE derivatives.

The substituent effect on the reaction energies of the phenoxy abstraction is remarkably large, even though we can deduce from the benzyl abstraction that the substituent effect on the PPE radical is small. Analysis of the substituent effects on phenol and phenoxy [91] shows that the phenoxy radical is stabilized and phenol is destabilized by the methoxy substituents. The phenoxy and phenol effects are additive, resulting in a large total substituent effect for the α- and β-channels.

The increase of the energy of the products for the phenoxy abstraction has important consequences. First, the lowest energy transition state for the β-hydrogen abstraction by phenoxy is lower than the energy of the products. This implies that the reverse reaction could have a negative Arrhenius activation energy. Such reactions have been observed in hydrogen atom transfers for highly polar reactants where pre-reactive interactions in the entrance channel (e.g., van der Waals complexes or hydrogen bonding) permit barriers to form below the reactant energies [96]. Second, because the energies of the transition states and products are similar or in opposite order for the methoxy-substituted PPE derivatives, reverse reactions will contribute to the overall pyrolysis mechanism. This violates one of the assumptions of Section 7.3.2.3 and the analytical kinetic model to predict α/β-selectivities is not applicable for PPE-*p*OCH_3, PPE-*o*OCH_3, and PPE-di-*o*OCH_3. Another aspect is the shortening of the kinetic chain length for PPE-*p*OCH_3, PPE-*o*OCH_3, and PPE-di-*o*OCH_3 due to the reduction of the oxygen–β-carbon BDE (see Section 7.4.2). Consequently, the rate constants of all reactions in Scheme 7.2, including initiation, migration, β-scission, reverse reactions, and termination, will need to be calculated before determining product distributions using numerical integration techniques as outlined in Section 7.3.3.

Still, we can calculate forward rate constants including the 117 transition states located for the hydrogen abstraction by phenoxy and benzyl using Eq. (7.30) and compare the influence of the substituents. The corresponding Arrhenius plots in the temperature range between 580 and 660K can be viewed in Figure 7.14. For the hydrogen abstraction by benzyl, α-abstraction is faster than β-abstraction and is accelerated by methoxy substituents, where the α-abstraction on PPE-di-*o*OCH_3 is fastest. Also, the β-abstraction is fastest for PPE-di-*o*OCH_3 but is decelerated by methoxy substituents in *ortho* and *para* positions. A different pattern emerges for the hydrogen abstraction by phenoxy, where abstraction rates are clustered according to the substitution distribution on PPE. The hydrogen abstraction for the α- and β-channels by the phenoxy radical is fastest for PPE, followed by PPE-*o*OCH_3 and PPE-di-*o*OCH_3. The abstraction on PPE-*p*OCH_3 is slowest. An exception is the β-abstraction on PPE-di-*o*OCH_3, which is slower than the rates for PPE-*p*OCH_3 and can be explained by steric hindrance in the β-transition states.

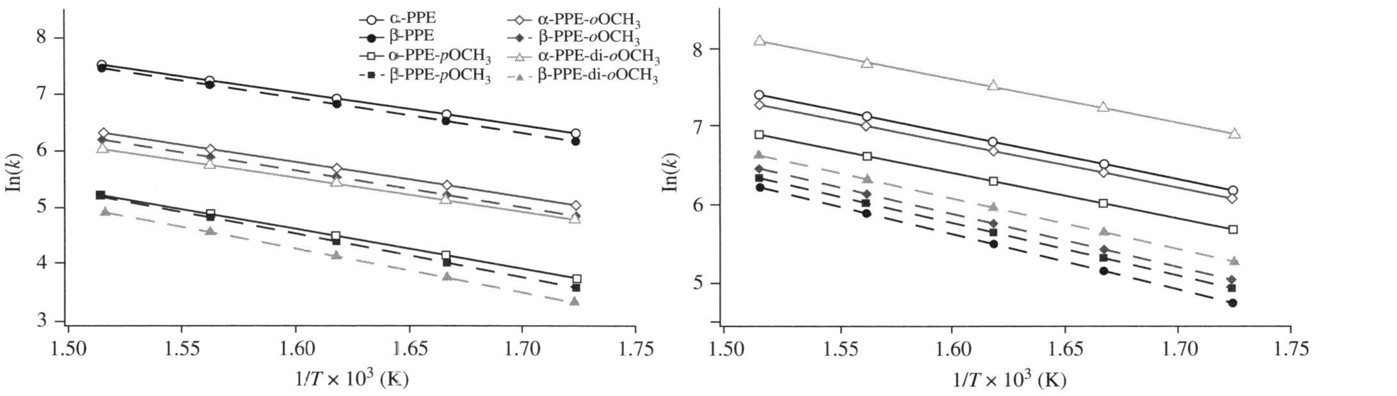

FIGURE 7.14 Arrhenius plots for the rate constants of the α- and β-pathways of the hydrogen abstraction on PPE, PPE-pOCH$_3$, PPE-oOCH$_3$, and PPE-di-oOCH$_3$ by the phenoxy radical (left) and the benzyl radical (right) in the temperature range 580–660K.

TABLE 7.6 Prefactors and Activation Energies Extracted from the Arrhenius Plots in Figure 7.14

	PPE		PPE-pOCH$_3$		PPE-oOCH$_3$		PPE-di-oOCH$_3$	
	ln(A)	E_a (kcal/mol)	ln(A)	E_a (kcal/mol)	ln(A)	E_a (kcal/mol)	ln(A)	E_a (kcal/mol)
Phenoxy								
α	16.4	11.7	15.9	14.0	15.7	12.3	15.1	11.9
β	16.9	12.4	16.6	15.0	16.1	13.0	16.5	15.2
Benzyl								
α	16.3	11.6	15.7	11.5	15.9	11.4	16.7	11.3
β	16.8	13.9	16.6	13.5	16.7	13.4	16.5	12.9

Temperature range 580–660K.

The activation energies and Arrhenius prefactors derived from the plots in Figure 7.14 are given in Table 7.6. Comparing the barriers for PPE of Section 7.4.3.1 with the barriers in Table 7.6 exposes a large discrepancy between the B3LYP and the M06-2X functionals. The experimental activation energies for benzyl abstraction are 14.4 kcal/mol for the α-channel and 17.6 kcal/mol for the β-channel with prefactors of 19.1 and 19.6, respectively, and relatively large error bars [97]. The B3LYP activation energies overestimate the experimental values by 4.4 and 3.6 kcal/mol (see Table 7.4). The M06-2X functional underestimates the experimental energies by 2.8 and 3.7 kcal/mol (see Table 7.6). However, the α/β-barrier difference, which is the determining factor for the α/β-selectivity, is very similar for the two functionals (B3LYP: 2.4 kcal/mol and M06-2X: 2.3 kcal/mol for the benzyl abstraction; B3LYP: 0.6 kcal/mol and M06-2X: 0.7 kcal/mol for the phenoxy abstraction), supporting the assumption of systematic error cancellation in the activation energy when calculating relative rate constants according to Section 7.3.2.3.

The rate constants for the hydrogen abstraction on PPE are used to calculate the α/β-selectivity for hydrogen abstraction (as has been done in Section 7.4.3.1 employing the B3LYP functional), which is determined to be 1.7. With the M06-2X functional we underestimate the experimental value of 3.8 [37] further than with the B3LYP functional. Since the α/β-barrier differences are predicted to be similar for the two functionals, we look for an alternative explanation. With the B3LYP functional we locate two transition states for the α-pathway and two transition states for the β-pathway of the hydrogen abstraction by benzyl, and three transition states for the α-pathway and two transition states for the β-pathway of the hydrogen abstraction by phenoxy. In contrast, with the M06-2X functional we find three transition states for the α-pathway and four transition states for the β-pathway of the hydrogen abstraction by benzyl, and three transition states for the α-pathway and four transition states for the β-pathway of the hydrogen abstraction by phenoxy. In Table 7.7, we list M06-2X α/β-selectivities for a varying number of transition states (for better comparison we omit tunneling corrections in the rate constants since they have not been included in the B3LYP values). When only the lowest transition states of each pathway are included (combination 1111 in Table 7.7), the total α/β-selectivity is low. For transition state combination 3434, the total and the individual α/β-selectivities are

TABLE 7.7 Total α/β-Selectivities and Individual α/β-Selectivities for the Benzyl and Phenoxy Hydrogen Abstraction on PPE at 618K Including a Varying Number of Transition States

Number of Transition States						
Benzyl		Phenoxy		Benzyl Selectivity	Phenoxy Selectivity	Total Selectivity
α	β	α	β			
1	1	1	1	4.8	1.2	1.9
2	*2*	*3*	*2*	*5.5*	*2.0*	*2.5*
3	4	3	4	3.7	1.1	1.7

Italic selectivities incorporate the same number of transition states as used to compute the B3LYP selectivities in Section 7.4.3.1.

even lower because for the benzyl and the phenoxy abstraction an additional reaction path is available for the β-channel compared to the α-channel. Transition state combination 2232 has the highest total α/β-selectivity, which can mainly be attributed to the higher selectivity for the phenoxy abstraction due to the larger number of α-transition states compared to β-transition states. In fact, the total M06-2X α/β-selectivity calculated with the same transition state combination as used for the B3LYP value gives nearly the same total α/β-selectivity, that is, 2.5 for M06-2X and 2.4 for B3LYP.

This shows the importance of an exhaustive conformational search. A perturbation (like switching from one DFT functional to another) can qualitatively alter the potential energy surface and cause the disappearance of a saddle point. This can have a large influence on the rate constant, particularly when transition states are close in energy. For example, the β-abstraction by the phenoxy radical on PPE can proceed through four different transition states (M06-2X functional); the lowest 0K barrier is 10.0 kcal/mol, followed by 10.9, 11.0, and 11.3 kcal/mol. The calculated rate constant at 618 K, when only the lowest transition state is used, is 256. When the second transition state is also included, the rate constant increases by a factor of 2.0; when the first three transition states are taken into account, the rate constant increases by a factor of 2.6; and when all four transition states are considered, the rate constant increases by a factor of 3.5. The appearance or disappearance of transition states with different computational levels becomes more likely the shallower the extreme point is. The transition states for hydrogen abstraction possess on the order of 10 frequencies below 100 cm^{-1}, which indicates that along these degrees of freedom the extreme points are in a shallow region of the potential.

7.4.4 Phenyl Rearrangement

We have discussed the hydrogen abstraction reactions occurring in the pyrolysis of PPE in Section 7.4.3. Because we expect the β-scission (reactions 3 and 5 in

Scheme 7.2) to be fast, our next focal point is the phenyl-shift reaction 4 in Scheme 7.2. Again, we start with PPE and PPE derivatives **2** (pHO-PPE) and **3** (pCH$_3$O-PPE) in Figure 7.4, where a hydroxy and a methoxy substituent are located in *para* position of the phenyl ring in the phenethyl group of PPE.

As we will see below, the phenyl migration is also fast compared to the hydrogen abstraction and we assume that major conformational changes do not occur in the β-PPE radical before reaction takes place. For PPE, the reactant of the phenyl rearrangement is derived from conformer (a) in Figure 7.12 by β-hydrogen abstraction. The hydroxy and methoxy substituent in pHO-PPE and pCH$_3$O-PPE can undergo a rotation, which leads to two minima, where the hydroxy and methoxy substituents are in plane with the aromatic ring of the phenethyl group. The minima are separated by a rotational barrier of 3.0 kcal/mol for both pHO-PPE and pCH$_3$O-PPE. To determine if inclusion of both rotational conformers is necessary, we choose a particular reaction path (conformer (c) in Figure 7.15) and calculate the energies of transition states and intermediates for both. For pHO-PPE the energy difference between the two rotational conformers in the transition states and intermediates does not exceed 0.2 kcal/mol and for pCH$_3$O-PPE 0.1 kcal/mol. We, therefore, did not treat the rotational conformers separately but used a symmetry number of 2 to account for two equivalent β-PPE radicals for PPE, pHO-PPE, and pCH$_3$O-PPE.

We determined that the phenyl migration in PPE proceeds through an intermediate. This has also been found for a similar, neophyl-like rearrangement [49,50]. For each model (PPE, pHO-PPE, and pCH$_3$O-PPE) we located three distinct intermediate conformers, which are displayed in Figure 7.15 for pHO-PPE as an example (PPE and pCH$_3$O-PPE yield equivalent conformers). Reactants and products are connected through a transition state leading to the intermediate and a transition state leading to the product. This is visualized in Figure 7.16 for conformer (a) of Figure 7.15.

The reaction profile for the phenyl migration in the β-radical of PPE, pHO-PPE, and pCH$_3$O-PPE is given in Figure 7.17. It is based on the lowest transition states leading to the intermediate, which corresponds to the rate-determining step of the rearrangement. There is only little influence of the substituents on the transition state energies; the largest effect is recorded for the intermediate, where the hydroxy and methoxy groups stabilize the intermediate by about 1 kcal/mol.

The phenyl rearrangement is a series of first-order reactions described by Eq. (7.31), for which the steady-state rate constant was derived in Section 7.3.2.2.

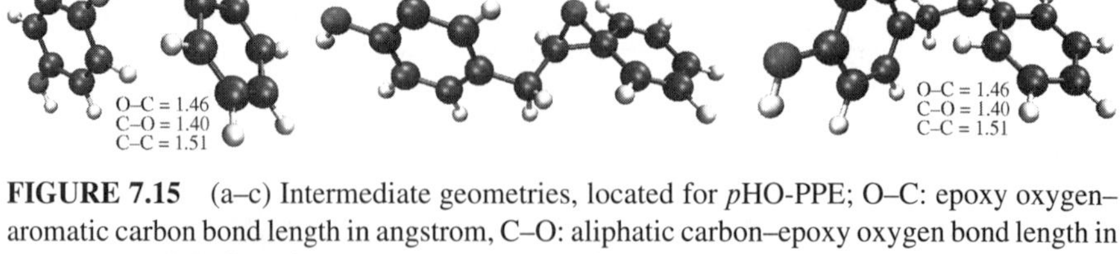

FIGURE 7.15 (a–c) Intermediate geometries, located for pHO-PPE; O–C: epoxy oxygen–aromatic carbon bond length in angstrom, C–O: aliphatic carbon–epoxy oxygen bond length in angstrom, and C–C: carbon–carbon bond length in epoxy group in angstrom.

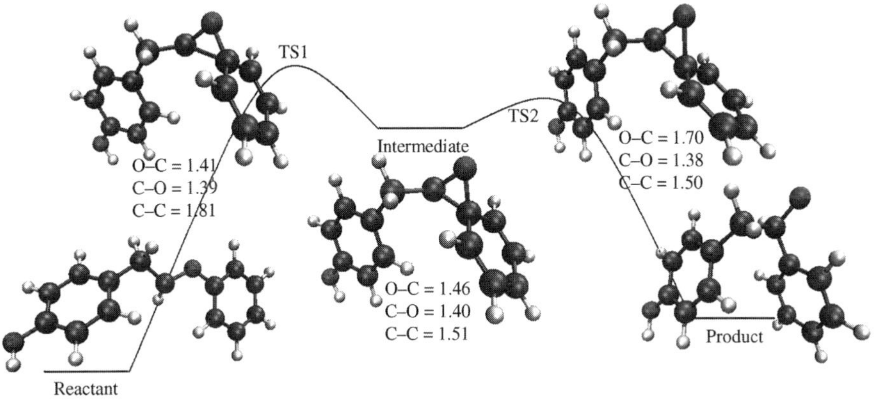

FIGURE 7.16 Transition state and intermediate geometries for conformer (a) of *p*HO-PPE in Figure 7.15; O–C: epoxy oxygen–aromatic carbon bond length in angstrom, C–O: aliphatic carbon–epoxy oxygen bond length in angstrom, and C–C: carbon–carbon bond length in epoxy group in angstrom.

The rate constants were calculated including three pathways for each PPE derivative. Figure 7.18 shows the Arrhenius plots for the phenyl migration in the β-radical of PPE, *p*HO-PPE, and *p*CH_3O-PPE in the temperature range 580–660 K. Table 7.8 contains the prefactors and activation energies extracted from Figure 7.18 and relative constants based on the rate constant for phenyl rearrangement in the β-radical of PPE at 618K. The phenyl-shift reaction is accelerated by the hydroxy and, more prominently, by the methoxy substituent. However, phenyl migration is between three and four orders of magnitude faster than hydrogen abstraction and relatively small

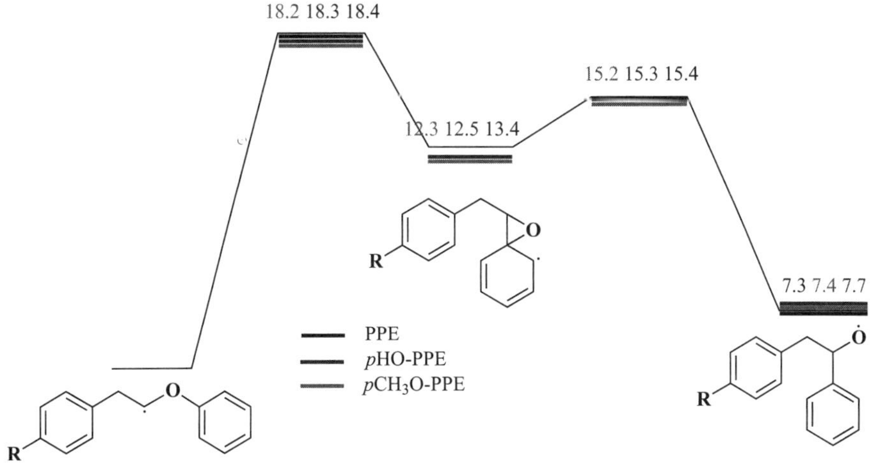

FIGURE 7.17 Reaction profile of the phenyl migration in PPE, *p*HO-PPE, and *p*CH_3O-PPE; only the pathways through the energetically lowest transition states leading to the intermediates are included.

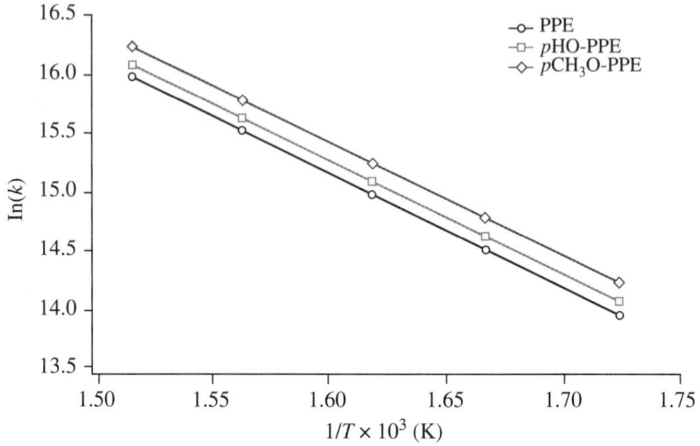

FIGURE 7.18 Arrhenius plots for the phenyl migration in the β-radical of PPE, *p*HO-PPE, and *p*CH_3O-PPE in the temperature range 580–660K.

TABLE 7.8 Prefactors and Activation Energies Extracted from the Arrhenius Plots in Figure 7.18

	PPE		*p*HO-PPE		*p*CH_3O-PPE	
	ln(A)	E_a (kcal/mol)	ln(A)	E_a (kcal/mol)	ln(A)	E_a (kcal/mol)
R-PPE	30.7	19.3	30.7	19.1	30.8	19.1
k_{R-PPE}/k_{PPE}	1		1.1		1.3	

Temperature range 580–660K. Also included are relative rate constants at 618K based on the rate constant for phenyl migration in the β-radical of PPE.

substituent effects on the rearrangement will be insignificant for the overall rates and product selectivities of PPE pyrolysis.

7.5 CONCLUSIONS AND OUTLOOK

Transition state theory in combination with density functional theory provides a framework for the calculation of rate constants for medium-sized molecules (on the order of 100 atoms). However, its routine and reliable application faces a multitude of challenges: a small error in the activation barrier causes a large deviation in the rate constant due to its exponential dependence on the activation energy; the determination of the prefactor, which represents entropic molecular contributions, is complicated by anharmonic effects in the vibrations; large organic molecules can have a shallow potential energy surface near extreme points, which makes the location of transition states difficult; the conformational space of reactants and transition states becomes large when low-energy transformations are possible; quantum effects are only taken into account through correction factors; recrossing of the barrier is neglected; and the separation of the reaction coordinate from all other degrees of freedom is an approximation.

Some of these difficulties have been addressed here. The calculation of relative rate constants allows for error cancellation, provided the errors are systematic. DFT and many-body activation energies tend to satisfy this condition; an example was given in Section 7.4.3.2. In contrast, the harmonic approximation for the vibrational contribution to the molecular entropy can produce large nonsystematic errors. Anharmonicity of low-frequency vibration was taken into account within the independent mode approximation using a semiclassical expression for the vibrational partition function (Section 7.3.1.3). Because of the exponential dependence of the vibrational partition function on the energy levels, the correction of low-frequency motion has the largest effect on the reduction of nonsystematic errors. In Section 7.3.2.3, a kinetic model for the pyrolysis of PPE was introduced, which explicitly exploits systematic error cancellation in relative rate constants of reactions with the same reactants and similar transition states and requires the computation of rate constants for hydrogen abstraction reactions only. The model was successfully applied in Section 7.4.3.1 for the pyrolysis of PPE and PPE derivatives, where a hydroxy or methoxy substituent is located at the phenyl ring of the phenethyl group. When a methoxy substituent is introduced at the phenyl ring adjacent to the ether oxygen, the energy of the products of hydrogen abstraction increases and the reverse reactions become important (Section 7.4.3.2). Because the analytical kinetic model assumes that reverse reactions are insignificant, the computation of product selectivities for the latter set of PPE derivatives involves the knowledge of rate constants for the entire pyrolysis mechanism and the application of a numerical integration technique, as outlined in Section 7.3.3. A first step toward a complete simulation of PPE pyrolysis was taken in Section 7.4.4, where rate constants were computed for the phenyl-shift reaction.

When reactions of different type are considered in a numerical integration scheme, error cancellation is less effective than in the calculation of relative rate constants for hydrogen abstraction by the same radical, merely on different carbon sides, and the requirements on the accuracy of rate constants increase. With rapidly evolving high-performance computing resources, the calculation of energy differences using high-level coupled-cluster methods for molecules of hundreds of atoms is already within reach. Developments to go beyond the independent mode approximation and to calculate molecular entropies more reliably are undertaken, sophisticated tunneling methods are available, and variational transition state theory provides a route to include barrier recrossing effects. Even the assumption of separability of the reaction coordinate can be relaxed by including a vibrational bath. New and automated techniques are investigated to sample conformational space, which can become large with increasing system size, as we have seen in Section 7.4.3.2, where 117 transition states were used to calculate the rate constants for 16 reactions.

Judging from the success of the analytical model in Section 7.4.3.1 to reproduce substituent effects on α/β-selectivities, we will soon be able to model the pyrolysis of lignin model compounds including a complete set of relevant reactions. In a future vision, the combination of hundreds of computational studies of model compounds in the kinetic simulation of biomass conversion will enable us to bridge the gap between molecular understanding and industrially applicable processing techniques.

ACKNOWLEDGMENTS

We would like to thank Robert Harrison who made this work possible. This research was sponsored by the Division of Chemical Sciences, Geosciences, and Biosciences, Office of Basic Energy Sciences, U.S. Department of Energy, and was performed in part using the resources of the Center for Computational Sciences at Oak Ridge National Laboratory under contract DE-AC05-00OR22725.

APPENDIX SUMMARY OF KINETIC PARAMETERS

Rate expressions are recommended to be used as relative quantities only. PPE, phenethyl phenyl ether; α-PPE$^•$, α-radical of PPE; β-PPE$^•$, β-radical of PPE; TST, transition state theory including anharmonic corrections; TST$^{\#}$, transition state theory including anharmonic corrections and Wigner tunneling correction; SS, steady-state approximation for intermediate.

Reaction	Temperature (K)	ln k	Method	Section
PPE + PhO$^•$ → α-PPE$^•$ + PhOH	580–660	$17.9–9.07 \times 10^3/T$	B3LYP/TST	7.4.3.1
PPE + PhO$^•$ → β-PPE$^•$ + PhOH	580–660	$17.8–9.36 \times 10^3/T$	B3LYP/TST	7.4.3.1
PPE + $PhCH_2^•$ → α-PPE$^•$ + $PhCH_3$	580–660	$17.1–9.41 \times 10^3/T$	B3LYP/TST	7.4.3.1
PPE + $PhCH_2^•$ → β-PPE$^•$ + $PhCH_3$	580–660	$17.1–1.06 \times 10^4/T$	B3LYP/TST	7.4.3.1
*p*HO-PPE + PhO$^•$ → α-*p*HO-PPE$^•$ + PhOH	580–660	$17.7–8.58 \times 10^3/T$	B3LYP/TST	7.4.3.1
*p*HO-PPE + PhO$^•$ → β-*p*HO-PPE$^•$ + PhOH	580–660	$17.3–9.14 \times 10^3/T$	B3LYP/TST	7.4.3.1
*p*HO-PPE + *p*HO-$PhCH_2^•$ → α-*p*HO-PPE$^•$ + *p*HO-$PhCH_3$	580–660	$16.8–9.65 \times 10^3/T$	B3LYP/TST	7.4.3.1
*p*HO-PPE + *p*HO-$PhCH_2^•$ → β-*p*HO-PPE$^•$ + *p*HO-$PhCH_3$	580–660	$16.9–1.09 \times 10^4/T$	B3LYP/TST	7.4.3.1
pCH_3O-PPE + PhO$^•$ → α-*p*CH_3O-PPE$^•$ + PhOH	580–660	$18.0–8.53 \times 10^3/T$	B3LYP/TST	7.4.3.1
pCH_3O-PPE + PhO$^•$ → β-*p*CH_3O-PPE$^•$ + PhOH	580–660	$17.3–9.10 \times 10^3/T$	B3LYP/TST	7.4.3.1
pCH_3O-PPE + *p*CH_3O-$PhCH_2^•$ → α-*p*CH_3O-PPE$^•$ + *p*CH_3O-$PhCH_3$	580–660	$17.1–9.64 \times 10^3/T$	B3LYP/TST	7.4.3.1
pCH_3O-PPE + *p*CH_3O-$PhCH_2^•$ → β-*p*CH_3O-PPE$^•$ + *p*CH_3O-$PhCH_3$	580–660	$17.0–1.10 \times 10^4/T$	B3LYP/TST	7.4.3.1

(*Continued*)

Reaction	Temperature (K)	ln *k*	Method	Section
PPE + PhO$^\bullet$ → α–PPE$^\bullet$ + PhOH	580–660	$16.4–5.89 \times 10^3/T$	M06-2X/TST$^\#$	7.4.3.2
PPE + PhO$^\bullet$ → β-PPE$^\bullet$ + PhOH	580–660	$16.9–6.25 \times 10^3/T$	M06-2X/TST$^\#$	7.4.3.2
PPE + $PhCH_2^\bullet$ → α–PPE$^\bullet$ + $PhCH_3$	580–660	$16.3–5.85 \times 10^3/T$	M06-2X/TST$^\#$	7.4.3.2
PPE + $PhCH_2^\bullet$ → β–PPE$^\bullet$ + $PhCH_3$	580–660	$16.8–7.01 \times 10^3/T$	M06-2X/TST$^\#$	7.4.3.2
PPE-$p$$OCH_3$ + $p$$CH_3O$-PhO$^\bullet$ → α-PPE$^\bullet$-$p$$OCH_3$ + $p$$CH_3O$-PhOH	580–660	$15.9–7.04 \times 10^3/T$	M06-2X/TST$^\#$	7.4.3.2
PPE-$p$$OCH_3$ + $p$$CH_3O$-PhO$^\bullet$ → β-PPE$^\bullet$-$p$$OCH_3$ + $p$$CH_3O$-PhOH	580–660	$16.6–7.55 \times 10^3/T$	M06-2X/TST$^\#$	7.4.3.2
PPE-$p$$OCH_3$ + $PhCH_2^\bullet$ → α-PPE$^\bullet$-$p$$OCH_3$ + $PhCH_3$	580–660	$15.7–5.81 \times 10^3/T$	M06-2X/TST$^\#$	7.4.3.2
PPE-$p$$OCH_3$ + $PhCH_2^\bullet$ → β-PPE$^\bullet$-$p$$OCH_3$ + $PhCH_3$	580–660	$16.6–6.80 \times 10^3/T$	M06-2X/TST$^\#$	7.4.3.2
PPE-$o$$OCH_3$ + $o$$CH_3O$-PhO$^\bullet$ → α-PPE$^\bullet$-$o$$OCH_3$ + $o$$CH_3O$-PhOH	580–660	$15.7–6.20 \times 10^3/T$	M06-2X/TST$^\#$	7.4.3.2
PPE-$o$$OCH_3$ + $o$$CH_3O$-PhO$^\bullet$ → β-PPE$^\bullet$-$o$$OCH_3$ + $o$$CH_3O$-PhOH	580–660	$16.1–6.53 \times 10^3/T$	M06-2X/TST$^\#$	7.4.3.2
PPE-$o$$OCH_3$ + $PhCH_2^\bullet$ → α-PPE$^\bullet$-$o$$OCH_3$ + $PhCH_3$	580–660	$15.9–5.72 \times 10^3/T$	M06-2X/TST$^\#$	7.4.3.2
PPE-$o$$OCH_3$ + $PhCH_2^\bullet$ → β-PPE$^\bullet$-$o$$OCH_3$ + $PhCH_3$	580–660	$16.7–6.74 \times 10^3/T$	M06-2X/TST$^\#$	7.4.3.2
PPE-di-$o$$OCH_3$ + di-$o$$CH_3O$-PhO$^\bullet$ → α-PPE$^\bullet$-di-$o$$OCH_3$ + di-$o$$CH_3O$-PhOH	580–660	$15.1–5.97 \times 10^3/T$	M06-2X/TST$^\#$	7.4.3.2
PPE-di-$o$$OCH_3$ + di-$o$$CH_3O$-PhO$^\bullet$ → β-PPE$^\bullet$-di-$o$$OCH_3$ + di-$o$$CH_3O$-PhOH	580–660	$16.5–7.64 \times 10^3/T$	M06-2X/TST$^\#$	7.4.3.2
PPE-di-$o$$OCH_3$ + $PhCH_2^\bullet$ → α-PPE$^\bullet$-di-$o$$OCH_3$ + $PhCH_3$	580–660	$16.7–5.71 \times 10^3/T$	M06-2X/TST$^\#$	7.4.3.2
PPE-di-$o$$OCH_3$ + $PhCH_2^\bullet$ → β-PPE$^\bullet$-di-$o$$OCH_3$ + $PhCH_3$	580–660	$16.5–6.50 \times 10^3/T$	M06-2X/TST$^\#$	7.4.3.2
β-PPE$^\bullet$ → β-$PhCH_2CH_2O^\bullet Ph$	580–660	$30.7–9.70 \times 10^3/T$	M06-2X/TST$^\#$, SS	7.4.4
β-pHO-PPE$^\bullet$ → β-pHO-$PhCH_2CH_2O^\bullet Ph$	580–660	$30.7–9.62 \times 10^3/T$	M06-2X/TST$^\#$, SS	7.4.4
β-$p$$CH_3O$-PPE$^\bullet$ → β-$p$$CH_3O$-$PhCH_2CH_2O^\bullet Ph$	580–660	$30.8–9.59 \times 10^3/T$	M06-2X/TST$^\#$, SS	7.4.4

REFERENCES

1. http://www.cfo.doe.gov/budget/10budget/Content/Highlights/FY2010Highlights.pdf.
2. R. Rinaldi and F. Schüth, *Energy Environ. Sci.* **2**, 610 (2009).
3. http://archive.greenpeace.org/climate/arctic99/reports/wildlife.html.

4. D. Mohan, C. U. Pittman, Jr., and P. H. Steele, *Energy Fuels* **20**, 848 (2006).
5. E. C. Petrou and C. P. Pappis, *Energy Fuels* **23**, 1055 (2009).
6. M. Stöcker, *Angew. Chem. Int. Ed.* **47**, 9200 (2008).
7. J. N. Chheda, G. W. Huber, and J. A. Dumesic, *Angew. Chem. Int. Ed.* **46**, 7164 (2007).
8. S. Czernik and A. V. Bridgwater, *Energy Fuels* **18**, 590 (2004).
9. G. W. Huber, S. Iborra, and A. Corma, *Chem. Rev.* **106**, 4044 (2006).
10. http://en.wikipedia.org/wiki/File:Lignin_structure.svg.
11. E. Dorrestijn, L. J. J. Laarhoven, I. W. C. E. Arends, and P. Mulder, *J. Anal. Appl. Pyrolysis* **54**, 153 (2000).
12. J. C. del Río, A. T. Martínez, and A. Gutiérrez, *J. Anal. Appl. Pyrolysis* **79**, 33 (2007).
13. M. Kleinert and T. Barth, *Chem. Eng. Technol.* **31**, 736 (2008).
14. P. J. M. Suhas Carrott and M. M. L. Ribeiro Carrott, *Bioresour. Technol.* **98**, 2301 (2007).
15. S. Baumlin, F. Broust, F. Bazer-Bachi, T. Bourdeaux, O. Herbinet, F. T. Ndiaye, M. Ferrer, and J. Lédé, *Int. J. Hydrogen Energy* **31**, 2179 (2006).
16. C. Fushimi, K. Araki, Y. Yamaguchi, and A. Tsutsumi, *Ind. Eng. Chem. Res.* **42**, 3929 (2003).
17. E. A. Borges da Silva, M. Zabkova, J. D. Araújo, C. A. Cateto, M. F. Barreiro, M. N. Belgacem, and A. E. Rodrigues, *Chem. Eng. Res. Des.* **87**, 1276 (2009).
18. H. Priefert, J. Rabenhorst, and A. Steinbüchel, *Appl. Microbiol. Biotechnol.* **56**, 296 (2001).
19. D. Ferdous, A. K. Dalai, S. K. Bej, and R. W. Thring, *Energy Fuels* **16**, 1405 (2002).
20. S. Wang, K. Wang, Q. Liu, Y. Gu, Z. Luo, K. Cen, and T. Fransson, *Biotechnol. Adv.* **27**, 562 (2009).
21. T. Faravelli, A. Frassoldati, G. Migliavacca, and E. Ranzi, *Biomass Bioenergy* **34**, 290 (2010).
22. C. Fargues, Á. Mathias, and A. Rodrigues, *Ind. Eng. Chem. Res.* **35**, 28 (1996).
23. G. X. Wu and M. Heitz, *J. Wood Chem. Technol.* **15**, 189 (1995).
24. A. Oasmaa, R. Alén, and D. Meier, *Bioresour. Technol.* **45**, 189 (1993).
25. G. Gellerstedt, J. Li, I. Eide, M. Kleinert, and T. Barth, *Energy Fuels* **22**, 4240 (2008).
26. K. Miura, *Energy and Fuels* **9**, 302 (1995).
27. M. T. Klein and P. S. Virk, *Energy Fuels* **22**, 2175 (2008).
28. Z. H. Craig, A. Bennett, M. T. Klein, and P. S. Virk, *Energy Fuels* **24**, 58 (2010).
29. J. B. Binder, M. J. Gray, J. F. White, C. Zhang, and J. E. Holladay, *Biomass Bioenergy* **33**, 1122 (2009).
30. K. Kuroda and A. Nakagawa-izumi, *Org. Geochem.* **37**, 665 (2006).
31. K. Kuroda, A. Nakagawa-izumi, T. Ashitani, and K. Fujita, *J. Anal. Appl. Pyrolysis* **86**, 185 (2009).
32. H. Kawamoto, S. Horigoshi, and S. Saka, *J. Wood Sci.* **53**, 168 (2007).
33. P. F. Britt, A. C. Buchanan, III, and E. A. Malcolm, *J. Org. Chem.* **60**, 6523 (1995).
34. P. F. Britt, A. C. Buchanan, III, M. J. Cooney, and D. R. Martineau, *J. Org. Chem.* **65**, 1376 (2000).
35. M. K. Kidder, P. F. Britt, A. L. Chaffee, and A. C. Buchanan, III, *Chem. Commun.* **1**, 52 (2007).
36. P. F. Britt, A. C. Buchanan, III, and E. A. Malcolm, *Energy Fuels* **14**, 1314 (2000).

37. P. F. Britt, M. K. Kidder, and A. C. Buchanan, III, *Energy Fuels* **21**, 3102 (2007).
38. P. H. Kandanarachchi, T. Autrey, and J. A. Franz, *J. Org. Chem.* **67**, 7937 (2002).
39. T. C. Drage, C. H. Vane, and G. D. Abbott, *Org. Geochem.* **33**, 1523 (2002).
40. R. Katahira, H. Kamitakahara, T. Takano, and F. Nakatsubo, *J. Wood Sci.* **52**, 255 (2006).
41. H. Kawamoto, M. Ryoritani, and S. Saka, *J. Anal. Appl. Pyrolysis* **81**, 88 (2008).
42. T. Watanabe, H. Kawamoto, and S. Saka, *Holzforschung* **63**, 424 (2009).
43. E. Dorrestijn, M. Kranenburg, M. V. Ciriano, and P. Mulder, *J. Org. Chem.* **64**, 3012 (1999).
44. K. E. Gilbert and J. J. Gajewski, *J. Org. Chem.* **47**, 4899 (1982).
45. M. T. Klein and P. S. Virk, *Ind. Eng. Chem. Fundam.* **22**, 35 (1983).
46. R. Brežný, V. Mihalov, and V. Kováčik, *Holzforschung* **37**, 199 (1983).
47. H. Kawamoto, S. Horigoshi, and S. Saka, *J. Wood Sci.* **53**, 268 (2007).
48. K. Kuroda, *J. Anal. Appl. Pyrolysis* **35**, 53 (1995).
49. M. Bietti, G. Ercolani, and M. Salamone, *J. Org. Chem.* **72**, 4515 (2007).
50. M. Smeu and G. A. DiLabio, *J. Org. Chem.* **72**, 4520 (2007).
51. T. Elder, *Biomacromolecules* **8**, 3619 (2007).
52. C. Agache and V. I. Popa, *Monatshefte Chem.* **137**, 55 (2006).
53. K. Kato, K. Endo, D. Matsumoto, T. Ida, K. Saito, K. Fukushima, and N. Kato, *Appl. Surf. Sci.* **255**, 1040 (2008).
54. L. Petridis and J. C. Smith, *J. Comp. Chem.* **30**, 457 (2008).
55. B. Durbeej, Y.-N. Wang, and L. A. Eriksson, *Lect. Notes Comput. Sci.* **2565**, 137 (2003).
56. M. Shigematsu and H. Masamoto, *J. Wood Sci.* **54**, 308 (2008).
57. C. Martínez, M. Sedano, J. Mendoza, R. Herrera, J. G. Rutiaga, and P. López, *J. Mol. Graph. Model.* **28**, 196 (2009).
58. C. Martínez, J. L. Rivera, R. Herrera, J. L. Rico, N. Flores, J. G. Rutiaga, and P. López, *J. Mol. Model.* **14**, 77 (2008).
59. G. da Silva and J. W. Bozzelli, *J. Phys. Chem. A* **111**, 7987 (2007).
60. I. Shavitt and R. J. Bartlett, *Many-Body Methods in Chemistry and Physics*, Cambridge University Press, New York (2009).
61. R. G. Parr and W. Yang, *Density-Functional Theory of Atoms and Molecules*, Oxford University Press, New York (1989).
62. A. G. Vandeputte, M. K. Sabbe, M.-F. Reyniers, V. Van Speybroeck, M. Waroquier, and G. B. Marin, *J. Chem. Phys. A* **111**, 11771 (2007).
63. J. Sun, Y. Tang, H. Sun, Y. Pan, X. Jia, X. Pan, and R. Wang, *Mol. Phys.* **106**, 1379 (2008).
64. X. Yang, *Annu. Rev. Phys. Chem.* **58**, 433 (2007).
65. D. G. Truhlar and B. C. Garrett, *Acc. Chem. Res.* **13**, 440 (1980).
66. D. G. Truhlar and B. C. Garrett, *Ann. Rev. Phys. Chem.* **35**, 159 (1984).
67. J. B. Anderson, *J. Chem. Phys.* **58**, 4684 (1973).
68. R. T. Skodje and D. G. Truhlar, *J. Phys. Chem.* **85**, 624 (1981).
69. B. C. Garrett, T. Joseph, T. N. Truong, and D. G. Truhlar, *Chem. Phys.* **136**, 271 (1989).
70. B. C. Garrett, D. G. Truhlar, R. S. Grev, and A. W. Magnuson, *J. Phys. Chem.* **84**, 1730 (1980).
71. W. H. Miller, *J. Chem. Phys.* **61**, 1823 (1974).

72. W. H. Miller, *J. Chem. Phys.* **62**, 1899 (1975).
73. M. Thoss and H. Wang, *Annu. Rev. Phys. Chem.* **55**, 299 (2004).
74. N. Makri, *J. Phys. Chem. A* **108**, 806 (2004).
75. N. E. Henriksen and F. Y. Hansen, *Theories of Molecular Reaction Dynamics*, Oxford University Press, New York (2008).
76. J. E. House, *Principles of Chemical Kinetics*, 2nd edition, Academic Press (2007).
77. D. A. McQuarrie and J. D. Simon, *Molecular Thermodynamics*, University Science Books, Sausalito, CA (1999).
78. A. Beste, *Chem. Phys. Lett.* **493**, 200 (2010).
79. A. Beste, A. C. Buchanan, III, P. F. Britt, B. C. Hathorn, and R. J. Harrison, *J. Phys. Chem. A* **111**, 12118 (2007).
80. A. D. Isaacson and D. G. Truhlar, *J. Chem. Phys.* **76**, 1380 (1982).
81. P. Y. Ayala and H. B. Schlegel, *J. Chem. Phys.* **108**, 2314 (1998).
82. A. Chakraborty and D. G. Truhlar, *J. Chem. Phys.* **124**, 184310 (2006).
83. D. G. Truhlar and A. D. Isaacson, *J. Chem. Phys.* **94**, 357 (1991).
84. R. Q. Topper and D. G. Truhlar, *J. Chem. Phys.* **97**, 3647 (1992).
85. M. B. Hansen, O. Christiansen, D. Toffoli, and J. Kongsted, *J. Chem. Phys.* **128**, 174106 (2008).
86. E. P. Wigner, *Phys. Rev.* **40**, 749 (1932).
87. J. G. Kirkwood, *Phys. Rev.* **44**, 31 (1933).
88. P. Linz and R. L. C. Wang, *Exploring Numerical Methods: An Introduction to Scientific Computing Using MATLAB*, Jones and Bartlett Publishers, Sudbury, MA (2003).
89. J. M. Younker, *Qinetic*, Pennsylvania State University, Pennsylvania (2010).
90. A. Beste, A. C. Buchanan, III, and R. J. Harrison, *J. Phys. Chem. A* **112**, 4982 (2008).
91. A. Beste and A. C. Buchanan, III, *Energy Fuels* **24**, 2857 (2010).
92. A. Beste and A. C. Buchanan, III, *J. Org. Chem.* **74**, 2837 (2090).
93. E. J. Bylaska et al., *NWChem, A Computational Chemistry Package for Parallel Computers, Version 5.1*, Pacific Northwest National Laboratory, Richland, WA (2007).
94. A. D. Becke, *J. Chem. Phys.* **98**, 5648 (1993).
95. C. Lee, W. Yang, and R. G. Parr, *Phys. Rev. B* **37**, 785 (1988).
96. Y. Zhao and D. G. Truhlar, *Acc. Chem. Res.* **41**, 157 (2008).
97. The best experimental rate constant for a reference reaction (2-allylbenzyl radical abstracting hydrogen from *meta*-xylene; J. A. Franz et al., *J. Org. Chem.* **51**, 1446 (1986)) has an error of 0.7 kcal/mol in the activation energy and 0.4 in log A. The group additivity parameters have errors associated with them also. An overall estimate of the errors in the experimental values is 1–2 kcal/mol for the activation energies and 2.5 for ln A.

CHAPTER 8

Quantum Chemistry Study on the Pyrolysis Mechanisms of Coal-Related Model Compounds

BAOJUN WANG and RIGUANG ZHANG

Key Laboratory of Coal Science and Technology, Taiyuan University of Technology, Ministry of Education and Shanxi Province, Taiyuan, China

LIXIA LING

Research Institute of Special Chemicals, Taiyuan University of Technology, Taiyuan, China

8.1 INTRODUCTION TO THE APPLICATION OF QUANTUM CHEMISTRY CALCULATION TO INVESTIGATION ON MODELS OF COAL STRUCTURE

The knowledge of coal structure and its reactivity has been the focus of investigations carried out by hundreds of researchers. Coal chemists have attempted to set up models of coal structure that represent a synthetic view of principal characteristics of coal organic matter. First, the strategies of constructing "average structural unit" of coal organic matter were proposed, and the structural units were, in fact, molecules described in a classically chemical way using atoms, chemical bonds, and some functional groups [1]. At a later time, with the advances of experimental means, molecular components of coal structure were investigated by systematic experimental studies including coal atomic composition, analysis of products from pyrolysis, and extraction or liquefaction treatments of coal samples, and the molecular masses of the products were determined by various experimental means and characterization instruments [2].

Recently, theoretical calculations have been used to investigate coal structure and its reactivity. For example, the partial structure of Shinn model [3], which represents

Rate Constant Calculation for Thermal Reactions: Methods and Applications, Edited by Herbert DaCosta and Maohong Fan.

the coal macromolecular model, was studied using quantum chemistry methods by Hou et al. [4]. Based on the analysis of partial coal structures, the reactivity of macromolecular coal has been well explained. Zhang [5] constructed the molecular structures of coal at different ranks, and the electric charge distribution on the surface of coal structure was calculated by means of quantum chemistry. Then, the mechanism by which polymers are adsorbed on coal surface was analyzed in detail, which successfully explains why anionic polymers are adsorbed on coal surfaces with negative charge. Takanohashi et al. [6] have simulated the interaction between coal and solvents by theoretical calculation, which indicates that "cooperative interactions through hydrogen bonds and aromatic–aromatic interactions at several sites of coal molecular structure will lead to a strong association structure i.e., even if one interaction is dissociated, many others can still hold the association structure." In our previous work [7–23], different coal structure models were chosen to describe gas adsorption on coal surfaces and the properties of coals at different ranks such as the stability, calorific power, and thermodynamic function change in combustion reaction. Aromatic plane models were used to, for instance, describe UV–Vis adsorption properties of coal. The studies discussed above show that quantum chemistry calculations can be used to analyze qualitatively and quantitatively the bond characteristics and molecular structure of coal and its change in reaction processes, which can provide theoretical guidance for the efficient utilization of coal. Calculations can also be used to illustrate experimental phenomena and to understand coal structure and its reactivity, namely, computational studies have gradually formed a new research field, and have provided feasible means for dealing with coal structure and its reactivity. However, the key step in applying quantum chemistry methods to the investigation of coal structure is the establishment of a reasonable initial model [24].

In this chapter, we will give some principles of choosing and building an appropriate coal structure model, and present an overview of recent advances that focus on the application of quantum chemistry methods to the study of the pyrolysis mechanisms of the coal-related model compounds, such as oxygen-, nitrogen-, and sulfur-containing compounds, which represent corresponding partial structures in coal.

8.2 THE MODEL FOR COAL STRUCTURE AND CALCULATION METHODS

8.2.1 The Proposal of Local Microstructure Model of Coal

It is key to choose or build an appropriate model when we study coal structure and its reactivity using quantum chemistry method. However, this entails many difficulties [25]. For a finite molecule, it can be built and described clearly, and for a periodic infinite molecule, such as crystal, it can also be built and studied accurately by quantum chemistry methods. However, the structure of coal is inherently complex and varies with origin, geographical area, and age or rank of the particular coal observed [26]. Therefore, it is difficult to obtain a clear structure model for this complex mixture. This is an inevitable problem for the research of coal chemistry.

In order to describe the chemical structure of coal, various kinds of coal molecular models have been proposed, but these models might not reflect the actual physical and chemical structure of coal. In fact, we think that it is impossible (and even unnecessary in practice) to precisely describe the microscopic structure of coal using a single model. Generally, any structural model is only a microscopic approximation for one or several macroscopic properties. Research results of our group show that it is feasible and reasonable for some model compounds to reflect specific properties of coal based on its local chemistry.

Therefore, for the molecular structure of coal, we should accept and adopt the view of an average molecular structure to build coal molecule structures, which might not represent the entire coal structure, but model one or several aspects of coal [27]. Furthermore, when we construct a suite of concrete models aiming at different coal ranks and coal types related to the practical reaction processes, we are convinced that the models constructed by the above principles are reasonable, exact, and scientific. They can give a reasonable microscopic description, reflect the properties of coal in one or several aspects, and provide reasonable microscopic explanation for the particular and local macroscopic property [28].

Based on the above knowledge of coal structure, the constructed model of coal is defined as local microstructure model of coal (LMSMC). The word "local" has two kinds of meanings: (1) the model might only be a local fragment of a coal macromolecule, which is a common and representative part, describing the chemical properties of neighboring atoms of a target atom or an organic group during pyrolysis, gasification, and so on; (2) the model might only be an abstract structure from a kind of property of coal, which is used to describe macrocharacteristics of coal, such as UV–Vis spectrum, reflectivity, and so on. LMSMC can be viewed as a methodology to study complex matter such as coal.

8.2.2 Coal-Related Model Compounds Describing the Properties of Coal Pyrolysis

A number of studies have focused on the molecular structure of coal [29–31]. The structures can be derived using data from a variety of sources, including coal atomic composition, analysis of pyrolysis products, extraction or liquefaction treatments of coal samples, and spectroscopic analyses using GC/MS, NMR, and IR. In our study, the method of Py–GC–MS, which consists of a CDS 5250 pyrolysis autosampler and focus GC-DSQII, has been used to study the pyrolysis of coal in order to obtain information about coal structure.

A series of oxygen-containing models such as phenol, furan, benzoic acid, benzaldehyde, and anisole, nitrogen-containing models such as pyrrole, indole, pyridine, 2-picoline, quinoline, and isoquinoline, and sulfur-containing models such as thiophene and benzenethiol, which represent oxygen-, nitrogen-, and sulfur-containing substructure in coal, were chosen as the coal-related heteroatom-containing model compounds. These models fall into the category of the LMSMC.

8.2.3 The Pyrolysis of Model Compounds Reflecting the Pyrolysis Phenomenon of Coal

The pyrolysis mechanisms of oxygen-, nitrogen-, and sulfur-containing models were proposed based on comparative experimental studies between coal model pyrolysis. For example, according to the correlations between the release of CO_2 and the loss of carboxyl groups, and the relationship between the release of CO and the loss of carbonyl groups in experiment [32], benzoic acid and benzaldehyde were chosen as the representatives of carboxyl and carbonyl groups in coal, and the release of CO and CO_2 during coal pyrolysis was investigated for these compounds.

The pyrolysis characteristics and products of some model compounds are the same as those of coal, for example, H_2S is detected during thiophene pyrolysis [33,34], which is in accordance with coal pyrolysis, namely, H_2S was formed via the evolution of sulfur in thiophenic compounds in coal at 1223 K [35]. Some pyrolysis aspects of model compounds are different to those of complex and real environment coal pyrolysis. For example, thiols in coal tend to decompose into H_2S and the amount of H_2S reaches a maximum around 973–1123 K. The reason is that the active sulfur-containing free radicals are formed first during coal pyrolysis, which can be stabilized by abstracting a hydrogen atom supplied by other hydrocarbon molecules [36] or which combine with H radicals formed by the homolysis of weak R–H (R may be alkyl) bonds or released in the process of polymerization during coal pyrolysis [37–39] resulting in the formation of H_2S. But during the pyrolysis of benzenethiol, besides benzene, a large number of polymerization products, diphenyl sulfide and diphenyl disulfide, were detected due to the lack of hydrogen sources [40]. In our study [23], benzenethiol has been chosen to investigate the pyrolysis of a typical thioalcohol compound in coal, but the polymerization process during the release process of small sulfur-containing molecules was not considered.

In general, the model compounds are chosen to reflect the pyrolysis process of complex coal. The pyrolysis mechanisms of oxygen-, nitrogen-, and sulfur-containing model compounds in coal will be introduced in detail in the following sections to illustrate some important pyrolysis reactions of coal.

8.2.4 The Calculation Methods

Density functional theory (DFT) methods, as implemented in the $DMol^3$ package, are used to investigate the pyrolysis mechanisms for the coal-related model compounds in this chapter. $DMol^3$ includes the local density approximation (LDA) and the generalized gradient approximation (GGA). All electron calculations are performed for all atoms; to determine accurate activation barriers in the pyrolysis mechanism of model compounds, transition states are searched by means of complete LST/QST for reactions [41]; starting from reactants and products, the linear synchronous transit (LST) method performs a single interpolation to a maximum energy, and the quadratic synchronous transit (QST) method alternates searches for an energy maximum with constrained minimizations in order to refine the transition state to a high degree.

$DMol^3$ is a DFT quantum chemistry software package that can perform accurate theoretical calculations on a wide range of compounds, including organic compounds, biological compounds, organometallics, and metal or metal oxides. The $DMol^3$ program has a long track record of successful commercial applications, with its unique approach to solving quantum mechanical equations. $DMol^3$ has long been one of the fastest methods available for performing calculations, an advantage that becomes apparent especially for larger systems with over 500 atoms. With its ability to model both molecular and solid state problems, the application provides a way to study the broadest possible range of problems in a single package. Finally, $DMol^3$ provides a user-friendly interface complying with Windows standards to allow even new users to use the program with confidence.

8.3 THE PYROLYSIS MECHANISMS OF COAL-RELATED MODEL COMPOUNDS

8.3.1 The Pyrolysis Mechanisms of Oxygen-Containing Model Compounds

Oxygen is one of the most important elements in coal. Its total content and chemical environment in coal have direct influence on pyrolysis, gasification, and so on. Model compounds with different oxygen-containing groups, which reflect substructure in coal, were chosen to study the migration and the release of oxygen during coal pyrolysis in detail.

8.3.1.1 Phenol and Furan Phenol is a hydroxyl group-containing aromatic compound in coal. The pyrolysis of phenol has been carried out by some researchers [42–45], and CO was detected as the main product. The formation of CO and cyclopentadiene via phenol pyrolysis was observed by experiment (Figure 8.1) [42]. The mechanism was further supported by labeling the phenol in different specific positions, and there has not been any theoretical study on this mechanism. However, the formation mechanism of CO via phenoxy radical decomposition has been studied by Olivella et al. [46] at the 3–21 G basis set used for the CASSCF geometry optimizations. To investigate the effect of the d polarization functions on the geometry of the calculated structures, additional geometry optimizations at the CASSCF level with the 6–31 G(d) basis set were performed for the most relevant stationary points, and the temperature correction to the sum of the CASPT2/6–31 G(d,p) energy and the

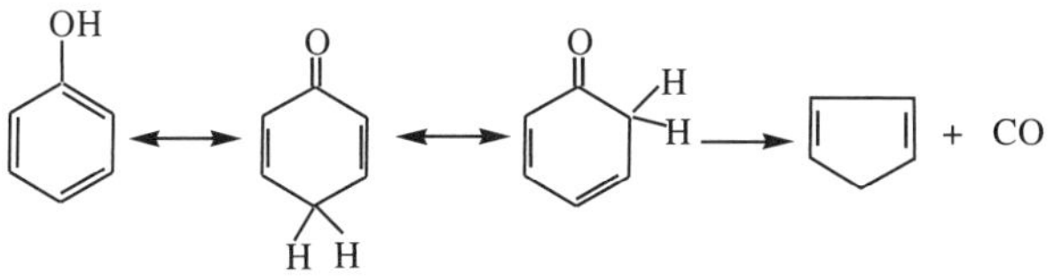

FIGURE 8.1 The formation of CO and cyclopentadiene from phenol.

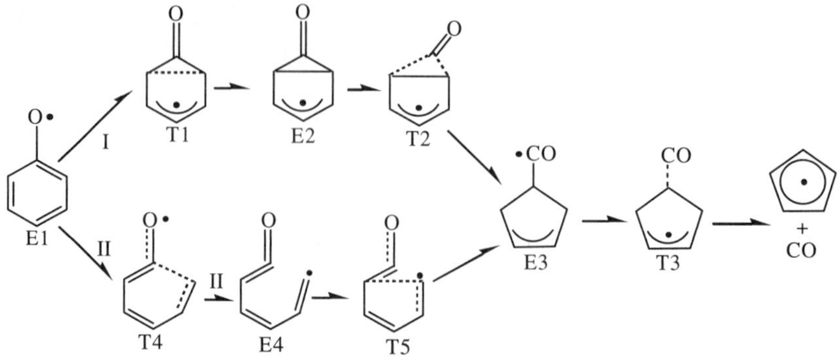

FIGURE 8.2 Pyrolysis mechanism of phenoxy.

ZPVE was evaluated. The pyrolysis mechanism is shown in Figure 8.2. Path I involves the formation of the bicyclic intermediate followed by an α-C–C bond cleavage of the three-membered ring to intermediate E3 and the elimination of the CO group from E3 leading to cyclopentadiene radical. Path II is a direct cleavage of the α-C–C bond followed by the cyclization to form the five-membered ring intermediate E3, and then the loss of CO. In path I, the relative energy of T2 is the highest, which is 53.9 kcal/mol higher than that of phenoxy radical. But the energy of T4 is 74.0 kcal/mol higher than that of phenoxy radical in path II, so path I is preferred. However, Liu et al. [47] believed that the activation energy, 53.9 kcal/mol at their best level, is about 10.0 kcal/mol higher than the value derived from experiment [48]. So the improved energy parameters were calculated; the result shows that the rate-determining step is E2 $\rightarrow$ E3 with transition state T2. The activation energy, 52 kcal/mol at their best G2M(rcc, MP2) level, is 8 kcal/mol higher than the value derived from experiment.

Due to the existence of five-membered furan rings in low-rank coals [49], the pyrolysis mechanism of furan has become the subject of several experimental studies [50–52]. Recently, some detailed theoretical studies have been carried out to understand the formation processes of CO, C_3H_4 (propyne and allene), ketene, acetylene, and so on.

The thermal decomposition of furan has been studied in a single shock tube over the temperature range 1050–1460 K. The major reaction products, C_3H_4 and CO, were detected and the formation mechanism was provided as shown in Figure 8.3 [51]. The other biradical mechanism of furan pyrolysis was also proposed by Organ and Mackie (see Figure 8.4) [52]. A study about the unimolecular decomposition channels was carried out to compute the geometries and the QCISD (T)/6–311 + + G** method to obtain energies using density functional (B3LYP) techniques by Liu et al. [53]. In this work, two mechanisms similar to that of isoxazole and one mechanism proposed by author were investigated. The results show that the highest activation energy is 77.0 kcal/mol, which occurs at TS9, and more than 77.0 kcal/mol is needed in other mechanisms. It can be concluded that the mechanism is more likely responsible for the formation of $CH_3C{\equiv}CH$ and CO (Figure 8.5). In this mechanism, intramolecular hydrogen migration plays an important role. Although this study

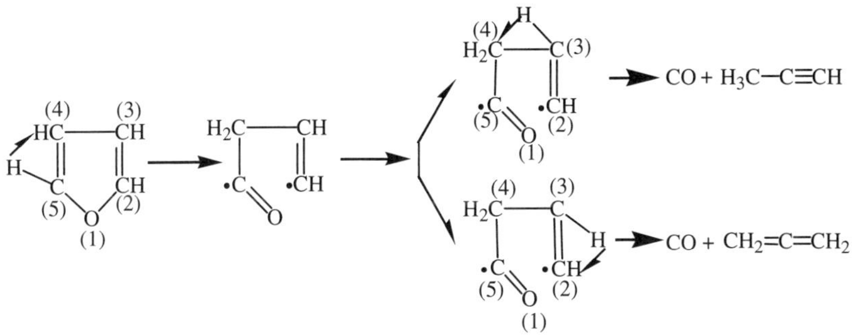

FIGURE 8.3 The formation mechanism of CO proposed by Lifshitz et al.

reported the energies of several decomposition pathways, no rate parameters were calculated and thus no comparison with experimental (kinetic modeling) studies was made, nor an investigation of the mechanisms was proposed in experiments [51,52]. Subsequently, Liu et al. [54] investigated the two biradical mechanisms proposed, which showed that the activation energies of the biradical mechanisms were about 40 kJ/mol higher than the experimental values [51,52]. It can be concluded that the biradical mechanisms are unlikely more competitive than the nonbiradical mechanism by kinetic analysis. A more comprehensive quantum chemical study in conjunction with kinetic modeling was carried out by Sendt et al. [55], and the calculations suggested that two-parallel paths initiated by 1,2-hydrogen migration, which results in the formation of cyclic carbine intermediates and leads to the decomposition products CO + propyne and C_2H_2 + ketene (as major and minor

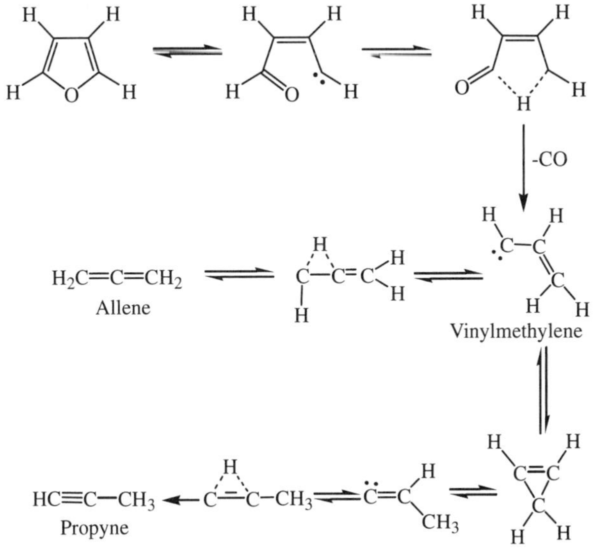

FIGURE 8.4 The formation mechanism of CO via furan pyrolysis proposed by Organ and Mackie.

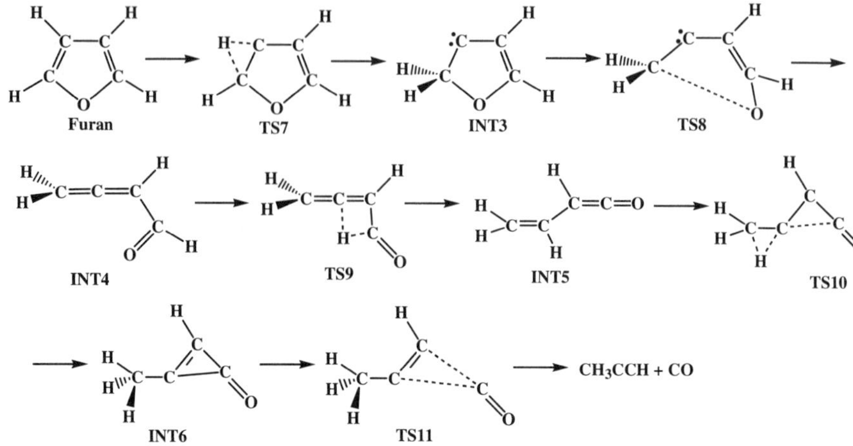

FIGURE 8.5 The feasible formation mechanism of CO during furan pyrolysis.

channels, respectively), are the dominant pathways, while the mechanism initiated by C–O bond scission to a biradical is unlikely.

8.3.1.2 Benzoic Acid and Benzaldehyde There are two contrary opinions about the correlation between low-temperature cross-linking reactions and decarboxylation with the release of CO_2 during coal pyrolysis. Suuberg et al. [56,57] observed that the low-temperature cross-linking associated with low-rank coals appeared to correlate with the release of CO_2. This correlation was confirmed by Solomon and coworkers [58,59]. However, the other researchers [60–62] believed that thermal decarboxylation of carboxylic acids and the release of CO_2 may not directly relate to cross-linking pathways because there is no evidence for aryl–aryl coupling (cross-linking) or other high molecular weight products that would be indicative of cross-linking reactions. Although the mechanism of decarboxylation reaction is not clearly understood, it is widely accepted that there is a correlation between the release of CO_2 and the loss of carboxyl groups, and the relationship between the release of CO and the loss of carbonyl groups [63].

The similar pyrolysis mechanisms of benzoic acid and benzaldehyde are proposed in Figures 8.6 and 8.7 [19];, the optimized structures and their atom numbers of reactants, intermediates, transition states, and products are shown in Figures 8.8 and 8.9; and the energy profiles of the stationary points for benzoic acid and benzaldehyde pyrolysis reactions are shown in Figures 8.10 and 8.11. According to the transition state theory [64], activation enthalpy $\Delta_r H_m^{\neq}$, activation entropy $\Delta_r S_m^{\neq}$, and activation energy E_a can be obtained from Eqs (8.1)–(8.3), respectively. The rate constant k can be expressed as shown in Eq. (8.4). These kinetic parameters are listed in Table 8.1.

$$\Delta_r H_m^{\neq} = H(TS) - H(R) \tag{8.1}$$

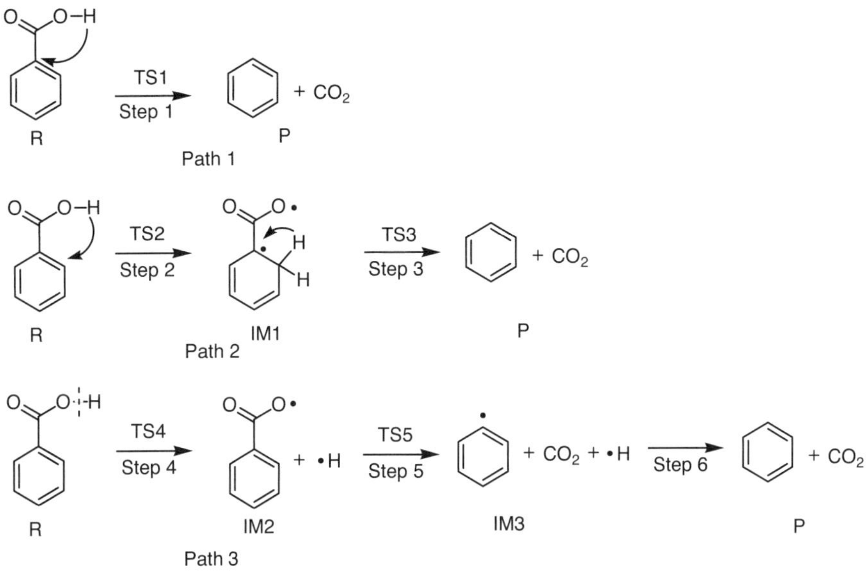

FIGURE 8.6 The pyrolysis mechanism of benzoic acid.

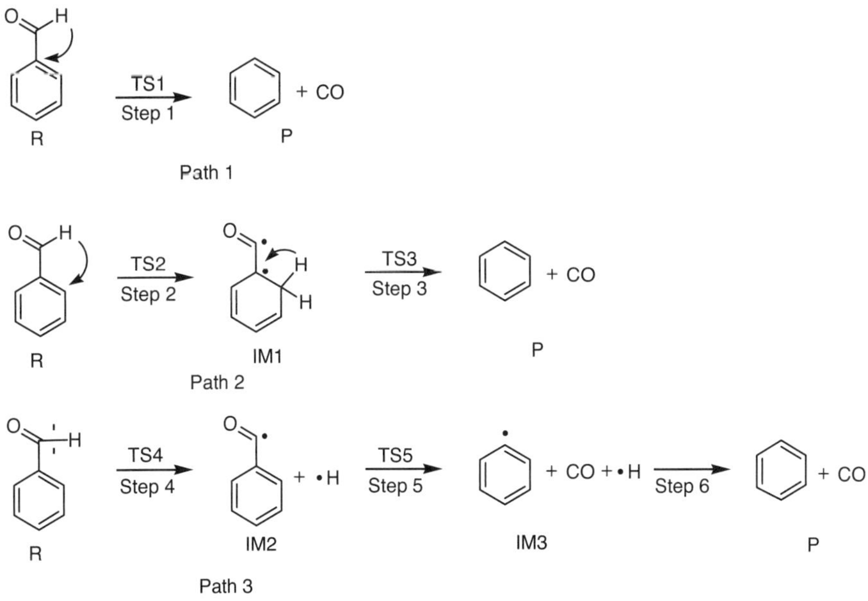

FIGURE 8.7 The pyrolysis mechanism of benzaldehyde.

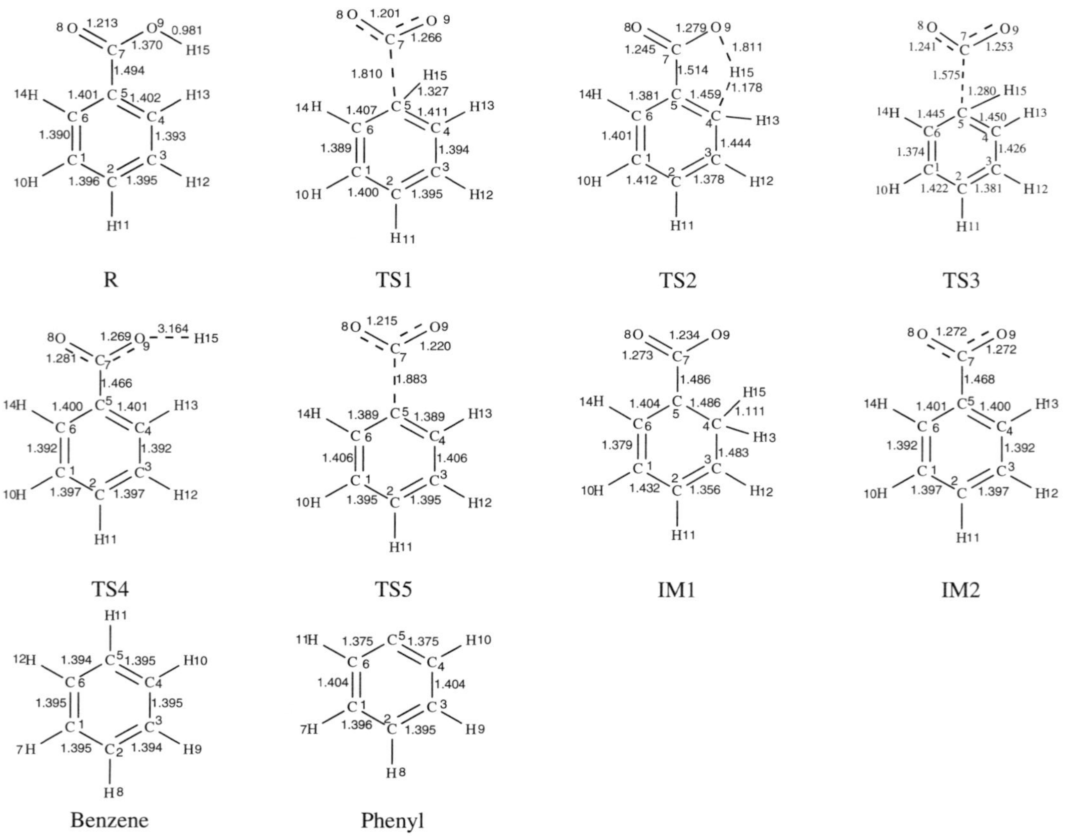

FIGURE 8.8 The optimized structures of all reactants, intermediates, transition states, and products during benzoic acid pyrolysis.

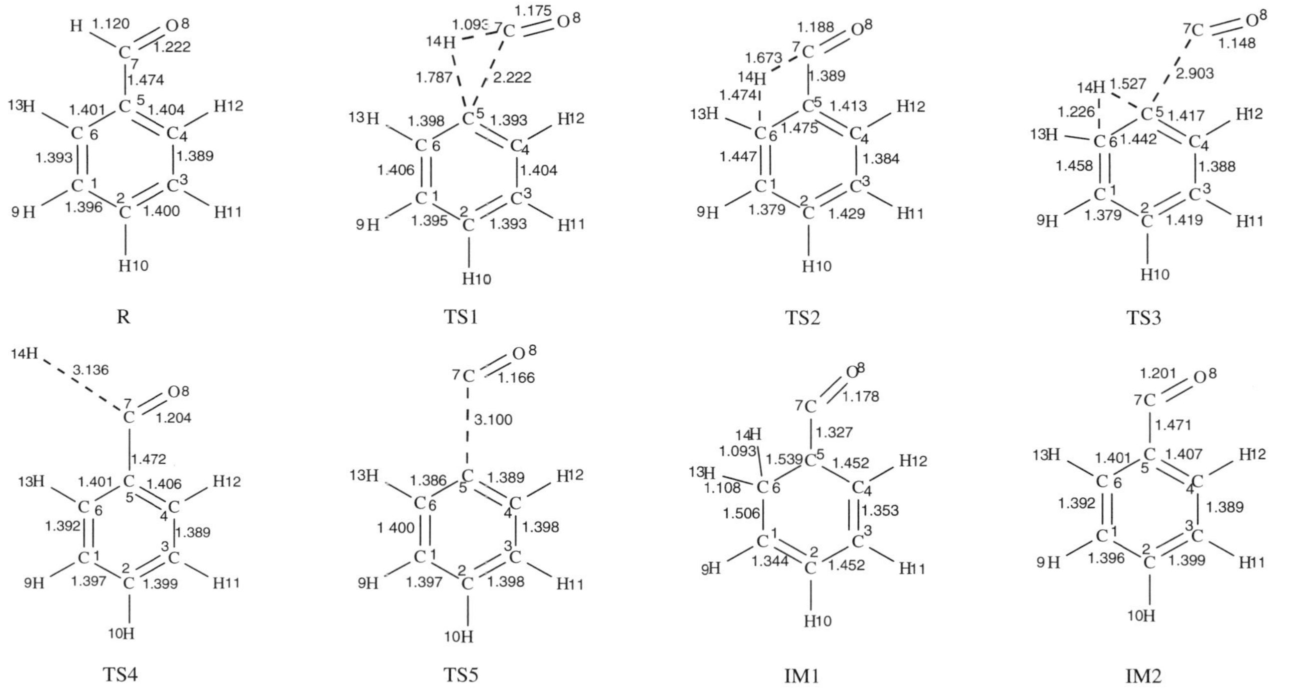

FIGURE 8.9 The optimized structures of all reactants, intermediates, transition states, and products during benzaldehyde pyrolysis.

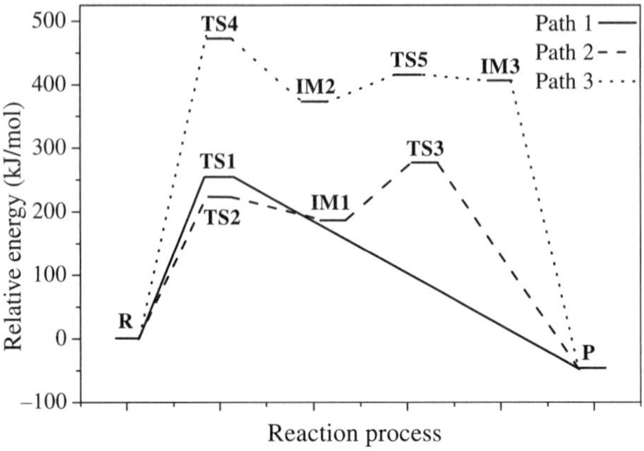

FIGURE 8.10 Energy profile of the stationary points for benzoic acid pyrolysis reaction.

$$\Delta_r S_m^{\neq} = S(TS) - S(R) \quad (8.2)$$

$$E_a = \Delta_r H_m^{\neq} + nRT \quad (8.3)$$

$$k = \frac{k_b T}{h}\left(\frac{p^0}{RT}\right)^{1-n} \exp\left(\frac{\Delta_r S_m^{\neq}}{R}\right) \exp\left(\frac{-\Delta_r H_m^{\neq}}{RT}\right) \quad (8.4)$$

where k is the rate constant, T is the reaction temperature, k_b is the Boltzmann constant, h is the Planck constant, p^0 is the standard atmospheric pressure, R is the fundamental gas constant, and n is the number of moles of reactant.

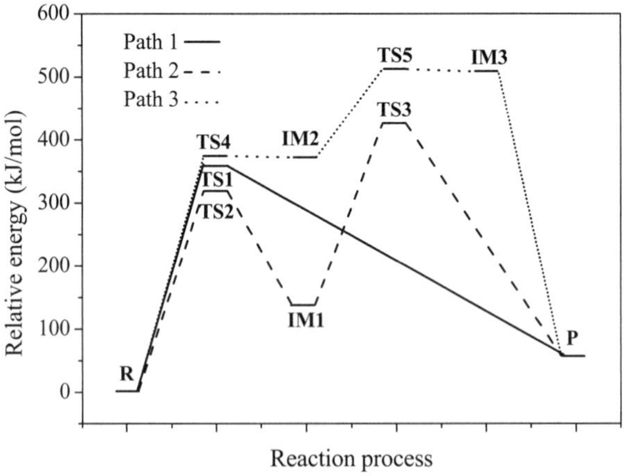

FIGURE 8.11 Energy profile of the stationary points for benzaldehyde pyrolysis reaction.

TABLE 8.1 The Activation Enthalpy, Activation Entropy, Activation Energy, and Rate Constant of Every Step During Benzoic Acid and Benzaldehyde Pyrolysis

Elementary Reaction	$\Delta_r H_m^{\neq}$ (kJ/mol)		$\Delta_r S_m^{\neq}$ (J/(mol K))		E_a (kJ/mol)		$\ln k$ (s^{-1})	
	298.15 K	1000 K	298.15 K	1000 K	298.15 K	1000 K	298.15 K	1000 K
Benzoic acid								
Step 1	256.73	255.24	10.26	7.96	259.21	263.56	−72.88	0.92
Step 2	221.78	219.01	−10.06	−14.74	224.26	227.33	−61.22	2.55
Step 3	90.48	87.68	−10.32	−15.06	92.95	96.00	−8.28	18.31
Step 4	478.46	482.22	20.06	27.67	480.94	490.54	−161.15	−24.01
Step 5	40.66	37.99	4.38	0.59	43.14	46.31	13.58	26.17
Benzaldehyde								
Step 1	362.49	363.09	18.73	21.34	364.97	371.41	−114.52	−10.44
Step 2	323.16	323.09	−0.79	−0.43	325.64	331.41	−101.01	−8.25
Step 3	290.48	291.73	25.12	28.51	292.96	300.04	−84.71	−0.99
Step 4	360.07	364.90	18.55	28.48	362.55	373.22	−113.57	−9.80
Step 5	145.34	143.27	19.44	16.83	147.82	151.58	−26.84	15.46

In the pyrolysis process of benzoic acid, paths 1 and 2 are the intramolecular hydrogen migration paths and path 3 is radical reaction path. Figure 8.10 shows that the highest energy barriers in paths 1 and 2 have little difference. The highest energy barrier occurs at TS4 in path 3, which has the highest relative energy in all species, and the activation energy of step 4 is also the largest in all elementary steps; the corresponding rate constant is the smallest. So the intramolecular hydrogen migration paths are more feasible than radical path via bond homolysis for decarboxylation of benzoic acid by kinetic analysis, and there is little possibility for cross-linking reaction to occur via radical combination. Above results show that decarboxylation process is not directly related to cross-linking reaction in coal, which gives a definite answer to the dispute whether decarboxylation and CO_2 evolution are directly related to cross-linking reactions in low-rank coals.

The characteristics of the energy profiles of the stationary points for benzoic acid and benzaldehyde pyrolysis (Figures 8.10 and 8.11) are quite similar. The deprivation of CO via intramolecular hydrogen migration paths (paths 1 and 2) is more feasible than the radical path (path 3). The activation energy of benzaldehyde pyrolysis is larger than that of benzoic acid, and the rate constant of benzaldehyde pyrolysis is smaller than that of benzoic acid, so the temperature of decarbonylation reaction is higher than that of CO loss, which is in agreement with the experimental results [65].

8.3.1.3 Anisole The formation of cresol via anisole pyrolysis isomerization is one of the important processes forming phenols during coal pyrolysis. O–CH_3 bond in anisole is weak, and it might cleave initially, leading to the formation of phenoxy radical and methyl radical during anisole pyrolysis [66–68]. There are three resonance structures in phenoxy radical [69], which are shown as IM1 in Figure 8.12 [20]. The

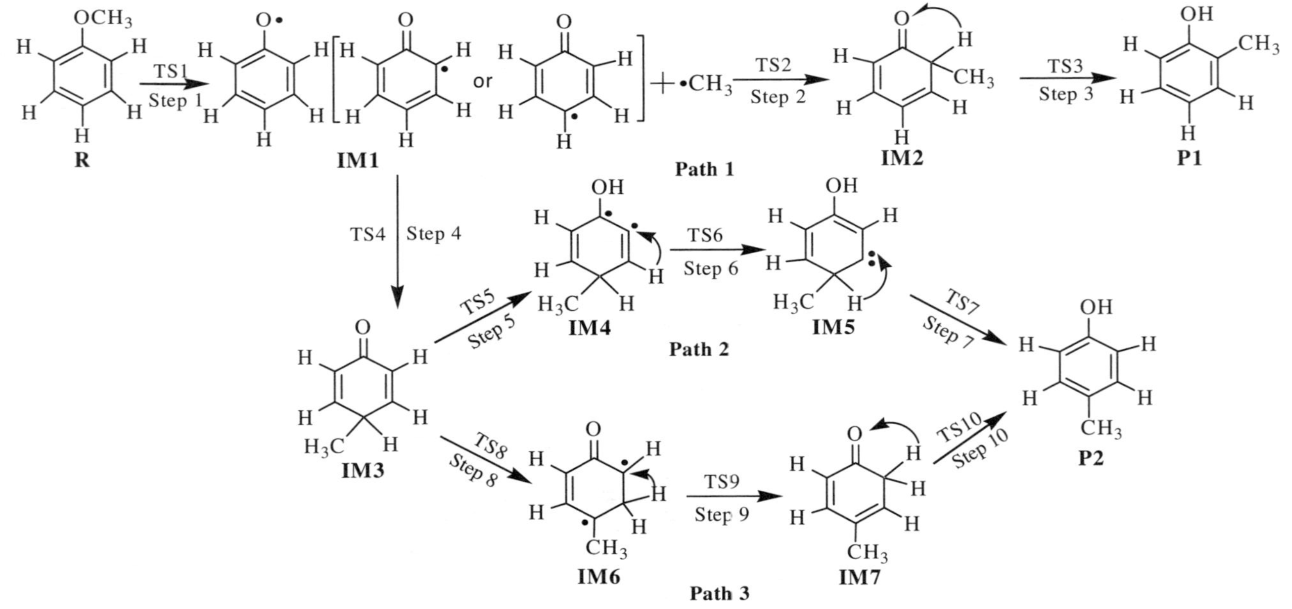

FIGURE 8.12 The pyrolysis mechanism of anisole.

resonance structures show that the *ortho* and *para* positions of O in IM1 are the combination sites with methyl except for oxygen position, which leads to the formation of *o*-methyl-cyclohexadienone and *p*-methyl-cyclohexadienone intermediates, which is shown in Figure 8.13. Besides, the electron spin densities of

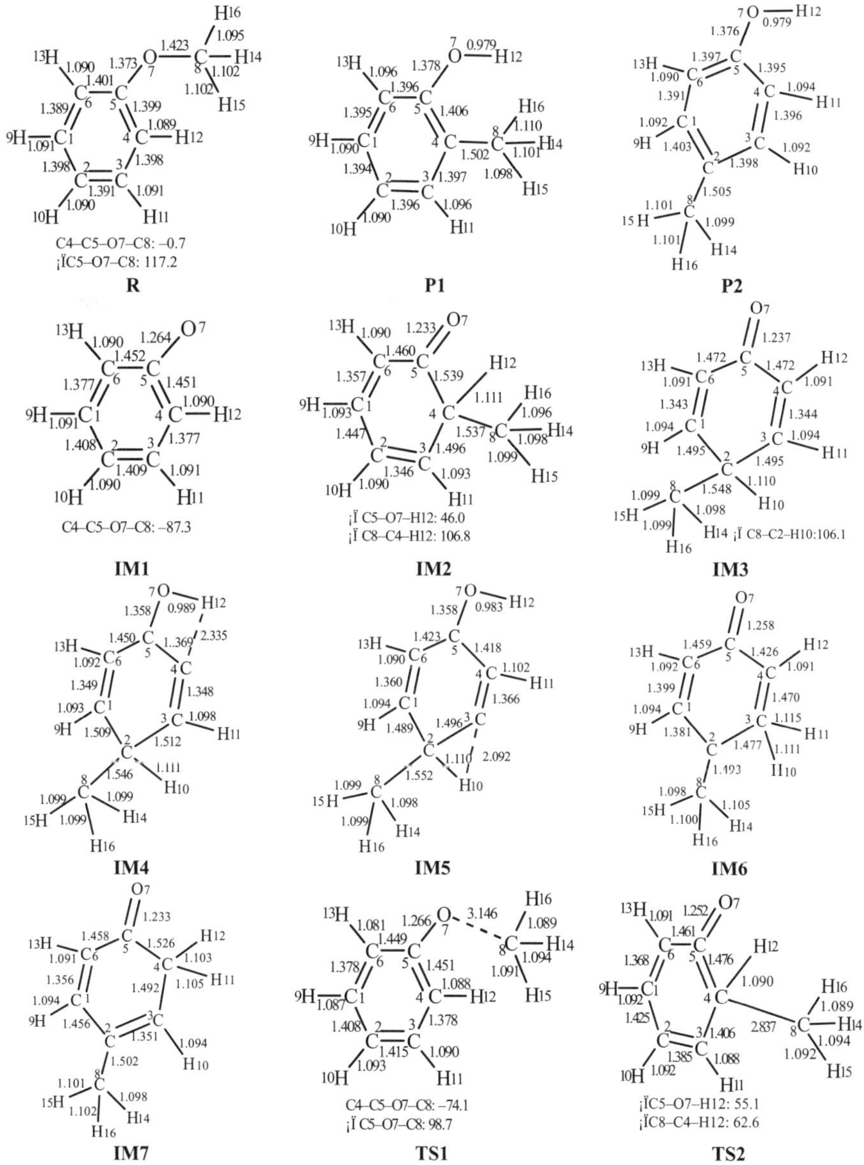

FIGURE 8.13 The optimized structures of all reactants, intermediates, transition states, and products.

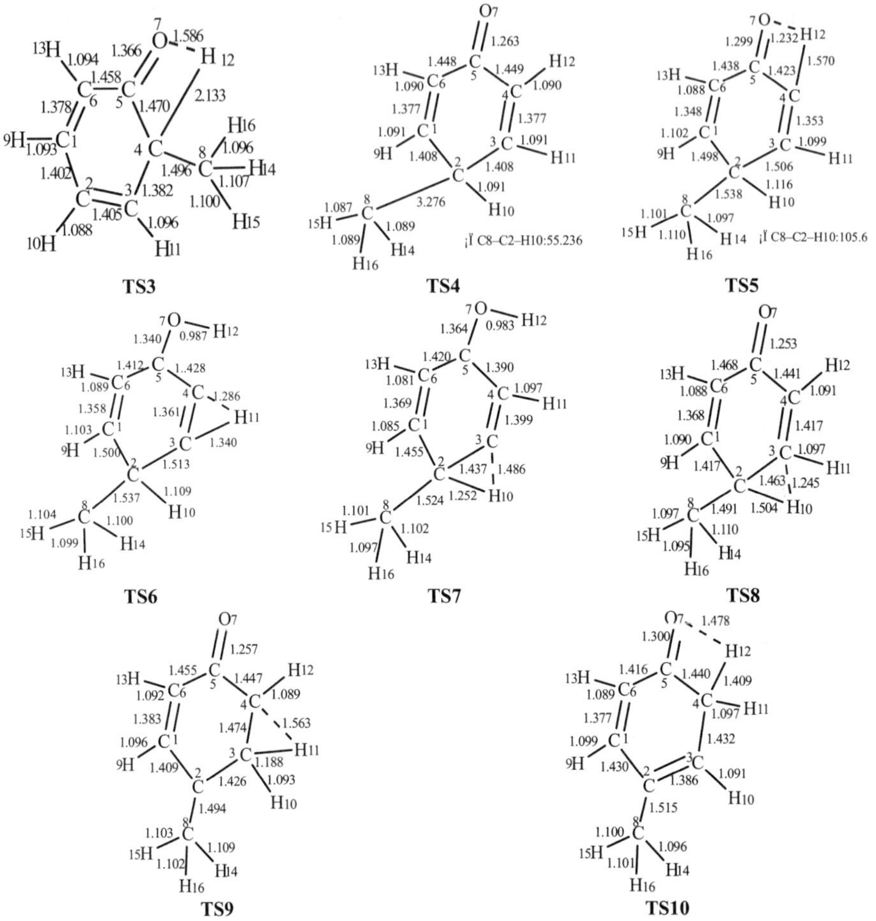

FIGURE 8.13 (*Continued*).

oxygen, *ortho*-carbon atom, and *para*-carbon atom in phenoxy radical are larger than other atoms, namely, the presence probability of unpaired electrons in these atoms is larger than that in other atoms, so the radical coupling reaction occurs by combination of a methyl radical with the phenoxy radical at the *ortho*- or *para*-position except for oxygen position. The resonance structures and the electron spin density analysis (Table 8.2) provide important basis for the existence of *o*-methyl-cyclohexadienone and *p*-methyl-cyclohexadienone intermediates. Then, these two intermediates lead to the formation of *o*- and *p*-cresol via different intramolecular hydrogen migrations. The energy profile of the stationary points for the isomerization reaction of anisole pyrolysis is shown in Figure 8.14. From this figure, we can see that the highest energy barrier of path 2 occurs at TS6, which is the largest in all species. The highest energy barriers of paths 1 and 3 occur at TS1, which is lower than that of TS6. From Table 8.3,

TABLE 8.2 Electron Spin Density Distribution of Phenoxy Radical

Atom	C1	C2	C3	C4	C5	C6	O7
Spin	−0.100	0.354	−0.099	0.261	−0.030	0.261	0.390

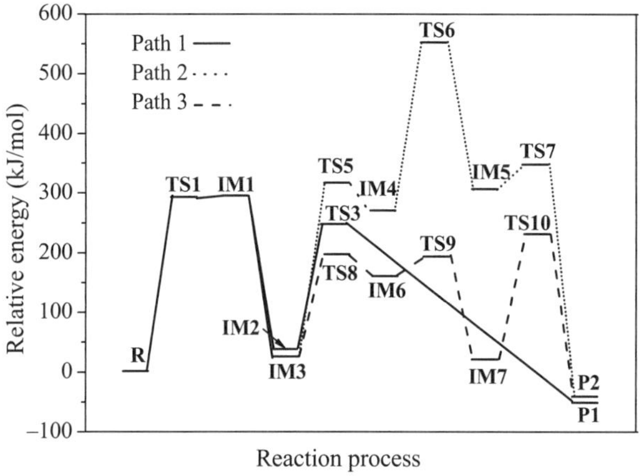

FIGURE 8.14 Energy profile of the stationary points for the isomerization reaction of anisole pyrolysis.

we can see that the activation energy of step 1 in paths 1 and 3 is the largest, and the corresponding rate constant is the smallest. The activation energy of step 5 in path 2 is similar to that of step 1, but the rate constant is smaller than that of step 1, so paths 1 and 3 are the favorable energy paths to form cresol. Besides, the same rate-determining step (step 1) is owned in paths 1 and 3, which shows that the rates of forming *o*- and *p*-cresol via anisole pyrolysis are the same. And the energies of *o*- and *p*-cresol are similar, which shows that the stabilities of two products are similar. So the yields of *o*- and *p*-cresol are similar according to the above analysis, which accords with the experimental result of Mulcahy and Williams [69].

8.3.2 The Pyrolysis Mechanisms of Nitrogen-Containing Model Compounds

Nitrogen oxides (NO_x) are environmental pollutants that lead to hazards such as acid rain and smog. As much as 75% of the total NO_x emissions comes from the process of coal utilization [70]. It has been an important problem in coal chemistry to understand the pyrolysis mechanism of nitrogen-containing structures. In addition, the role of nitrogen migration in coal is very complex and has not been understood at the molecular level [71,72]. The formation of NO_x is considered to occur in two steps: the

TABLE 8.3 The Activation Enthalpy, Activation Entropy, Activation Energy, and Rate Constant of Three Pyrolysis Paths

Elementary Reaction	$\Delta_r H^{\neq}_m$ (kJ/mol)		$\Delta_r S^{\neq}_m$ (J/(mol K))		E_a (kJ/mol)		$\ln k$ (s^{-1})	
	298.15 K	1000 K	298.15 K	1000 K	298.15 K	1000 K	298.15 K	1000 K
Path 1								
Step 1	272.64	277.59	46.62	57.79	275.11	285.90	−74.92	4.23
Step 3	198.77	198.64	−5.83	−5.23	201.25	206.95	−51.43	6.15
Path 2								
Step 1	272.64	277.59	46.62	57.79	275.11	285.90	−74.92	4.23
Step 5	270.92	268.99	−2.26	−5.51	273.40	277.30	−80.11	−2.35
Step 6	262.13	261.54	6.66	6.32	264.61	269.85	−75.49	−0.03
Step 7	34.22	30.43	−8.22	−14.54	36.70	38.74	14.66	25.26
Path 3								
Step 1	272.64	277.59	46.62	57.79	275.11	285.90	−74.92	4.23
Step 8	154.57	152.93	−0.33	−2.74	157.05	161.24	−32.94	11.94
Step 9	27.37	22.04	−22.07	−31.37	29.85	30.35	15.76	24.24
Step 10	195.10	193.92	−11.81	−13.77	197.58	202.23	−50.67	5.69

first step includes the conversion of nitrogen species in coal to NO_x precursors such as HCN and NH_3 and the following step is the oxidation under combustion conditions to form NO_x. The first step occurs usually in the primary pyrolysis of coal and is very complex [37]. Thus, a series of nitrogen-containing model compounds was chosen to investigate in further detail the formation of HCN and NH_3, the nitrogen migration processes during coal pyrolysis, and the release of polluting agents.

8.3.2.1 Pyrrole and Indole The five-membered heterocyclic pyrrole is an important structure in coals, and the thermal decomposition reaction of pyrrole is a key issue for the production of NO_x precursors. The pyrolysis of pyrrole has been studied under continuous flow (CFP) condition at 1123 K and about 40 products were identified by Patterson et al. [73]. Axworthy et al. [70] confirmed these results. Besides these brief studies of thermal decomposition of pyrrole, the detailed kinetic study for the pyrolysis of pyrrole was also carried out by Lifshitz et al. [74,75]. They studied the thermal decomposition of 1% mixture of pyrrole in argon in a single pulse shock tube (SPST) over the temperature range of 1050–1450 K and at pressure of about 3 atm. The major products *cis*-crotonitrile ($CH_3CH{=}CHCN$), HCN + C_3H_4 (propyne and allene), and allyl cyanide ($CH_2{=}CHCH_2CN$) with a branching ratio of approximately 3.5:1.5:1 were detected by the SPST technique. And the formation mechanism of these products via a common open-chain biradical intermediate was proposed (Figure 8.15). The initial step of this mechanism is a concerted N1−C5 bond cleavage and 1,2-hydrogen migration from the nitrogen to C5 to a biradical. This

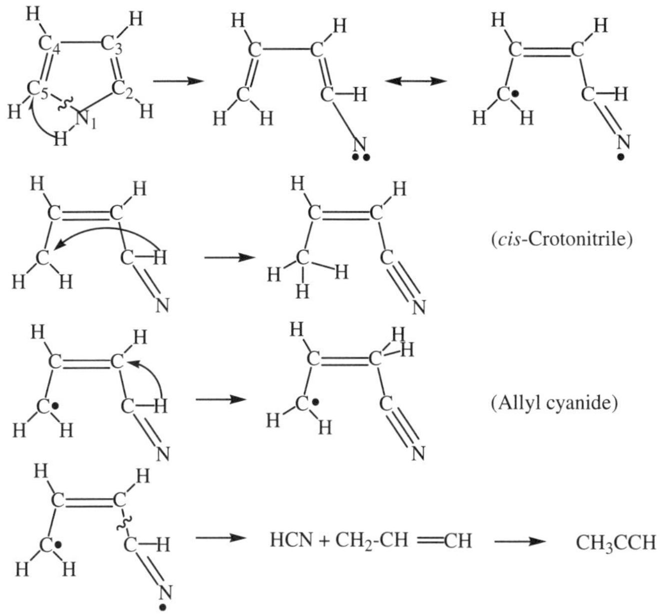

FIGURE 8.15 Pyrolysis mechanism of pyrrole proposed by Lifshitz et al.

biradical can undergo a 1,4-hydrogen migration from C2 to C5 to yield *cis*-crotonitrile. This biradical can also go through a 1,2-hydrogen migration from C2 to C3, leading to allyl cyanide, or a C2–C3 bond cleavage to form HCN and CH_2–CH=CH. The latter can rearrange to yield propyne and allene.

Mackie et al. [76,77] have used the same approach as Lifshitz et al. to study the pyrolysis of pyrrole over the temperature range of 1200–1700 K. In these studies, *cis*-crotonitrile, allyl cyanide, hydrogen cyanide, and propyne were found as main products. Further products are acetylene, ketene imine, and allene, which are in agreement with Lifshitz et al.'s proposal that the major products were formed via the same biradical intermediate, but they argued that a direct C–N bond cleavage is unfavorable because the C–N bond was very strong due to the aromatic nature of the pyrrole ring. They estimated a C–N bond dissociation energy of 377 kJ/mol, which was significantly higher than that of 310–314 kJ/mol for overall disappearance of pyrrole derived from rate constants. The difference suggested that the observed barrier may not correspond to a direct ring scission, so they proposed that the first step of pyrrole pyrolysis is a 1,2-hydrogen migration from nitrogen to C2 to form 2*H*-pyrrole (pyrrolenine), followed by C–N bond cleavage to yield the biradical intermediate, as described in Figure 8.16.

Zhai et al. [78] approved the opinion of Mackie et al.'s, and they investigated the pyrolysis mechanisms of pyrrole by density functional theory calculations. The results show that the initial step of pyrrole pyrolysis, via a 1,2-hydrogen migration from nitrogen to an adjacent carbon, leads to 2*H*-pyrrole. The activation barrier was

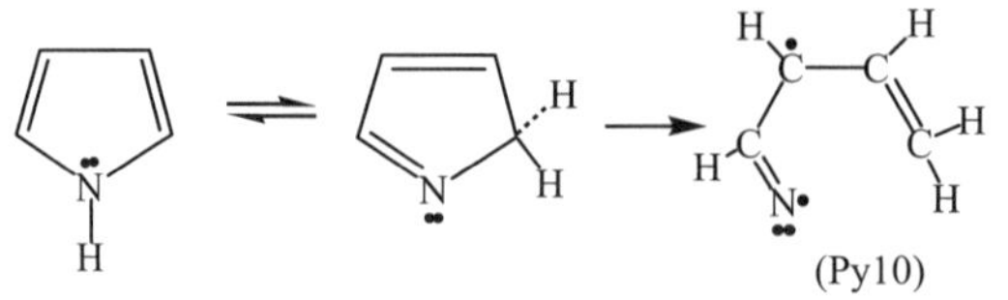

FIGURE 8.16 Mechanism of biradical intermediate formation proposed by Mackie et al.

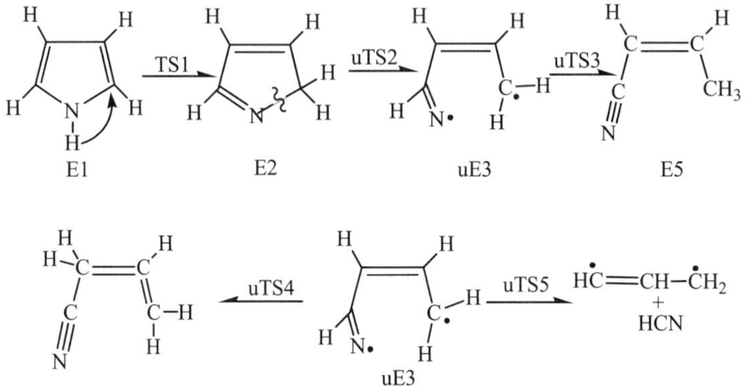

FIGURE 8.17 Pyrolysis mechanism of pyrrole proposed by Lifshitz et al. with slight modification by Mackie et al.

predicted to be 184.1 kJ/mol. The product, 2*H*-pyrrole, was predicted to be 44.4 kJ/mol higher in energy than pyrrole. Then, the ring-opening reaction of 2*H*-pyrrole and subsequent isomerization and decomposition process of the open-shell biradical were investigated, as described in Figure 8.17.

Indole contains a pyrrole ring fused to benzene. Laskin and Lifshitz [79] have investigated the thermal reaction of indole over the temperature range 1050–1650 K. Similar to pyrrole, the main thermal reactions of indole are isomerizations. Three products were found as the result of the opening of the pyrrole ring: benzyl cyanide and *o*- and *m*-tolunitriles. And the isomerization mechanism of indole was proposed as shown in Figure 8.18, which was investigated in detail by using Becke's three-parameter hybrid density functional method in conjunction with Lee–Yang–Parr's correlation functional (B3LYP) by Zhou and Liu [80]. The results showed that two tautomers of indole via hydrogen migration were predicted to be important intermediates in the reaction yielding benzyl cyanide and *o*- and *m*-tolunitriles. Different paths of hydrogen migration may lead to different products.

8.3.2.2 Pyridine Pyridine is the main form of six-membered heterocycles, which exists in chemically bound nitrogen in coal and coal-derived liquids. And above researches about pyrrole and indole show that hydrogen migration plays an important role in the pyrolysis mechanism. Therefore, a detailed theoretical study of the pyrolysis processes of pyridine via hydrogen migration was carried out in our

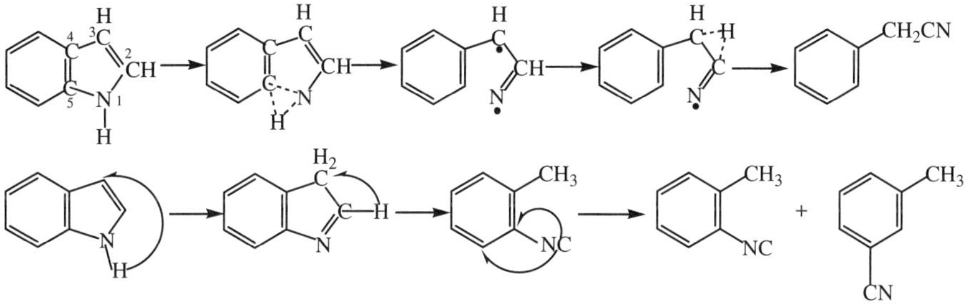

FIGURE 8.18 Pyrolysis mechanism of indole proposed by Laskin and Lifshitz.

work. Migration of hydrogen atom from pyridine ring was proposed as the first step of the pyridine pyrolytic decomposition; then, different hydrogen migrations will happen. The following possible two-paralleled reaction paths are described schematically in Figure 8.19.

The calculated results show that the 1,2-hydrogen migration of pyridine via TS1 leading to IM1 is thought to be preferred reaction path. TS1 is predicted to be 337.85 kJ/mol higher than pyridine. The 1,3- and 1,4-hydrogen migration of pyridine leads to the formation of IM2 and IM3 via TS2 and TS3, respectively, which are predicted to be 491.23 and 486.55 kJ/mol higher than pyridine. Therefore, IM2 and IM3 cannot be produced due to their relatively higher energy barrier.

Starting from IM1, the results indicate that the initial step is 1,2-hydrogen migration to yield an intermediate, 2*H*-pyridine. The formation of N$-$$CH_2$ bond goes through a transition state TS4. 2*H*-pyridine in energy is predicted to be

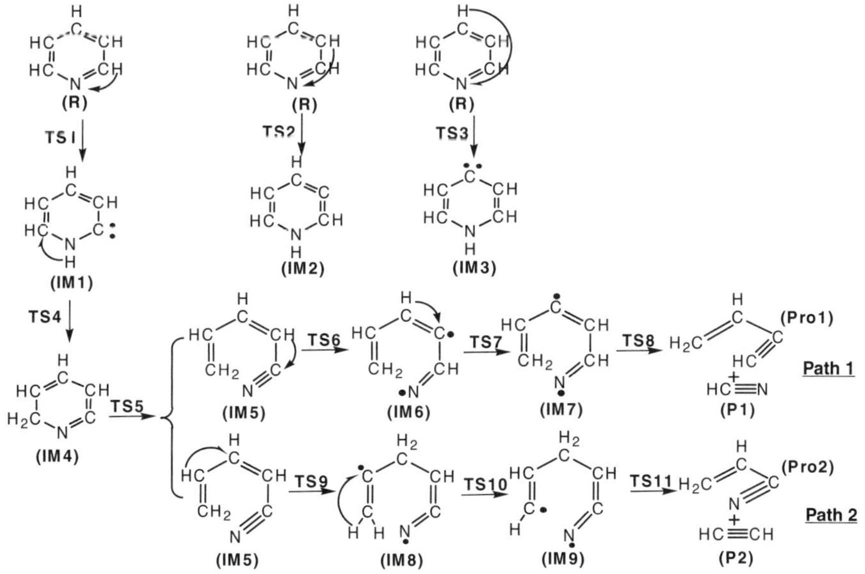

FIGURE 8.19 The decomposition mechanism of pyridine via hydrogen migration.

327.81 kJ/mol higher than pyridine with an activation barrier of 131.19 kJ/mol. Subsequent C_6–N_1 cleavage results in the formation of an important intermediate IM5, and the transition state of this C–N cleavage, TS5, is only 8.67 kJ/mol higher than 2*H*-pyridine. Then, the successive hydrogen migration from intermediate IM5 through different pathways gave finally hydrogen cyanide, acetylene, vinylacetylene, and acrylonitrile. In the two-parallel paths, R $\rightarrow$ IM5 is the common reaction step. From the intermediate IM5 beginning, path 1 is a three-step reaction with two intermediates (IM6, IM7) and three transition states (TS6, TS7, and TS8). Comparing the activation energy between three-step reactions, it can be found that IM5 $\rightarrow$ TS6 is the rate-controlling step of path 1 (R $\rightarrow$ P1) and corresponding energy barrier is *E* (TS6) – *E*(IM5) = 389.28 kJ/mol. Path 2 (R $\rightarrow$ P2) is the same result as path 1. IM5 $\rightarrow$ TS9 is the rate-controlling step of path 2 and corresponding energy barrier is *E*(TS9) – *E*(IM5) = 427.02 kJ/mol. Moreover, analyzing and comparing the activation energies of the two-parallel pathways, on the whole, it is found that the potential highest energy barrier of path 1 is smaller than that of path 2. Obviously, path 1 is the main reaction path for the initial pyrolysis process of pyridine and path 2 is the competitive pathway due to the small difference between two activation barriers of two-parallel paths.

Ninomiya et al. [81] investigated the formation of 2-, 3-, and 4-pyridyl radicals by C–H bond fission from pyridine. The calculation results show that the heats of formation are 532.20, 551.87, and 550.61 kJ/mol, which correspond to 2-, 3-, and 4-pyridyl radicals, respectively. It can be seen that the energy of hydrogen migration reaction is smaller than that of C–H bond dissociation in pyridine and the sequence of energy does not change.

The above researches show that intramolecular hydrogen migration in the decomposition routes of pyridine plays an important role to make the reaction go easily and adjust different distribution of products in pyrolysis reaction.

8.3.2.3 2-Picoline Considering that 2-picoline not only contains conjugated six-membered ring representing the aromatic structure of coal but also includes methyl next to nitrogen as bridge bond and aliphatic branched chain in coal, 2-picoline is chosen as the nitrogen-containing model compound for describing coal pyrolysis.

There have been several previous investigations about the kinetics of 2-picoline pyrolysis. The pyrolysis of 2-picoline in dilute mixtures with argon using single pulse shock was investigated by Doughty et al. [82,83], and it was found that the major products were acetylene, methane, hydrogen, HCN, and cyanoacetylene. They also successfully modeled the pyrolysis process with 70 free radical reactions so as to provide three mechanisms, including the mechanism via a bicyclic norbornadiene-type intermediate (N-containing analogue of norbornadiene), the mechanism via a seven-membered ring intermediate, and three direct ring-opening mechanisms, in which the mechanism via N-containing analogue of norbornadiene (Figure 8.20) was widely authorized. Ikeda and Mackie [84] also investigated the pyrolysis of 2-picoline; however, the kinetic data were not the same as those obtained from the experiment of Doughty et al., which shows that the pyrolysis is a complex process, and the pyrolysis mechanism needs to be understood. However, we are not aware of any

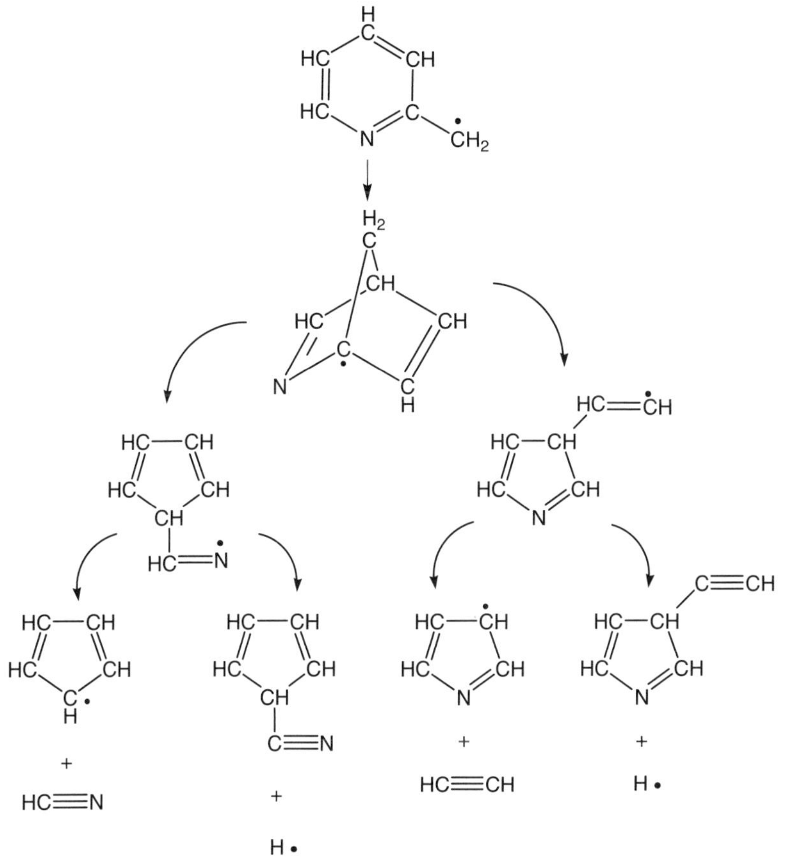

FIGURE 8.20 The pyrolysis mechanism of 2-picoline via the intermediate of N-containing analogue of norbornadiene proposed by Doughty et al.

detailed quantum mechanism study to examine the validity of the mechanism because of the lack of reliable thermochemical data for many species involved, and the arguments supporting the above mechanism were usually based on crude analogy and estimation.

The formation process of the main product HCN during 2-picoline pyrolysis was investigated using detailed DFT method, as shown in Scheme 8.1 [85]. Five transition states (TS1, TS2, TS3, TS4, and TS5) and four intermediates (IM1, IM2, IM3, and •HCNH) were determined during the pyrolysis of 2-picoline via the intermediate of N-containing analogue of norbornadiene, and these structures are shown in Figure 8.21.

$$\mathrm{R} \xrightarrow[\text{Step1}]{\text{TS1}} \mathrm{IM1} \xrightarrow[\text{Step 2}]{\text{TS2}} \mathrm{IM2} \xrightarrow[\text{Step 3}]{\text{TS3}} \mathrm{IM3} \xrightarrow[\text{Step 4}]{\text{TS4}} \mathrm{C_5H_5{\cdot}+HCNH{\cdot}} \xrightarrow[\text{Step 5}]{\mathrm{C_5H_5{\cdot}+TS5}} \mathrm{C_5H_5{\cdot}+HCN+H{\cdot}} \xrightarrow[\text{Step 6}]{} \mathrm{C_5H_6+HCN}$$

SCHEME 8.1 The formation process of HCN during 2-picoline pyrolysis.

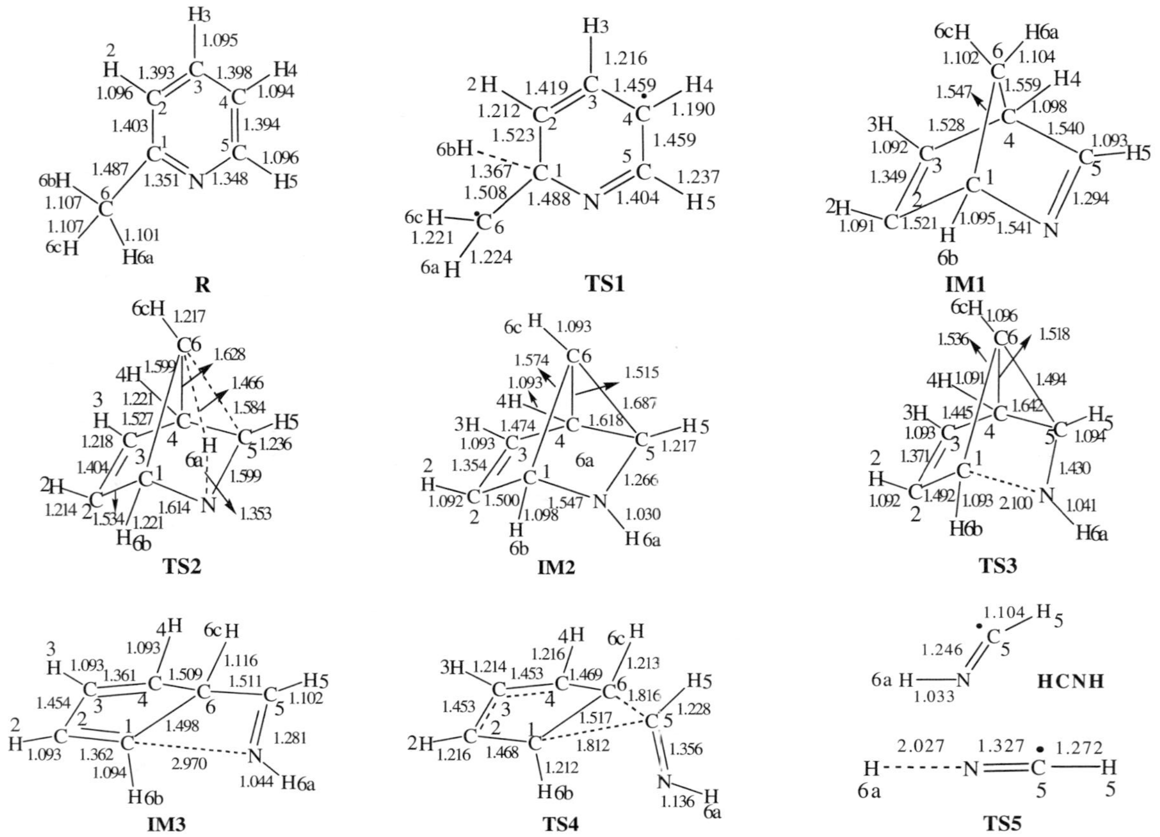

FIGURE 8.21 The optimized equilibrium structures, transition states, and their atomic numbers during 2-picoline pyrolysis.

TABLE 8.4 The Activation Energy and Rate Constant of Every Elementary Reaction During 2-Picoline Pyrolysis

Elementary Reaction	E_a		ln k	
	298.15 K	875 K	298.15 K	875 K
Step 1	428.66	436.88	−144.354	−29.877
Step 2	571.23	579.78	−199.289	−46.720
Step 3	51.49	28.84	9.944	27.377
Step 4	220.70	221.91	−61.507	−2.918
Step 5	269.99	277.74	−77.325	−4.885

In step 1, N-containing analogue of norbornadiene intermediate is formed followed by H6b migrating from C6 to C1 and C6 connecting with C4 via TS1 with an energy barrier of 428.66 kJ/mol. Then, H6a in IM1 migrates from C6 to N leading to the formation of IM2 via TS2. A higher energy barrier of 571.23 kJ/mol is need in this step because N-containing analogue of norbornadiene intermediate is very stable. In step 3, an energy barrier of 51.49 kJ/mol is needed, which shows that this step occurs easily. Subsequently, C5–C6 bond is cleaved via TS4 resulting in the formation of HCNH radical and cyclopentadiene radical, and an energy barrier of 220.70 kJ/mol is needed. Finally, H6a in HCNH radical is abstracted, resulting in the formation of HCN—an energy of 269.99 kJ/mol is needed in this step—and cyclopentadiene is formed via cyclopentadiene radical combining with H radical.

The activation energy and rate constant of every elementary reaction during 2-picoline pyrolysis are listed in Table 8.4. We can see that the activation energy of step 2 (IM1 → IM2) is the largest and the rate constant is the smallest in all reactions. So it is confirmed that hydrogen migration in step 2 is the rate-determining step during the pyrolysis of 2-picoline.

8.3.2.4 *Quinoline and Isoquinoline* The pyrolysis of quinoline and isoquinoline was studied at 923–1223 K by Patterson et al. [86] to evaluate the relative stabilities. Bruinsma et al. [87] investigated the pyrolysis of quinoline in a flow reactor, and it was concluded that quinoline was more stable and benzene was identified as one of pyrolysis products. Furthermore, benzonitrile was observed as a pyrolysis product of quinoline [70]. However, the kinetic data for distribution of pyrolysis products or for the pyrolysis mechanism were absent in their experiments. Laskin and Lifshitz [88] investigated the pyrolysis of quinoline and isoquinoline in a single pulse shock tube over the temperature range of 1275–1700 K in detail. The experimental results showed that pyrolysis products were $HC{\equiv}CCN$, C_6H_6, C_2H_2, and C_6H_5CN, and the total disappearance rates of quinoline and isoquinoline were the same whether the reactant was quinoline or isoquinoline. The tautomerism between quinoline and isoquinoline did not occur; neither traces of isoquinoline in shock-heated mixture of quinoline nor traces of quinoline in shock-heated mixture of isoquinoline were found. To illustrate this fact, 1-indene imine radical was assumed as an important intermediate, and the pyrolysis mechanisms of quinoline and isoquinoline via the intermediate were proposed. Winkler et al. [89] investigated the continuous flow pyrolysis of quinoline and isoquinoline at 1173 K; the same major products as those of Laskin and Lifshitz [88] were identified by GC–MS, and they also agreed with the mechanism proposed by Laskin and Lifshitz.

The presence of 1-indene imine has not been identified experimentally, although it is widely believed to be an important intermediate during pyrolysis of quinoline and isoquinoline. In order to better understand the fact that the total disappearance rates of quinoline and isoquinoline are the same and the pyrolysis products are also the same for both reactants. In order to verify the importance of 1-indene imine intermediate, a study on the pyrolysis mechanism of quinoline and isoquinoline was done. The pyrolysis mechanisms of quinoline and isoquinoline leading to P1 ($HC{\equiv}CCN$, C_6H_6) and P2 (C_2H_2, C_6H_5CN) via 1-indene imine intermediate are proposed in Figure 8.22 [21].

The Paths of Quinoline Pyrolysis Yielding C_6H_6 and $HC{\equiv}CCN$ Two possible reaction paths (paths 1A and 1B) are found for the pyrolysis: quinoline (R) → C_6H_6 + $HC{\equiv}CCN$ (P1), as shown in Scheme 8.2.

First, α-H in quinoline (R) migrates to N to yield IM1. The calculated activation energy of the step via TS1 is 344.99 kJ/mol, which is close to the experimental activation energy of 326.00 kJ/mol [87] and 316.47 kJ/mol [88]. Thus, the theoretical calculation seems to support the proposed hydrogen migration but not the scheme of direct bond cleavage for the initial step of quinoline pyrolysis. Thus, the reaction undergoes a series of hydrogen migration leading to P1. The calculated results show that the rate-determining steps of paths 1A and 1B are steps 1 and 6, with activation energies of 344.99 and 361.93 kJ/mol, respectively. There is only a difference of 16.93 kJ/mol in the activation energy, so paths 1A and 1B may be considered to be the coexistent competitive reaction paths.

The Paths of Isoquinoline Pyrolysis Yielding C_6H_5CN and $HC{\equiv}CH$ Two possible reaction paths (paths 2A and 2B) are also found for the pyrolysis: isoquinoline (Ri) → C_6H_5CN + $HC{\equiv}CH$ (P2), as shown in Scheme 8.3.

Isoquinoline has two α-H next to the N atom, but their chemical surroundings are different. In the formation process of IM1i, hydrogen migration destroys the conjugate structure of the N-containing six-membered-ring, so it needs to overcome the higher energy barrier of 338.94 kJ/mol. Then the reaction undergoes a series of hydrogen migration leading to P2. The calculated results show that the rate-determining steps of paths 2A and 2B are steps 1i and 7i, with activation energies of 338.94 and 370.76 kJ/mol, respectively. There is only a difference of 31.82 kJ/mol in activation energy, so paths 2A and 2B may be considered to be the coexistent competitive reaction paths like the case of paths 1A and 1B.

The Paths of Quinoline Pyrolysis Yielding C_6H_5CN and $HC{\equiv}CH$ The formation process of P2 (C_6H_5CN, $HC{\equiv}CH$) during quinoline pyrolysis has two different paths (paths 3A and 3B), which is shown in Scheme 8.4.

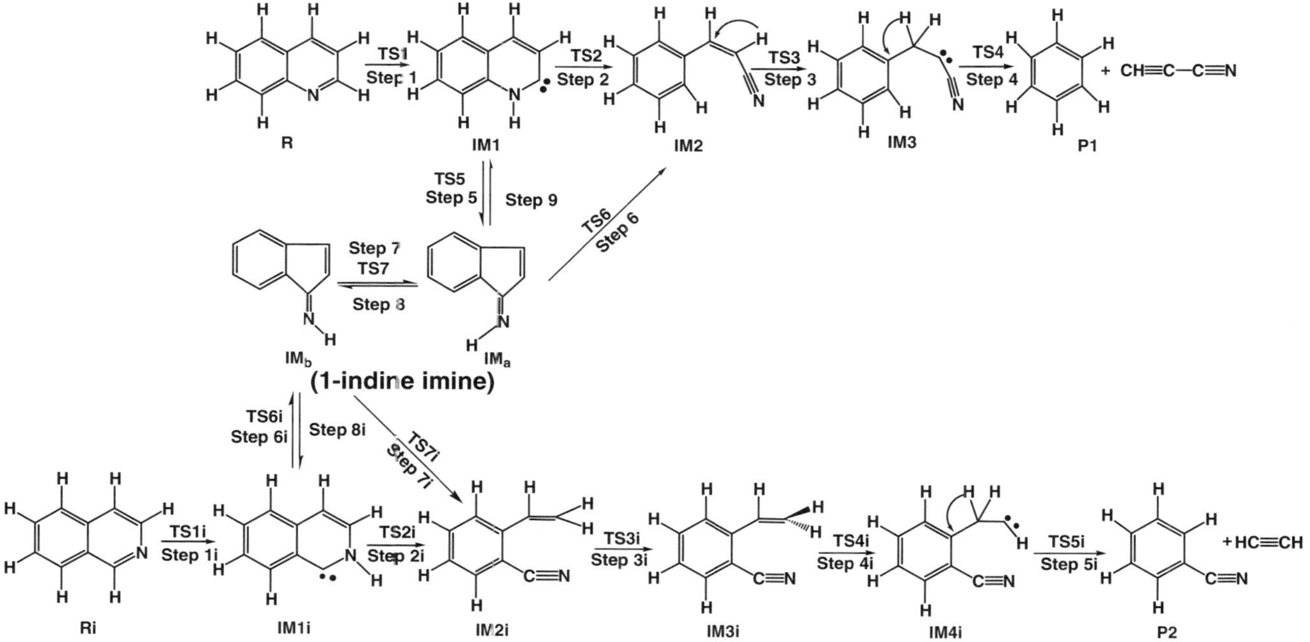

FIGURE 8.22 The pyrolysis mechanisms of quinoline (R) and isoquinoline (Ri).

$$\text{Path 1A: } R \xrightarrow[\text{Step 1}]{\text{TS1}} IM1 \xrightarrow[\text{Step 2}]{\text{TS2}} IM2 \xrightarrow[\text{Step 3}]{\text{TS3}} IM3 \xrightarrow[\text{Step 4}]{\text{TS4}} P1$$

$$\text{Path 1B: } R \xrightarrow[\text{Step 1}]{\text{TS1}} IM1 \xrightarrow[\text{Step 5}]{\text{TS5}} IM_a \xrightarrow[\text{Step 6}]{\text{TS6}} IM2 \xrightarrow[\text{Step 3}]{\text{TS3}} IM3 \xrightarrow[\text{Step 4}]{\text{TS4}} P1$$

SCHEME 8.2 The reaction schemes for pyrolysis of quinoline (R) leading to P1 (C_6H_6, $HC{\equiv}CCN$).

$$\text{Path 2A: } Ri \xrightarrow[\text{Step 1i}]{\text{TS1i}} IM1i \xrightarrow[\text{Step 2i}]{\text{TS2i}} IM2i \xrightarrow[\text{Step 3i}]{\text{TS3i}} IM3i \xrightarrow[\text{Step 4i}]{\text{TS4i}} IM4i \xrightarrow[\text{Step 5i}]{\text{TS5i}} P2$$

$$\text{Path 2B: } Ri \xrightarrow[\text{Step 1i}]{\text{TS1i}} IM1i \xrightarrow[\text{Step 6i}]{\text{TS6i}} IM_b \xrightarrow[\text{Step 7i}]{\text{TS7i}} IM2i \xrightarrow[\text{Step 3i}]{\text{TS3i}} IM3i \xrightarrow[\text{Step 4i}]{\text{TS4i}} IM4i \xrightarrow[\text{Step 5i}]{\text{TS5i}} P2$$

SCHEME 8.3 The reaction schemes for pyrolysis of isoquinoline (Ri) leading to P2 (C_6H_5CN, $HC{\equiv}CH$).

$$\text{Path 3A: } R \xrightarrow[\text{Step 1}]{\text{TS1}} IM1 \xrightarrow[\text{Step 5}]{\text{TS5}} IM_a \xrightarrow[\text{Step 8}]{\text{TS7}} IM_b \xrightarrow[\text{Step 7i}]{\text{TS7i}} IM2i \xrightarrow[\text{Step 3i}]{\text{TS3i}} IM3i \xrightarrow[\text{Step 4i}]{\text{TS4i}} IM4i \xrightarrow[\text{Step 5i}]{\text{TS5i}} P2$$

$$\text{Path 3B: } R \xrightarrow[\text{Step 1}]{\text{TS1}} IM1 \xrightarrow[\text{Step 5}]{\text{TS5}} IM_a \xrightarrow[\text{Step 8}]{\text{TS7}} IM_b \xrightarrow[\text{Step 8i}]{\text{TS6i}} IM1i \xrightarrow[\text{Step 2i}]{\text{TS2i}} IM2i \xrightarrow[\text{Step 3i}]{\text{TS3i}} IM3i \xrightarrow[\text{Step 4i}]{\text{TS4i}} IM4i \xrightarrow[\text{Step 5i}]{\text{TS5i}} P2$$

SCHEME 8.4 The reaction schemes for pyrolysis of quinoline (R) leading to P2 (C_6H_5CN, $HC{\equiv}CH$).

Paths 3A and 3B have the same steps as path 1B from quinoline (R) to IM_a. Then IM_a goes through a quick conformational tautomerization to IM_b with an activation energy of 113.98 kJ/mol. The calculated results show that the activation energies of the rate-determining steps are 370.76 kJ/mol in path 3A and 371.37 kJ/mol in path 3B, so paths 3A and 3B are considered to be the coexistent competitive paths.

The Paths of Isoquinoline Pyrolysis Yielding C_6H_6 and $HC{\equiv}CCN$ The formation process of P1 (C_6H_6, $HC{\equiv}CCN$) during isoquinoline pyrolysis also has two different paths (paths 4A and 4B), which is shown in Scheme 8.5.

Path 4A has the same steps as path 2B from isoquinoline (Ri) to IM_b, and then goes through a quick conformational tautomerization to IM_a with an activation energy of 113.96 kJ/mol. In step 6, H17 in IM_a migrates from N10 to C4 followed by a concerted C4–C9 bond cleavage to form IM2 with an activation energy of 361.93 kJ/mol. From IM2 to P1, paths 4A and 1A have the same steps.

$$\text{Path 4A: } Ri \xrightarrow[\text{Step1i}]{\text{TS1i}} IM1i \xrightarrow[\text{Step 6i}]{\text{TS6i}} IM_b \xrightarrow[\text{Step 7}]{\text{TS7}} IM_a \xrightarrow[\text{Step 6}]{\text{TS6}} IM2 \xrightarrow[\text{Step 3}]{\text{TS3}} IM3 \xrightarrow[\text{Step 4}]{\text{TS4}} P1$$

$$\text{Path 4B: } Ri \xrightarrow[\text{Step 1i}]{\text{TS1i}} IM1i \xrightarrow[\text{Step 6i}]{\text{TS6i}} IM_b \xrightarrow[\text{Step 7}]{\text{TS7}} IM_a \xrightarrow[\text{Step 9}]{\text{TS5}} IM1 \xrightarrow[\text{Step 2}]{\text{TS2}} IM2 \xrightarrow[\text{Step 3}]{\text{TS3}} IM3 \xrightarrow[\text{Step 4}]{\text{TS4}} P1$$

SCHEME 8.5 The reaction schemes for pyrolysis of isoquinoline (Ri) leading to P1 (C_6H_6, $HC{\equiv}CCN$).

Path 4B has the same steps as path 4A from isoquinoline (Ri) to IM_a. Subsequently, in step 9, C4 in IM_a links to N10 followed by a concerted C4–C9 bond cleavage to form IM1, which is the reverse reaction of step 5 and has an activation energy of 383.48 kJ/mol. From IM1 to P1, paths 4B and 1A have the same steps.

The activation energies of rate-determining steps are 361.93 kJ/mol in path 4A and 383.48 kJ/mol in path 4B, so paths 4A and 4B may be considered to be the coexistent competitive paths.

The Analysis of Total Reaction Paths According to the above analysis of each path, it can be concluded that the most feasible formation processes of P1 (C_6H_6, $HC{\equiv}C{-}C{\equiv}N$) and P2 ($C_6H_5C{\equiv}N$, $HC{\equiv}CH$) during quinoline and isoquinoline pyrolysis should involve two parts: leading to the intermediate 1-indene imine with two conformational tautomers and yielding P1 and P2 from the common intermediate.

Quinoline goes through three paths (paths 1B, 3A, and 3B) toward IM_a and isoquinoline goes through other three paths (paths 2B, 4A, and 4B) toward IM_b by the kinetic mechanism involving the intramolecular hydrogen migration, in which the highest energy barriers of 344.99 kJ/mol in step 1 and 338.94 kJ/mol in step 1i have to be overcome, which are close to the experimental activation energies 326.00 kJ/mol [87] and 316.47 kJ/mol [88]. Because of the similar activation energies, the production rates of 1-indene imine are almost the same, whether the original reactant is quinoline or isoquinoline. Then there is a quick conformational tautomeric equilibrium between IM_a and IM_b with activation energies of 113.98 and 113.96 kJ/mol.

The common intermediate 1-indene imine goes through three paths (paths 1B, 4A, and 4B) toward P1, and through other three paths (paths 2B, 3A, and 3B) toward P2. The calculated results show that the highest energy barriers yielding P1 and P2 are 361.93 kJ/mol in step 6 and 370.76 kJ/mol in step 7i. The highest energy barriers are almost the same, slightly higher than the experimental results [87,88]. Thus, it is concluded that the total disappearance rates of quinoline and isoquinoline are the same and the composition of the pyrolysis products is also the same whether the original reactant is quinoline or isoquinoline, which is in agreement with the experimental results [88,89].

The above researches show that the conformational tautomerism of 1-indene imine intermediate plays an important role in the kinetic mechanisms of pyrolysis of quinoline and isoquionline. 1-indene imine intermediate determines the composition of the pyrolysis products to be the same, and also determines the total disappearance rates of the reactants to be the same whether the original reactant is quinoline or isoquinoline. The intramolecular hydrogen migration is an important reaction step, which appears widely in the paths of the pyrolysis mechanism.

8.3.3 The Pyrolysis Mechanisms of Sulfur-Containing Model Compounds

The presence of sulfur in coal makes its extensive utilization more challenging. The pyrolysis of coal is the first step in the coal gasification, coal liquefaction, coal combustion, and so on. A significant amount of the sulfur is released with these products as H_2S, COS, and CS_2 during coal pyrolysis. Other forms include substituted and unsubstituted thiophenes, benzothiophenes, and dibenzothiophenes. Of all the volatile sulfur species, H_2S is the most abundant. These sulfur-containing products not only pollute the environment but also have serious influences on subsequent production processes, such as leading to the poisoning of catalysts, the corrosion of equipment, and so on. Therefore, to solve these problems, the desulfurization of coal has long been an important topic of investigation [90]. A number of methods have been developed to study the desulfurization, such as biodesulfurization [91], electrochemistry desulfurization [92], and chemistry desulfurization [93]. However, due to the fact that the formation mechanism of sulfur-containing compounds during coal pyrolysis is still not perfectly understood, an effective and economic process to remove sulfur has not been found to date. Therefore, understanding the role of sulfur migration and release during coal pyrolysis remains highly relevant.

8.3.3.1 Thiophene In order to study the formation of H_2S from the pyrolysis of sulfur-containing in coal, thiophene (C_4H_4S) is often used as an organosulfur model [94]. Cullis and Norris [95] found that H_2S was formed in large quantities at 1323 K during thiophene pyrolysis. Winkler et al. [33] found that the amount of H_2S increased strongly above 1123 K during the continuous flow pyrolysis of thiophene, and fragmental products are also formed from thiophene, such as C_4H_2. Memon et al. [34] also detected major gaseous product H_2S when studying the pyrolysis of thiophene over the temperature range 1598–2022 K. All of these show that H_2S was the major product of thiophene pyrolysis, but no kinetics information was obtained for the formation of H_2S. Recently, a study of the unimolecular decomposition of thiophene was carried out to investigate the formation of H_2S [96]. Energies and the bond Mulliken population of several decomposition pathways were analyzed, but no reaction rate constants were calculated and thus no comparison with different paths was made. So it is necessary to make a comprehensive quantum chemical study in conjunction with kinetic modeling in order to obtain rate constants for the decomposition reactions and to elucidate the important pathways in the thermal decomposition of thiophene.

Some experiments [34,97] supported that the C–S bond fission preceded other reactions and formed biradical during thiophene pyrolysis. The work done by Martoprawiro et al. [77] indicates that the bond between carbon and heteroatom is strong. Indeed, the calculated C–S bond energy in thiophene was 560.00 kJ/mol [98]. Other experiments [87,99] supported that the C–H bond was weak, and might cleave initially. So an α-H atom abstracted from thiophene was investigated first in our study, and the calculated bond dissociation energy was about 476.64 kJ/mol, which was in agreement with the C–H bond energy of 488.18 kJ/mol [100], but significantly higher than the activation energy of the reaction derived from kinetic data [76]. So the mechanism of α-H migration to S as the first step was proposed in paths 1 and 2. Besides, Hore and Russell's research showed that 1,2-hydrogen migration was the most probable step (activation energy 300.0 kJ/mol) before bond fission of thiophene [98], so path 3 was proposed. The possible reaction mechanism of thiophene

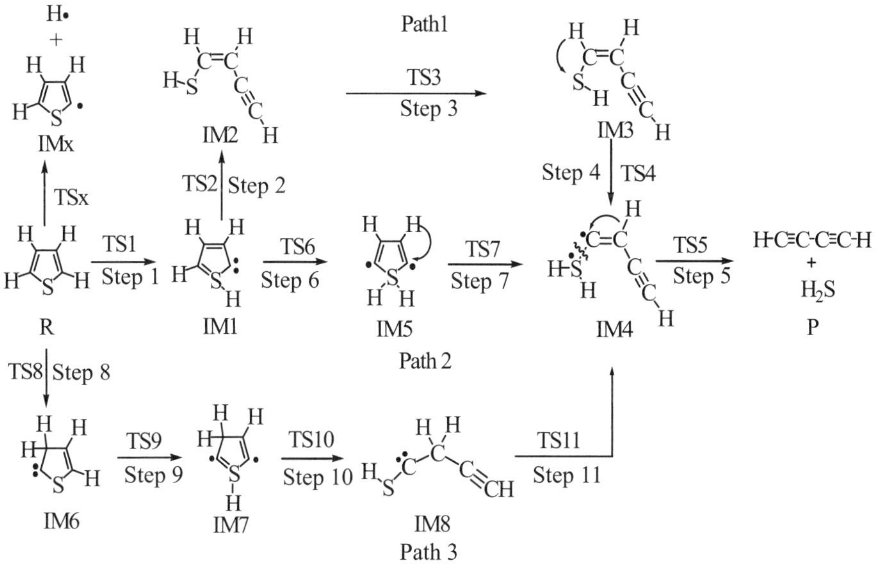

FIGURE 8.23 The proposed reaction mechanism of thiophene pyrolysis.

pyrolysis is shown in Figure 8.23 [22]; the optimized structures and their atom numbers of reactants, intermediates, transition states, and products are shown in Figure 8.24. The kinetic parameters of every elementary step are shown in Table 8.5, and the energy profile of the stationary points for the three paths of thiophene pyrolysis is shown in Figure 8.25.

In path 1, it can be seen that the activation energy of step 3 (IM2 $\longrightarrow$ IM3) is the smallest, and the rate constant $\ln k_3$(298.15 K) is higher than others. It shows that step 3 occurs easily. Meanwhile, from the molecular structures of transition state and the reactant, we can see that this step is a *cis–trans*-isomerization process, and neither bond cleavage nor new bond formation, which facilitates the reaction of step 4. The activation energy of step 1 (R $\longrightarrow$ IM1) is the highest, and the rate constant $\ln k_1$ is the smallest, so it can be concluded that step 1 is the rate-determining step in path 1. It is an H migration process, and the aromaticity of thiophene destroyed. The increase in reaction temperature is in favor of the pyrolysis reaction of thiophene under isothermal and isotonic conditions by comparing to the rate constants at different temperatures.

In path 2, from the activation energy and rate constant, we can see that the activation energy of step 6 is the highest, and the rate constant $\ln k_6$ is the smallest, so it can be concluded that the step 6 is the rate-determining step in path 2. In path 3, the activation energy of step 9 is the highest.

Comparing the three paths, it is concluded that the activation energy of rate-determining step (step 1) of path 1 is 52.09 kJ/mol smaller than that of step 6 in path 2, and is 80.72 kJ/mol smaller than that of step 9 in path 3; both $\ln k_6$ and $\ln k_9$ are smaller than $\ln k_1$, so we conclude that path 1 is the favorable energy path, namely, α-H migrates to S first; then β-H migrates to α-C followed by a concerted C–S bond cleavage resulting in the cyclic structure of thiophene turning into a chain structure. Finally, H_2S and butadiyne are formed via hydrogen migration.

8.3.3.2 Benzenethiol The organic sulfur in coal is eliminated as H_2S mainly during coal pyrolysis [101], especially the unstable thiol compounds, which eliminate H_2S at 973–1123 K [35]. Benzenethiol (C_6H_5SH) is the simplest aromatic thiol [102], which is often used to investigate the desulfurization of a typical thioalcohol compound [103–105]. It was investigated by flow pyrolysis at 973 K and vacuum pyrolysis at 1073 K, and benzene was identified as one of the major products [40]. Benzene as the major product during benzenethiol decomposition in the presence of CO or CO/H_2O was also investigated by Van Buren et al. [106]. The decomposition of benzenethiol was studied in a flow reactor by Bruinsma et al. [107], and it can be concluded that the intramolecular hydrogen migration in phenol resulting in the formation of cyclopentadiene demonstrated by Cypres [108] was not observed for benzenethiol. The mass spectrum of the molecular ions of benzenethiol produced by

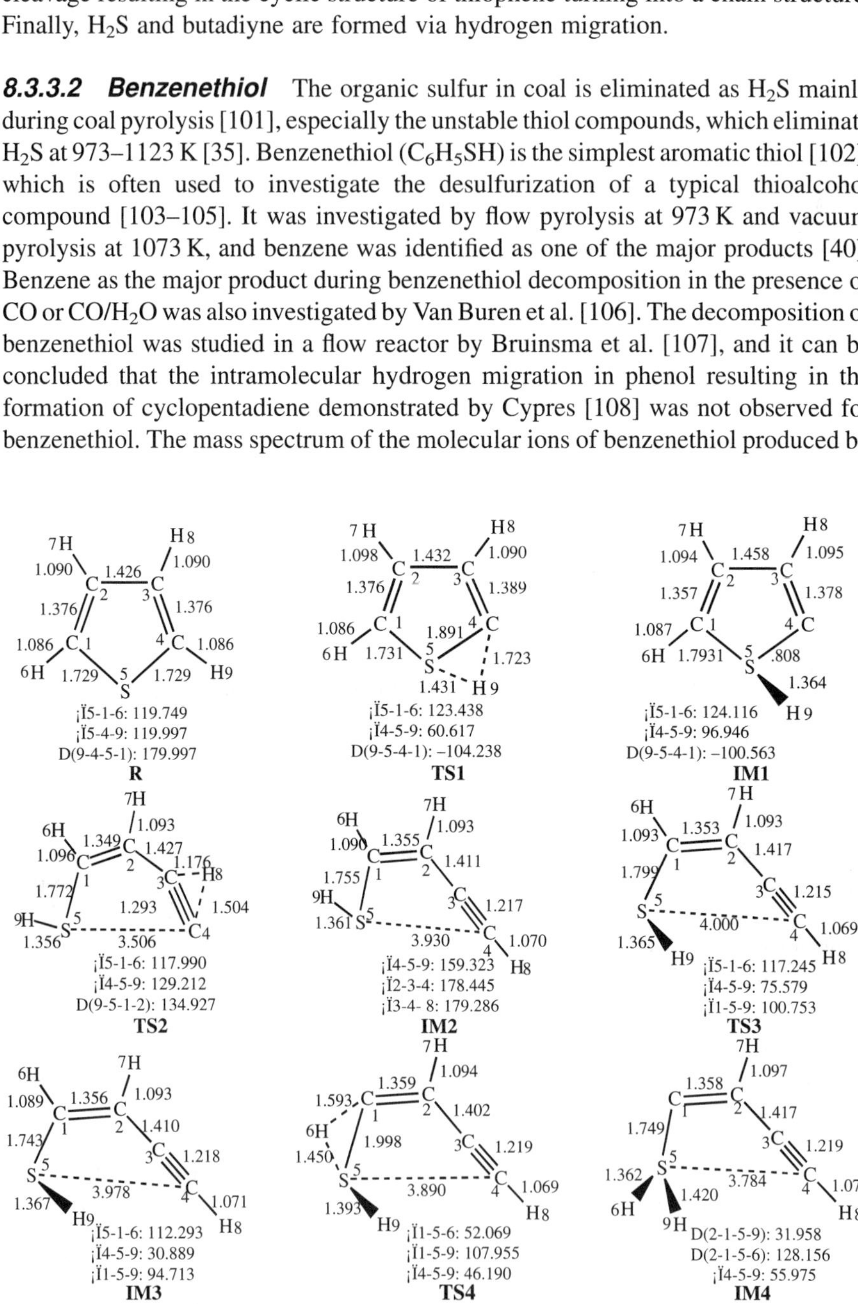

FIGURE 8.24 The optimized structures of all reactants, intermediates, transition states, and products.

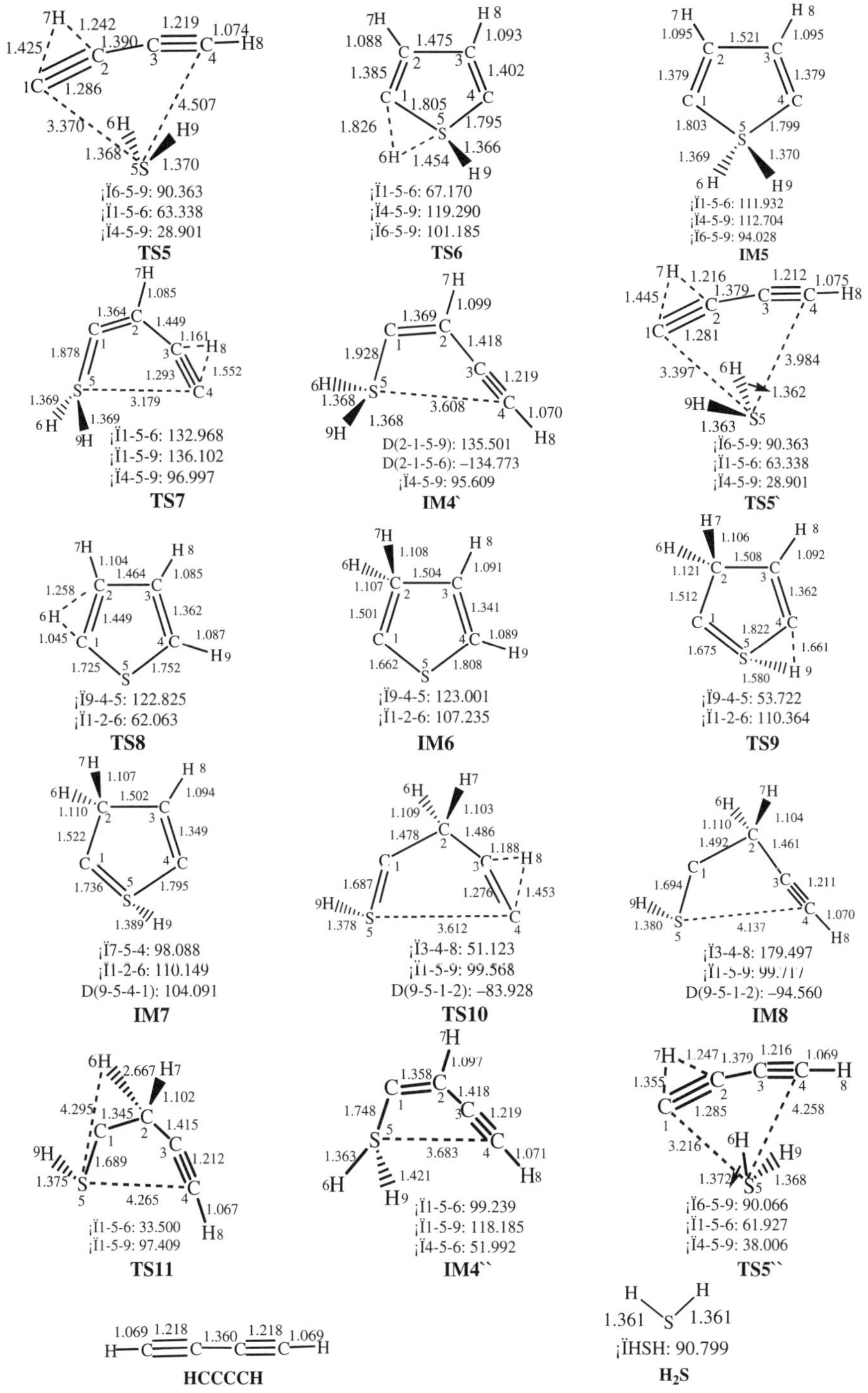

FIGURE 8.24 *(Continued)*

TABLE 8.5 The Activation Enthalpy, Activation Entropy, Activation Energy, and Rate Constants of Elementary Reaction During Thiophene Pyrolysis

Elementary Reaction	$\Delta_r H^{\neq}_m$ (kJ/mol)		$\Delta_r S^{\neq}_m$ (J/(mol K))		E_a (kJ/mol)		ln k (s^{-1})	
	298.15 K	875 K	298.15 K	875 K	298.15 K	875 K	298.15 K	875 K
Path 1								
Step 1	349.15	348.67	4.30	3.97	351.63	355.95	−110.88	−16.92
Step 2	82.91	81.88	27.92	27.29	85.39	89.15	−0.63	22.56
Step 3	26.07	21.11	−13.37	−22.64	28.55	28.39	17.33	24.91
Step 4	345.88	347.14	18.63	21.83	348.36	354.42	−107.84	−14.56
Step 5	94.06	95.54	43.39	47.82	96.54	102.82	−3.27	23.15
Path 2								
Step 1	349.15	348.67	4.30	3.97	351.63	355.95	−110.88	−16.92
Step 6	401.24	400.37	9.42	8.22	403.72	407.64	−131.28	−23.51
Step 7	5.10	4.12	23.98	23.38	7.58	11.40	30.28	32.78
Step 5′	60.89	62.00	35.08	38.69	63.37	69.26	9.11	26.67
Path 3								
Step 8	276.82	274.88	−0.30	−3.68	279.29	282.15	−82.25	−7.69
Step 9	429.87	430.88	2.83	5.16	432.35	438.16	−143.62	−28.08
Step 10	74.20	73.90	48.35	49.29	76.68	81.17	5.34	26.30
Step 11	136.38	139.92	22.98	30.61	138.86	147.20	−22.79	14.98
Step 5″	97.04	99.30	43.23	49.42	99.52	106.58	−4.49	22.83

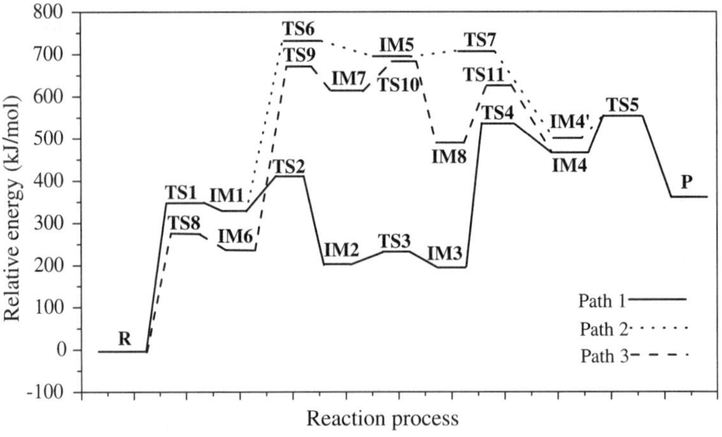

FIGURE 8.25 Energy profile of the stationary points for the three paths of thiophene pyrolysis.

electron ionization of neutral benzenethiol showed that the most intense peaks were observed at *m/z* 84 and 66 (corresponding to the losses of ethyne and carbon monosulfide, respectively). These are similar to the reactions already observed without the presence of a collision gas, indicating that unimolecular rearrangements of the metastable molecular ions, thiophene, and cyclopentadiene ions (*m/z* 84:*m/z* 66 = 0.83) were likely to be generated [109], which showed that cyclopentadiene and thiophene might be also the decomposition products of benzenethiol. Some of the thiols being transformed into thiophenes during coal pyrolysis have been investigated in early studies [36]. We can see from the above experiments that the sulfur in benzenethiol may migrate to the products CS, H_2S, and thiophene. A detailed theoretical study of the sulfur migration during benzenethiol pyrolysis was carried out in our work [23].

Four different paths were proposed to describe the migration process of sulfur (see Figure 8.26), two of which were responsible for the sulfur migration eventually resulting in H_2S. The optimized structures and their atom numbers of reactants, intermediates, transition states, and products are shown in Figure 8.27, and the energy

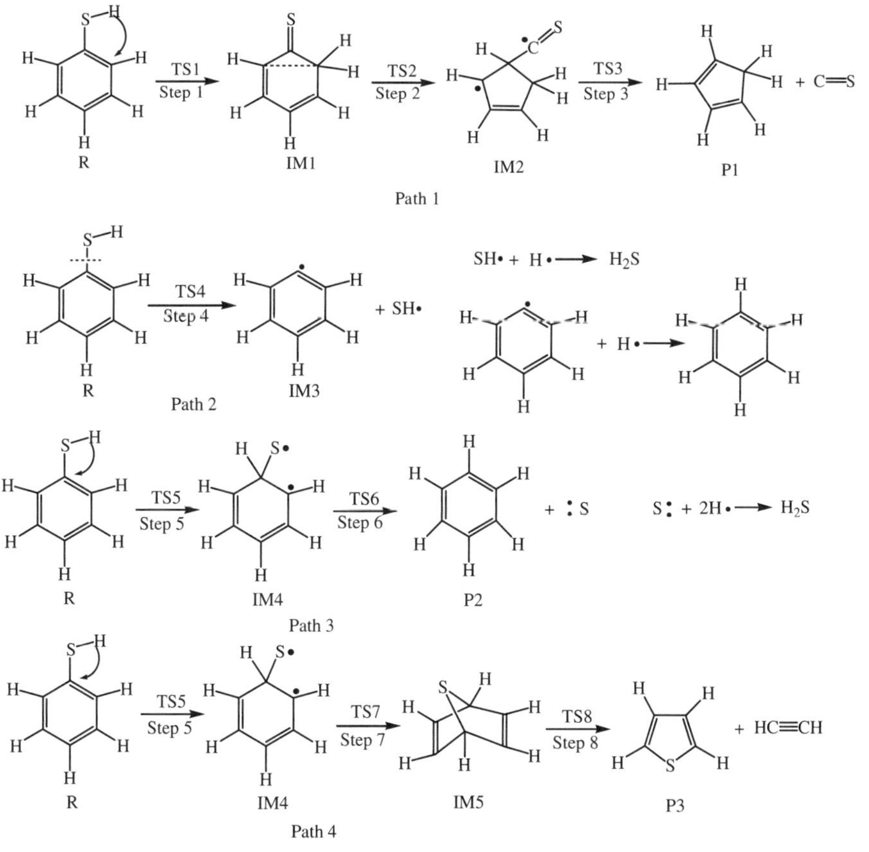

FIGURE 8.26 The proposed sulfur migration mechanisms during benzenethiol pyrolysis.

profile of the stationary points for benzenethiol pyrolysis is shown in Figure 8.28. We can see that the highest energy barrier along the sulfur migration yielding CS occurs at TS2. In the two paths of H_2S formation, the highest energy barrier occurs at TS4. The second path to H_2S formation is path 3, and the highest energy barrier occurs at TS6.

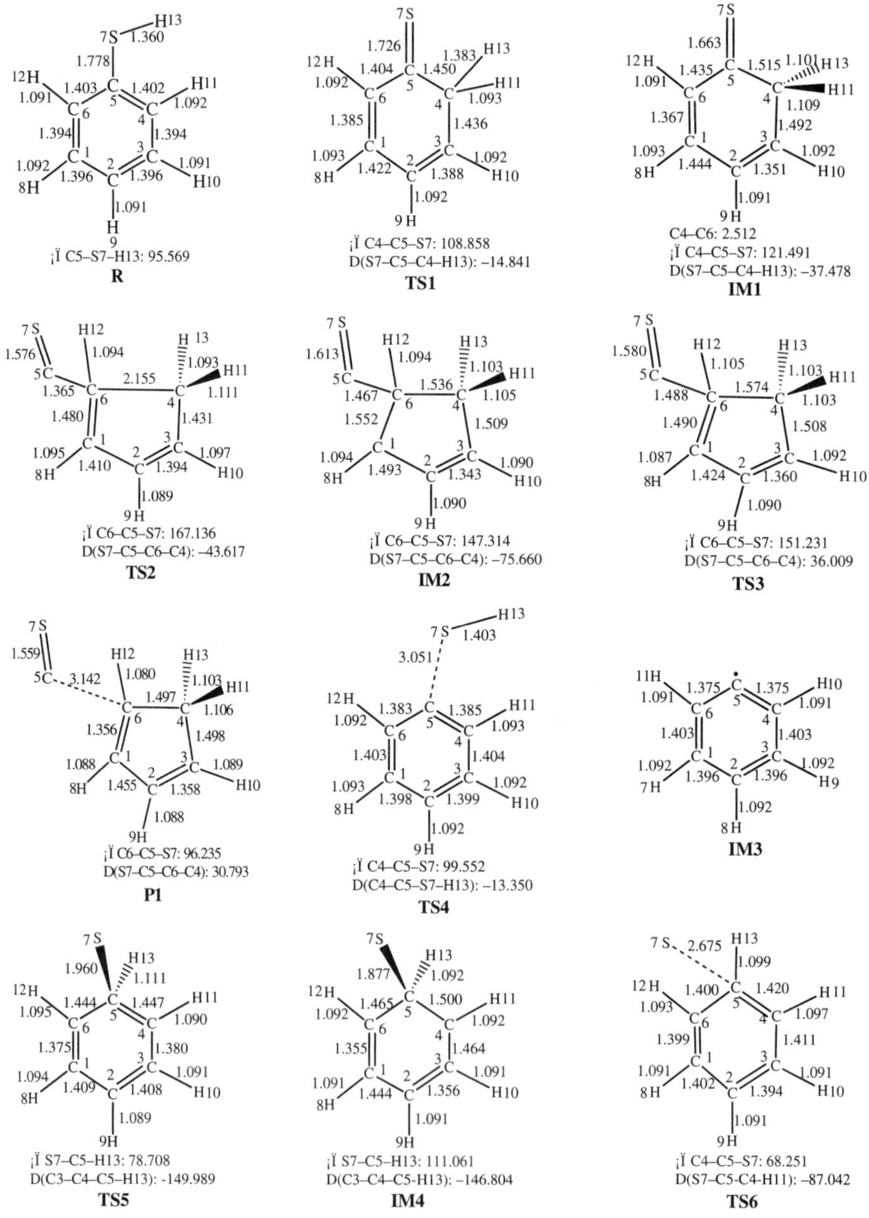

FIGURE 8.27 The optimized geometries of all reactants, intermediates, transition states, and products during benzenethiol pyrolysis.

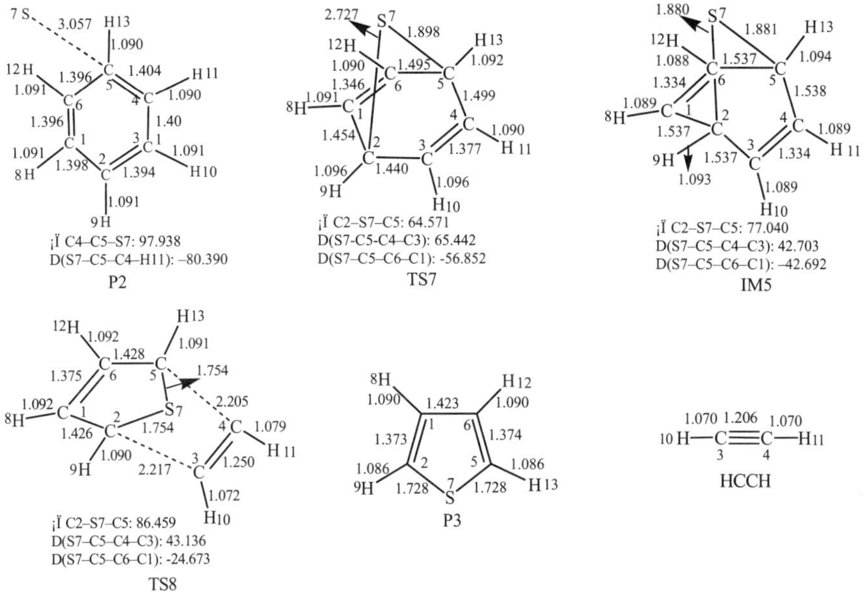

FIGURE 8.27 *(Continued)*

The highest energy barrier along the sulfur migration yielding thiophene occurs at TS8. Comparing the four paths, the lowest energy path is path 3, that is, H of the thiol group migrates to the ipso C and S radical is eliminated by β-scission reaction. The S radical combining with H radicals formed during coal pyrolysis results in the formation of H_2S. The sulfur migrations leading to the formation of CS and thiophene are more difficult. The products benzene, cyclopentadiene, and acetylene are also formed in the process of the sulfur migration. Benzene is formed easily, which is the reason that a large amount of benzene is found by flow pyrolysis of benzenethiol at 973 K [40].

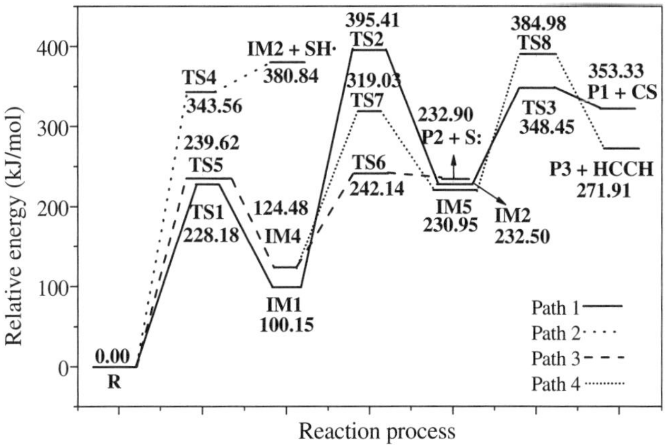

FIGURE 8.28 Energy profile of the stationary points for benzenethiol pyrolysis reaction.

Comparing the activation energy for the formation of H_2S via benzenethiol and thiophene pyrolysis, it can be concluded that the former is easier than the latter, which explains the experimental fact that thiophene was pyrolyzed at a higher temperature [34] than that of benzenethiol [40] from a kinetic point of view.

8.4 CONCLUSION

The pyrolysis mechanisms of a series of oxygen-, nitrogen-, and sulfur-containing model compounds are summarized, and the main results can be listed as following:

1. The pyrolyses of oxygen-, nitrogen-, and sulfur-containing model compounds are the important reactions in coal chemistry. The proposed reaction mechanisms are based on the experimental results and the clues to the intermediates and products. The pyrolysis of these models can reflect well the pyrolysis process of coal.
2. Through the researches on pyrolysis mechanisms of a series of nitrogen- and sulfur-containing models by quantum chemistry calculation, the migration of nitrogen and sulfur during coal pyrolysis has been investigated, and the production of pollution agents such as HCN, NH_3, and H_2S has been analyzed from a point of view of chemical reaction mechanism.
3. The methodology of LMSMC can be used to reasonably describe the pyrolysis property of coal, and reflect the pyrolysis process of coal aiming at a certain property.
4. The intermediates in some reactions play an important role in coal pyrolysis, such as the three resonance structures of phenoxy radical and the isomerization structures of 1-indene imine, which can be found by quantum chemistry calculation but can rarely be detected by pure experiment.
5. The intramolecular hydrogen migration appears often in the pyrolysis mechanism of oxygen-, nitrogen-, and sulfur-containing model compounds, and plays an important role, which can be summarized as follows: (a) it offers more reasonable reaction paths for the formation of products; (b) it allows for reasonable interpretation of experimental results; (c) it needs less activation energy; (d) it explains the distribution of products.

REFERENCES

1. A. Marzec, Towards an understanding of the coal structure: a review, *Fuel Process. Technol.* **77–78** 25–32 (2002).
2. K. C. Xie, *Coal Structure and its Reactivity*. Science Press, Beijing (2002), pp. 115–207.
3. J. H. Shinn, From coal to single-stage and two-stage products: a reactive model of coal structure, *Fuel* **63**, 1187–1196 (1984).

4. X. J. Hou, J. L. Yang, and Y. W. Li, Ab initio study of coal structure, *J. Fuel Chem. Technol.* **27**(Suppl.) 142–148 (1999).
5. J. L. Zhang, Quantum chemistry study on effect mechanism of polymer on coal surface, *J. Univ. Sci. Technol. Beijing* **23**(1), 6–8 (2001).
6. T. Takanohashi, M. Iino, and K. Nakamura, Simulation of interaction of coal associates with solvents using the molecular dynamics calculation, *Energy Fuels* **12** (6), 1168–1173 (1998).
7. B. J. Wang, *Study on Quantum Chemistry of Coal Structure and Reactivity*, Taiyuan University of Technology, Taiyuan, Shanxi (2006).
8. M. Li, *Research on Oxygen-Containing Functional Groups on Coal Surface*, Taiyuan University of Technology, Taiyuan, Shanxi (2004).
9. Q. Y. Zhao, *Quantum Chemistry Calculation of Effect of Gases on Coal Surfaces and Carbon Nanotubes*, Taiyuan University of Technology, Taiyuan, Shanxi (2005).
10. R. G. Zhang, *Study on the Formation and Structure of Aromatic Plane in Coal Using Quantum Chemistry Calculation*, Taiyuan University of Technology, Taiyuan, Shanxi (2006).
11. L. J. Zhao, *Quantum Chemistry Study on the Mechanism for the Pyrolysis of Coal-Related Oxygen-Containing Model Compounds*, Taiyuan University of Technology, Taiyuan, Shanxi (2008).
12. L. X. Ling, *Quantum Chemistry Study on Pyrolysis of Coal-Related Heteroatom-Containing Model Compounds and Removal of Sulfur-Containing Compounds*, Taiyuan University of Technology, Taiyuan, Shanxi (2010).
13. R. G. Zhang, B. J. Wang, Y. J. Tian, and L. X. Ling, Quantum chemistry studies on the growth mechanism of polycyclic arenes from benzene precursors, *Chin. J. Chem. Eng.* **17**(3), 394–400 (2009).
14. B. J. Wang, L. X. Ling, Q. Y. Zhao, R. G. Zhang, and K. C. Xie, Quantum chemistry calculation on thermochemical properties of coal, *J. Chin. Coal Soc.* **34**(9), 1239–1243 (2009).
15. B. J. Wang, L. X. Ling, R. G. Zhang, and K. C. Xie, Quantum chemistry study on adsorption of gases on coal surface, *J. Chem. Ind. Eng. Chin.* **60**(4), 995–1000 (2009).
16. H. P. Meng, W. Zhao, R. G. Zhang, and B. J. Wang, Effect of char on the conversion of methane to produce syngas. Part IV The theoretical studies on catalysis mechanism of oxygen-containing groups in char, *Coal Conversion* **31**(3), 31–35 (2008).
17. B. J. Wang, M. Li, Q. Y. Zhao, Y. H. Qin, and K. C. Xie, Relationship between surface potential and functional groups of coals, *J. Chem. Ind. Eng. Chin.* **55**(8), 1229–1334 (2004).
18. B. J. Wang, W. Y. Li, and R. G. A. Zhang, Density Functional Theory Study on the Pyrolysis Mechanism of 2-Picoline, *The 9th China–Japan Symposium on Coal and C1 Chemistry Proceedings*, Chengdu City, China, October 2006, pp. 59–60.
19. L. X. Ling, L. J. Zhao, R. G. Zhang, and B. J. Wang, Pyrolysis mechanisms of benzoic acid and benzaldehyde based on quantum chemistry, *Chin. J. Chem. Eng.* **60**(5), 1224–1230 (2009).
20. L. J. Zhao, L. X. Ling, R. G. Zhang, X. F. Liu, and B. J. Wang, Theoretical study on pyrolysis mechanism of O-containing model compound anisole in coal, *Chin. J. Chem. Eng.* **59**(8), 2095–2102 (2008)
21. L. X. Ling, R. G. Zhang, B. J. Wang, and K. C. Xie, Pyrolysis mechanisms of quinoline and isoquinoline with density functional theory, *Chin. J. Chem. Eng.* **17**(5), 805–813 (2009).

22. L. X. Ling, R. G. Zhang, B. J. Wang, and K. C. Xie, Density functional theory study on the pyrolysis mechanism of thiophene in coal, *J. Mol. Struct. (Theochem)* **905**, 8–12 (2009).

23. L. X. Ling, R. G. Zhang, B. J. Wang, and K. C. Xie, DFT study on the sulfur migration during benzenethiol pyrolysis in coal, *J. Mol. Struct. (Theochem)* **952**, 31–35 (2010).

24. B. J. Wang, Y. G. Zhang, and K. C. Xie, Application of quantum chemistry calculation to investigation on coal structure and reactivity, *J. Chem. Ind. Eng. Chin.* **54**(4), 477–488 (2003).

25. K. C. Xie, *Coal Structure and Its Reactivity*, Science Press, Beijing (2002), pp. 181–184.

26. G. A. Carlson, Computer simulation of the molecular structure of bituminous coal, *Energy Fuels* **6**(6), 771–778 (1992).

27. K. C. Xie, *Coal Structure and its Reactivity*, Science Press, Beijing (2002), pp. 85–86.

28. B. J. Wang, *Study on Quantum Chemistry of Coal Structure and Reactivity*, Taiyuan University of Technology, Taiyuan, Shanxi (2006), pp. 25–28.

29. F. Li, Y. F. Zhang, and K. C. Xie, Study on small molecular phase in macromolecules of Pingshuo bituminous coal, *J. Fuel Chem. Technol.* **21**(3), 293–297 (1993).

30. F. Li, Y. F. Zhang, and K. C. Xie, Characterization of the macromolecular structure of Pingshuo coal macerals using ^{13}C-NMR, XPS, FTIR and XRD, *Fuel Sci. Tech. Int.* **11**(8), 1113–1131 (1993).

31. B. J. Wang, Z. M. Wu, F. Li, and K. C. Xie,*The Study on the Electronic Structures and Molecular Models of Coal Macerals,* Vol. I, ICCS (1999), pp. 271–274.

32. X. D. Zhu, Z. B. Zhu, C. J. Han, and L. H. Tang, Fundamental study of coal pyrolysis: III. Functional group and pyrolysis products, *J. East Chin. Univ. Sci. Technol.* **26**(1), 14–17 (2000).

33. J. K. Winkler, W. Karow, and P. Rademcher, Gas-phase pyrolysis of heterocyclic compounds, part 1 and 2: flow pyrolysis and annulation reactions of some sulfur heterocycles: thiophene, benzo[*b*]thiophene, and dibenzothiophene. A product-oriented study, *J. Anal. Appl. Pyrolysis* **62**(1), 123–141 (2002).

34. H. U. R. Memon, A. Williams, and P. T. Williams, Shock tube pyrolysis of thiophene, *Int. J. Energy Res.* **27**(3), 225–239 (2003).

35. L. B. Sun, Z. H. Ni, L. F. Zhang, Z. M. Zong, and X. Y. Wei, Research advancement of nitrogen and sulfur separate out form in coal pyrolysis process, *Clean Coal Technol.* **8**(3), 47–50 (2002).

36. A. Attar, Chemistry, thermodynamics and kinetics of reactions of sulphur in coal-gas reactions: a review, *Fuel* **57**(14), 201–212 (1978).

37. L. L. Tan and C. Z. Li, Formation of NO_x and SO_x precursors during the pyrolysis of coal and biomass. Part I. Effects of reactor configuration on the determined yields of HCN and NH_3 during pyrolysis, *Fuel* **79**(15), 1883–1889 (2000).

38. J. Feng, W. Zhao, and C. Z. Li, Formation of NO_x and SO_x precursors during the pyrolysis of coal and biomass: Part IV. Pyrolysis of a set of Australian and Chinese coals, *Fuel* **80** (15), 2131–2138 (2001).

39. L. P. Chang, Z. L. Xie, K. C. Xie, K. C. Pratt, J. Hayashi, T. Chiba, and C. Z. Li, Formation of NO_x precursors during the pyrolysis of coal and biomass: Part VI. Effects of gas atmosphere on the formation of NH_3 and HCN, *Fuel* **82**(10), 1159–1166 (2003).

40. D. E. Johnson, Pyrolysis of benzenethiol, *Fuel* **66**(2), 255–260 (1987).

41. T. A. Halgren and W. N. Lipscomb, The synchronous transit method for determining reaction pathways and locating transition states, *Chem. Phys. Lett.* **49**, 225–232 (1977).

42. A. J. Colussi, F. Zabel, and S. W. Benson, The very low-pressure pyrolysis of phenyl ethyl ether, phenyl allyl ether, and benzyl methyl ether and the enthalpy of formation of the phenoxy radical, *Int. J. Chem. Kinet.* **IX** 161–178 (1977).

43. K. Brezinsky, M. Pecullan, and I. Glassman, Pyrolysis and oxidation of phenol, *J. Phys. Chem. A* **102**(4), 8614–8619 (1998).

44. C. Horn, K. Roy, P. Frank, and T. Just,*Shock-Tube Study on the High-Temperature Pyrolysis of Phenol,* Twenty-Seventh Symposium (International) on Combustion, The Combustion Institute, 1998, pp. 321–328.

45. A. B. Lovell, K. Brezinsky, and I. Glassman, The gas phase pyrolysis of phenol, *Int. J. Chem. Kinet.* **21**(7), 547–560 (1989).

46. S. Olivella, A. Sole, and A. Garcia-Raso, Ab Initio calculations of the potential surface for the thermal decomposition of the phenoxyl radical, *J. Phys. Chem.* **99**(26), 10549–10556 (1995).

47. R. F. Liu, K. Morokuma, M. M. Alexander, and M. C. Lin, Ab initio study of the mechanism for the thermal decomposition of the phenoxy radical, *J. Phys. Chem.* **100**(22), 9314–9322 (1996).

48. C. Y. Lin and M. C. Lin, Unimolecular decomposition of the phenoxy radical in shock waves, *Int. J. Chem. Kinet.* **17**, 1025–1028 (1985)

49. J. H. Lee and I. N. Tang, Absolute rate constants for the hydroxyl radical reactions with ethane, furan, and thiophene at room temperature, *J. Chem. Phys.* **77**(9), 4459–4463 (1982).

50. M. A. Grela, V. T. Amorebieta, and A. J. Colussi, Very low pressure pyrolysis of furan, 2-methylfuran, and 2,5-dimethylfuran. The stability of the furan ring, *J. Phys. Chem.* **89**(1), 38–41 (1985).

51. A. Lifshitz, M. Bidani, and S. Bidani, Thermal reactions of cyclic ethers at high temperatures. 3. Pyrolysis of furan behind reflected shocks, *J. Phys. Chem.* **90**(21), 5373–5377 (1986).

52. P. P. Organ and J. C. Mackie, Kinetics of pyrolysis of furan, *J. Chem. Soc. Faraday Trans.* **87**(6), 815–823 (1991).

53. R. F. Liu, X. F. Zhou, and L. Zhai, Theoretical investigation of unimolecular decomposition channels of furan, *J. Comput. Chem.* **19**(2), 240–249 (1998).

54. R. F. Liu, X. F. Zhou, and T. M. Zuo, The pyrolysis mechanism of furan revisited, *Chem. Phys. Lett.* **325**(4), 457–464 (2000).

55. K. Sendt, G. B. Bacskay, and J. C. Mackie, Pyrolysis of furan: ab initio quantum chemical and kinetic modeling studies, *J. Phys. Chem. A* **104**(9), 1861–1875 (2000)

56. E. M. Suuberg, D. Lee, and J. W. Larsen, Temperature dependence of crosslinking processes in pyrolysing coals, *Fuel* **64**(12), 1668–1671 (1985).

57. E. M. Suuberg, P. E. Unger, and J. W. Larsen, Relation between tar and extractables formation and cross-linking during coal pyrolysis, *Energy Fuels* **1**(3), 305–308 (1987).

58. G. V. Deshpande, P. R. Solomon, and M. A. Serio, Crosslinking reactions in coal pyrolysis, *Am. Chem. Soc. Div. Fuel Chem. Prepr.* **33**(2), 310–321 (1988).

59. P. R. Solomon, M. A. Serio, G. V. Despande, and E. Kroo, Cross-linking reactions during coal conversion, *Energy Fuels* **4**(1), 42–54 (1990).

60. T. P. Eskay, P. F. Britt, and A. C. Buchanan, III, Does decarboxylation lead to cross-linking in low-rank coals, *Energy Fuels* **10**(6), 1257–1261 (1996).
61. T. P. Eskay, P. F. Britt, and A. C. Buchanan, III, Pyrolysis of aromatic carboxylic acid: potential involvement of anhydrides in retrograde reactions in low-rank coal, *Energy Fuels* **11**(6), 1278–1287 (1997).
62. J. A. Manion, D. F. McMillen, and R. Malhotra, Decarboxylation and coupling reactions of aromatic acids under coal-liquefaction conditions, *Energy Fuels* **10**(3), 776–788 (1996).
63. X. D. Zhu, Z. B. Zhu, C. F. Zhang, and L. H. Tang, Fundamental study of coal pyrolysis: IV. Functional group pyrolysis model, *J. East Chin. Univ. Sci. Technol.* **27**(2), 113–120 (2001).
64. X. C. Fu, W. X. Shen, and T. Y. Yao, *Physical Chemistry*, 5th edition, Higher Education Press, Beijing (2006), p. 244.
65. L. H. Zhao, H. Q. Guo, and Q. L. Ma, Study on gaseous products distributions during coal pyrolysis, *Coal Conversion* **30**(1), 5–8 (2007).
66. C. Y. Lin and M. C. Lin, Thermal decomposition of methyl phenyl ether in shock waves: the kinetics of phenoxy reactions, *J. Phys. Chem.* **90**(3), 425–431 (1986).
67. R. H. Schlosberg, P. F. Szajowski, G. D. Dupre, J. A. Danik, A. Kurs, T. R. Ashe, and W. I. Olmstead, Pyrolysis studies of organic oxygenates: 3. High temperature rearrangement of aryl alkyl ethers, *Fuel* **62**(6), 690–694 (1983).
68. W. C. E. Isabel, R. L. Arends, and M. Peter, Kinetic study of the thermolysis of anisole in a hydrogen atmosphere, *J. Phys. Chem.* **97**(30), 7914–7925 (1993).
69. M. F. R. Mulcahy and D. J. Williams, Reaction of phenoxy radicals with methyl radicals in the gaseous phase, *Nature* **199**, 761–762 (1963).
70. A. E. Axworthy, V. H. Dayan, and G. B. Martin, Reactions of fuel-nitrogen compounds under conditions of inert pyrolysis, *Fuel* **57**(1), 29–35 (1978).
71. X. Y. Cao, Z. G. Niu, L. Q. Ying, Z. H. Wang, J. H. Zhou, J. Z. Liu, and K. F. Cen, Releasing of fuel-nitrogen during blind coal pyrolysis, *J. Fuel Chem. Technol.* **3**(16), 538–542 (2003).
72. J. X. Mao, J. Q. Mao, and S. M. Zhao, *Clean Burning of Coal*, Science Press, Beijing (1998), pp. 209–210.
73. J. M. Patterson, A. Tsamasfyros, and W. T. Smith, Pyrolysis of pyrrole, *J. Heterocycl. Chem.* **5**(5), 727–729 (1968).
74. A. Lifshitz, C. Tamburu, and A. Suslensky, Isomerization and decomposition of pyrrole at elevated temperatures: studies with a single-pulse shock tube, *J. Phys. Chem.* **93**(15), 5802–5808 (1989).
75. F. Dubnikova and A. Lifshitz, Isomerization of pyrrole. Quantum chemical calculations and kinetic modeling, *J. Phys. Chem. A* **102**(52), 10880–10888 (1998).
76. J. C. Mackie, M. B. Colket, P. F. Nelson, and M. Esler, Shock tube pyrolysis of pyrrole and kinetic modeling, *Int. J. Chem. Kinet.* **23**, 733–759 (1991).
77. M. Martoprawiro, G. B. Bacskay, and J. C. Mackie, Ab initio quantum chemical and kinetic modeling study of the pyrolysis kinetics of pyrrole, *J. Phys. Chem. A* **103**(20), 3923–3934 (1999).
78. L. Zhai, X. F. Zhou, and R. F. Liu, A theoretical study of pyrolysis mechanisms of pyrrole, *J. Phys. Chem. A* **103**(20), 3917–3922 (1999).

79. A. Laskin and A. Lifshitz, Isomerization and decomposition of indole. experimental result and kinetic modeling, *J. Phys. Chem.* **101**(42), 7787–7801 (1997).

80. X. F. Zhou and R. F. Liu, A density functional theory study of the pyrolysis mechanisms of indole, *J. Mol. Struct. (Theochem)* **461–462** 569–579 (1999).

81. Y. Ninomiya, Z. B. Dong, Y. Suzuki, and J. Koketsu, Theoretical study on the thermal decomposition of pyridine, *Fuel* **79**, 449–457 (2000).

82. A. Terentis, A. Doughty, and J. C. Mackle, Kinetics of pyrolysis of a coal model compound, 2-picoline, the nitrogen heteroaromatic analog of toluene. 1. Product distributions, *J. Phys. Chem.* **96**(25), 10334–10339 (1992).

83. A. Doughty and J. C. Mackle, Kinetics of pyrolysis of a coal model compound, 2-picoline, the nitrogen heteroaromatic analogue of toluene. 2. The 2-picolyl radical and kinetic modeling, *J. Phys. Chem.* **96**, 10339–10348 (1992).

84. E. Ikeda and J. C. Mackie, Thermal decomposition of two coal model compounds—pyridine and 2-picoline: kinetics and product distributions, *J. Anal. Appl. Pyrol.* **34**(1), 47–63 (1995).

85. B. J. Wang, L. X. Ling, and R. G. Zhang,*2-Picoline Pyrolysis Mechanism and its Molecular Catastrophe Phenomenon,* BIT's Annual World Congress of Well Stimulation and EOR, Southwest Petroleum University, China, 2010, p. 097.

86. J. M. Patterson, C. H. Issidorides, E. P. Papadopoulos, and W. T. Smith, The thermal interconversion of quinoline and isoquinoline, *Tetrahedron Lett.* **15**, 1247–1250 (1970).

87. O. S. L. Bruinsma, P. J. J. Tromp, H. J. J. de Sauvage Nolting, and J. A. Moulijn, Gas phase pyrolysis of coal-related aromatic compounds in a coiled tube flow reactor: 2. Heterocyclic compounds, their benzo and dibenzo derivatives, *Fuel* **67**(3), 334–340 (1988).

88. A. Laskin and A. Lifshitz, Thermal decomposition of quinoline and isoquinoline: the role of 1-indene imine radical, *J. Phys. Chem. A* **102**, 928–946 (1998).

89. J. K. Winkler, W. Karow, and P. Rademacher, Gas phase pyrolysis of heterocyclic compounds: Part 3. Flow pyrolysis and annulation reaction of some nitrogen heterocycles: a product oriented study, *Arkivoc* **1**(4), 576–602 (2000).

90. H. K. Chen, B. Q. Li, J. L. Yang, and B. J. Zhang, Transformation of sulfur during pyrolysis and hydropyrolysis of coal, *Fuel* **77**(6), 487–493 (1998).

91. J. Klein, Technological and economic aspects of coal biodesulfurisation, *Biodegradation* **9**(3–4) 293–300 (1998).

92. Z. C. Wang, W. R. Yao, P. Cui, and T. Song, Organic sulphur removal from coal by electrolysis, *J. Fuel Chem. Technol.* **31**(5), 509–512 (2003).

93. Z. Q. Gao, X. D. Hu, Y. Q. Wu, and J. S. Gao, Study on chemical desulfurization of coal, *Environ. Chem.* **13**(6), 486–491 (1994).

94. P. Mills, S. Korlann, M. E. Bussell, M. A. Reynolds, M. V. Ovchinnikov, R. J. Angelici, C. Stinner, T. Weber, and R. Prins, Vibrational study of organometallic complexes with thiophene ligands: models for adsorbed thiophene on hydrodesulfurization catalysts, *J. Phys. Chem. A* **105**(18), 4418–4427 (2001).

95. C. F. Cullis and A. C. Norris, The pyrolysis of organic compounds under conditions of carbon formation, *Carbon* **10**(5), 525–537 (1972).

96. C. Huang, J. Y. Zhang, J. Chen, and C. G. Zheng, Quantum chemistry study on the pyrolysis of thiophene functionalities in coal, *Coal Conversion* **28**(20), 33–35 (2005).

97. M. Bajus, V. Veselý, and J. Baxa, Steam cracking of hydrocarbons: 5. Effect of thiophene on reaction kinetics and coking, *Ind. Eng. Chem. Prod. Res. Dev.* **20**(4), 741–745 (1981).

98. N. R. Hore and D. K. Russell, The thermal decomposition of 5-membered rings: a laser pyrolysis study, *New J. Chem.* **28**, 606–613 (2004).

99. C. D. Hurd, R. V. Levetan, and A. R. Macon, Pyrolytic formation of arenes. II. Benzene and other arenes from thiophene, 2-methylthiophene and 2- (methyl-^{14}C)-thiophene, *J. Am. Chem. Soc.* **84**(23), 4515–4519 (1962).

100. C. Barckholtz, T. A. Barckholtz, and C. M. Hadad, C–H and N–H bond dissociation energies of small aromatic hydrocarbons, *J. Am. Chem. Soc.* **121**(3), 491–500 (1999).

101. J. C. Zhao, X. D. Hu, and J. S. Gao, Study on the variations of organic sulfur in coal by pyrolysis, *Coal Conversion* **16**(2), 77–81 (1993).

102. J. Nara, S. Higai, Y. Morikawa, and T. Ohno, Density functional theory investigation of benzenethiol adsorption on Au (111), *J. Chem. Phys.* **120**(14), 6705–6711 (2004).

103. D. A. Chen, C. M. Friend, and H. Xu, Desulfurization of benzenethiol on Co–Mo (110) phases, *Surf. Sci.* **395**, L221–L226 (1998).

104. D. R. Huntley, The mechanism of the desulfurization of benzenethiol by nickel (110), *J. Phys. Chem.* **96**(11), 4550–4558 (1992).

105. T. S. Rufael, D. R. Huntley, D. R. Mullins, and J. L. Gland, Adsorption and reactions of benzenethiol on the Ni (111) surface, *J. Phys. Chem.* **98**(49), 13022–13027 (1994).

106. R. L. Van Buren, R. J. Baltisberger, N. F. Woolsey, and V. I. Stenberg, Formic acid and the high-temperature reductive desulfurization of aromatic sulfides, *J. Org. Chem.* **47**(21), 4107–4110 (1982).

107. O. S. L. Bruinsma, R. S. Geertsma, P. Bank, and J. A. Moulijn, Gas phase pyrolysis of coal-related aromatic compounds in a coiled tube flow reactor: 1. Benzene and derivatives, *Fuel* **67**(3), 327–333 (1988).

108. R. Cypres, Aromatic hydrocarbons formation during coal pyrolysis, *Fuel Process. Technol.* **15**, 1–15 (1987).

109. P. C. Nam, R. Flammang, H. T. Le, P. Gerbaux, and M. T. Nguyen, Protonation and methylation of thiophenol, thioanisole and their halogenated derivatives: mass spectrometric and computational study, *Int. J. Mass Spectrom.* **228**(2–3), 151–165 (2003).

CHAPTER 9

Ab Initio Kinetic Modeling of Free-Radical Polymerization

MICHELLE L. COOTE

ARC Centre of Excellence in Free-Radical Chemistry and Biotechnology, Research School of Chemistry, Australian National University, Canberra, Australia

9.1 INTRODUCTION

9.1.1 Free-Radical Polymerization Kinetics

Free-radical polymerization is an important industrial process, responsible for the production of around 50% of all synthetic polymers worldwide (about 100 million tons per annum) [1]. As in all radical chain processes, polymerization proceeds via a series of initiation, propagation, and termination reactions (see Scheme 9.1). Free radicals are typically produced *in situ* via thermal, photolytic, or electrochemical decomposition of an initiator. The resulting radicals then attack the π-bond of alkene monomer via a free-radical addition reaction that results in formation of a new σ-bond between the attacking radical and monomer, and a new radical center, which can then undergo further radical addition reactions with further monomer molecules. This step is usually highly regioselective, with addition occurring at the least substituted carbon center of the π-bond. As the radical center is sp^2 hybridized, attack can occur from either face, and control over the stereochemistry of the polymer resulting from *conventional* free-radical polymerization is therefore limited. However, in recent years, successful *stereocontrolled* radical polymerization has been achieved for some classes of monomer (vinyl esters, acrylamides, methacrylates) via complexation of the growing radical with Lewis acids [2].

Termination of the growing radical occurs via bimolecular radical–radical termination reactions, which, depending on the system, occur via combination and/or disproportionation. These reactions tend to be inherently much faster than propagation, to the extent that they are usually diffusion limited. In conventional

Rate Constant Calculation for Thermal Reactions: Methods and Applications, Edited by Herbert DaCosta and Maohong Fan.

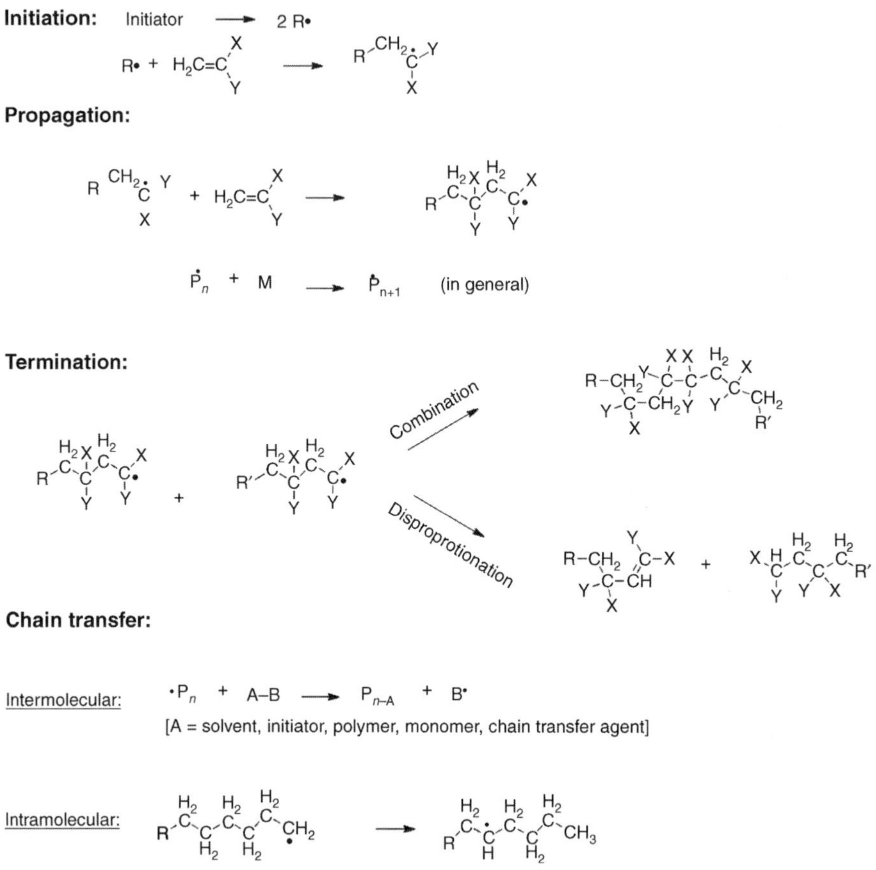

SCHEME 9.1 Conventional free-radical polymerization.

free-radical polymerization, bimolecular termination occurs randomly throughout the process leading to polymers that have a broad molecular weight distribution and end groups that are not normally suitable for further modification. The propagating radical is also highly susceptible to side reactions, such as hydrogen or halogen atom abstraction from species such as the monomer, solvent, initiator, or added chain transfer agent. These reactions, which are typically followed by further propagation of the new radical center, can be exploited to limit the molecular weight of the resulting polymer and confer specific end groups, but are not always desirable. Intermolecular and intramolecular chain transfer to polymer can also occur in some systems. When followed by subsequent propagation, this leads, respectively, to long- and short-chain branchings of the polymer, which may or may not be desirable, depending on the application.

Until recently, free-radical polymerization was thought to offer only limited control over the molecular weight distribution, chain end composition, and other aspects of the polymer microstructure. However, in the last couple of decades, the

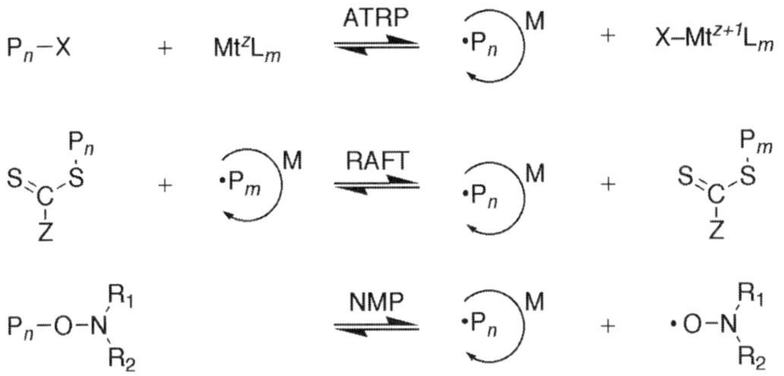

SCHEME 9.2 Main control equilibria in atom transfer radical polymerization (ATRP), reversible addition–fragmentation chain transfer (RAFT) polymerization, and nitroxide-mediated polymerization (NMP).

field has been transformed by the development of controlled/living radical polymerization techniques such as reversible addition–fragmentation chain transfer (RAFT) polymerization [3], atom transfer radical polymerization (ATRP) [4], and nitroxide-mediated polymerization (NMP) [5]. The key feature of these methods is their protection of the growing polymer chains from the bimolecular termination reactions that normally occur in conventional free-radical polymerization through their reversible trapping as a dormant species. The chemical nature of the control equilibrium varies according to the process, with some of the leading examples provided in Scheme 9.2. Through this equilibrium, the termination rate (which depends on the square of the radical concentration) is minimized with respect to the propagation rate (which only depends on the first order of the radical concentration). Some bimolecular termination is inevitable but, provided the process is correctly optimized, most of the chains survive throughout the polymerization, and therefore have a narrow molecular weight distribution. They can then be isolated in their "dormant" form and their active end groups used for further polymerization or functionalization.

Controlled radical polymerization combines the advantages of conventional radical polymerization with the ability to produce polymers with narrow molecular weight distributions, designer end groups, and special architectures such as stars, blocks, and grafts for use in bioengineering and nanotechnology applications [1a]. However, its development has also given rise to new technical challenges. In particular, its success is crucially dependent on choosing control agents and reaction conditions that strike an optimal balance between the rates of several competing reactions, such that concentration of the dormant species is orders of magnitude greater than the free species, and the exchange between the two forms is rapid. It has therefore become more important than ever to build accurate kinetic models for predicting the molecular weight, composition, and microstructure of the resulting (co)polymer as a function of the reaction conditions and the rate coefficients of the individual reactions for use in process optimization and control. At a deeper level, it is also important to understand the links between the chemical structures of the reagents and the individual rate coefficients themselves so as to aid the selection and/or design of new and improved control agents and catalysts.

Unfortunately, the increasing complexity of radical polymerization processes (which may contain hundreds or thousands of kinetically distinct reactions) can significantly hinder experimental efforts to extract this information for all but the simplest systems. The fundamental problem is that experimental techniques can only measure the "observables" of a process—typically the time-dependent concentrations of some of the major species or (more often) some of the major functional groups. Linking this *macroscopic* information to the *microscopic* properties of the process (i.e., the rate coefficients of the individual reactions) requires model-based assumptions, which can be subject to significant errors [6].

It is therefore not surprising that there is growing interest in the application of quantum chemistry to these systems [7]. Unlike experimental techniques, quantum chemistry allows one to study the microscopic chemical behavior directly, free from model-based assumptions other than the laws of quantum mechanics. Its main disadvantage is that the accuracy of the calculations can be affected by the numerical approximations inherent in the level of theory used. While high levels of theory can deliver chemically accurate results, such methods tend to be very computationally intensive and their cost scales exponentially with the size of the molecular system—a major problem when attempting to study polymers. Nonetheless, aided by rapid and continuing advances to computer power and the development of cost-effective strategies for studying for larger molecules, these problems are being overcome to the extent that chemically accurate predictions of rate coefficients are now possible [8]. Indeed, computational chemistry has already helped to clarify the mechanism and kinetics of various controlled [9–11] and conventional [8,12–14] free-radical polymerization processes, has helped to design optimal control agents [15], and has contributed to the discovery of new types of radical polymerization process [16,17].

9.1.2 Scope of this Chapter

This chapter illustrates how quantum chemistry can be taken one step further and be used to build and implement accurate kinetic models of complicated processes such as radical polymerization. The cornerstone of this technique, termed *ab initio* kinetic modeling, is the use of quantum chemistry to clarify the reaction scheme for a given process and predict from first principles the necessary rate and equilibrium constants of the individual reactions. From this microscopic information, it is then possible to build and implement accurate kinetic models that can be used to predict the macroscopic outcome of the chemical process—predictions that can then be tested directly against experiment. As will be discussed below, this approach offers several significant advantages over conventional kinetic modeling, while at the same time expanding the scope of the quantum chemical predictions from the microscopic to the macroscopic.

In what follows, we first outline conventional kinetic modeling before comparing and contrasting with its quantum chemical equivalent. We then describe the quantum chemical methodology required to obtain accurate predictions of the various individual rate coefficients for radical polymerization processes. We conclude with a case study drawn from some recent work on modeling RAFT polymerization [18]. Finally, it should be noted that *ab initio* kinetic modeling is not by any means unique to the polymer field; various versions of this approach to kinetic modeling have been adopted for studying other complicated multistep processes, particularly catalysis and combustion [19]. However, in this chapter we focus solely on the special problems and strategies relevant to modeling free-radical polymerization kinetics.

9.2 *AB INITIO* KINETIC MODELING

Kinetic models are equations (or, more often, sets of simultaneous equations) that, when solved, can be used to predict the macroscopic outcome of a chemical process as a function of the reaction conditions. That is, they can allow one to predict the time-dependent product distributions of a chemical process as a function of the initial concentrations of the reagents and the set of relevant parameter values (typically the rate coefficients of the individual reactions that comprise the process). Kinetic models are particularly valuable for complicated processes such as free-radical polymerization, where the link between the microscopic behavior of the individual reactions and the macroscopic outcome of the process is neither simple nor obvious. In such cases, kinetic models make it possible to identify the optimal reaction conditions that, for example, target a specific molecular weight distribution or minimize the occurrence of unwanted structural defects. They can also be used to identify optimal values of the rate coefficients for the various individual reactions and in this way provide the link between structure–reactivity studies on the individual reactions and optimal reagent design. This is particularly important for controlled radical polymerization processes since, as noted above, the degree of control is highly sensitive to the rates of individual reactions, which are in turn dependent on both the concentrations of reagents and their chemical structures. Not surprisingly, the development and implementation of accurate kinetic models for such processes has been a major target of research in this field.

9.2.1 Conventional Kinetic Modeling

The traditional approach to modeling complicated multistep processes such as free-radical polymerization is to begin by assuming a complete reaction scheme for the process, one that includes every kinetically distinct reaction that may potentially take place. In doing this it is necessary to identify any possible side reactions and, where polymeric species are involved, decide the extent to which the rate coefficients depend on the chain length. For example, if propagation is assumed to be chain length independent, it is sufficient to include just one generic propagation reaction in the reaction scheme (i.e., of the form $P_n^\bullet + M \rightarrow P_{n+1}^\bullet$) with one unique rate coefficient; if it is assumed to be chain length dependent up to the trimer stage, three separate reactions and rate coefficients are required (for propagation of the unimer, dimer, trimer, and above), and so forth. In each case it is also necessary to determine whether reactions are likely to be reversible under the reaction conditions; in those cases both the forward and reverse reactions are then considered in the reaction scheme.

For each reaction in the scheme, the rate expression is then written out, thereby forming a set of coupled differential equations. These equations are then solved simultaneously so as to obtain the time-dependent concentrations of the various species as a function of the initial conditions and the rate parameters of the various individual reactions. For very simple systems, it is possible to solve such equations analytically and thereby obtain closed expressions for time-dependent concentrations of the various species. However, for most real polymerizations this is not possible, particularly when modeling the full molecular weight distribution of the resulting polymer, as this typically entails solving high-dimensional systems of differential equations with millions of individual components (e.g., corresponding to the reactions of polymer radicals of various chain lengths). In such cases, even standard numerical algorithms for solving differential equations are impractical and specialized methods are required. The most widely used method for polymerization kinetic modeling is the Galerkin hp method that is implemented into the commercial software PREDICI® [20]. The theoretical details of this method are beyond the scope of this chapter; it is merely worth noting that, using PREDICI® software, the solution of the coupled differential equations is straightforward, is relatively computationally inexpensive, and does not entail model reductions such as limits on the number of reactions or chain lengths, or the assumption of steady state.

Having designed a kinetic model and implemented it using software such as PREDICI®, it is necessary to obtain accurate and precise values for all of the model parameters. As explained above, these are the set of rate coefficients for the individual reactions within the assumed reaction scheme. *Unfortunately this requirement creates a major bottleneck for conventional kinetic modeling*. This is because these individual reactions typically only take place within the context of a complicated polymerization process, and (in most cases) there are no experimental methods for studying them in isolation. Instead, one has to "measure" the individual rate coefficients by fitting the kinetic model itself to experimental data for the overall process under various initial conditions.

This model fitting-based approach to the measurement of the model parameters raises two very significant problems. The first is that, by fitting the model to the data, it becomes very difficult, often impossible, to test its validity. In most cases, provided there are sufficient adjustable parameters, any number of different models can be made to fit the same data, irrespective of the physical validity of their underlying assumptions. The second problem is that only a small number of unknown parameters (usually less than 10) can be estimated by model fitting to typically available polymerization data. Beyond this number, one tends to obtain large and highly correlated parameter uncertainties, often with multiple sets of parameter estimates providing equally good fits to the data. To avoid this situation, it becomes necessary to eliminate some of the unknown parameters by making simplifying assumptions (such as assuming side reactions do not happen or ignoring possible chain length dependencies of various rate parameters). These assumptions (which, due to problem one, are difficult to test) can severely compromise the validity of the model.

Why are these problems significant? Provided a model, or indeed any random polynomial, can provide a good fit to the available data, it is capable of reproducing those same data under the same range of reaction conditions and can therefore be very useful. However, if the underlying physical assumptions of the model are invalid, then the model is likely to lack predictive value when the reaction conditions are altered and/or it is used to predict other observables for the same process. For example, for most copolymerizations, one may successfully fit the terminal model to measured composition data; however, when the same model is used to predict the average propagation rate coefficients for the same system, it fails dramatically [21].

This lack of predictive capacity can extend to the model parameters as well. As explained above, these model parameters ostensibly correspond to the rate coefficients of the individual reactions within the reaction scheme. However, if these model parameters are "measured" by fitting an invalid model to available experimental data, they will not necessarily bear any direct relation to these rate coefficients but will instead be some complex and unknown function of the rates of several different chemical reactions. This in turn will compromise any predictions and/or measurements that are made when these parameters are used as rate coefficients for the same individual reaction when modeling other chemical processes.

At a more fundamental level, invalid kinetic models can hamper the development of better methods for controlling the products of free-radical polymerization. To state the obvious, if side reactions are giving rise to unwanted defect structures or loss of control over the molecular weight distribution, it will be difficult to develop strategies for minimizing these side reactions (e.g., by changes to the reaction conditions) if the kinetic model assumes that these particular side reactions are not occurring. Likewise, it is difficult to model and explain the effects of chemical structure on the values of model parameters when they do not directly correspond to the rate coefficients of the assumed chemical reactions. The development of accurate kinetic models is essential if such structure–reactivity studies are to be physically meaningful and of assistance in the development of optimal reagents.

9.2.2 *Ab Initio* Kinetic Modeling

Ab initio kinetic modeling aims to address the principal problem with conventional kinetic modeling: its reliance on model fitting for parameter estimation. Instead, the required rate coefficients for the various individual reactions are obtained using high-level *ab initio* molecular orbital theory calculations. Like in conventional kinetic modeling, one first assumes a complete reaction scheme for the process, containing all kinetically distinct individual reactions. However, unlike conventional modeling, one then proceeds to calculate the forward (and, if relevant, reverse) rate coefficients for each of these individual reactions using quantum chemistry. In doing this, it actually becomes possible to build the correct reaction scheme from the ground up—additional reactions can be ruled in or out of the scheme on the basis of their calculated rate coefficients. Having assembled the set of

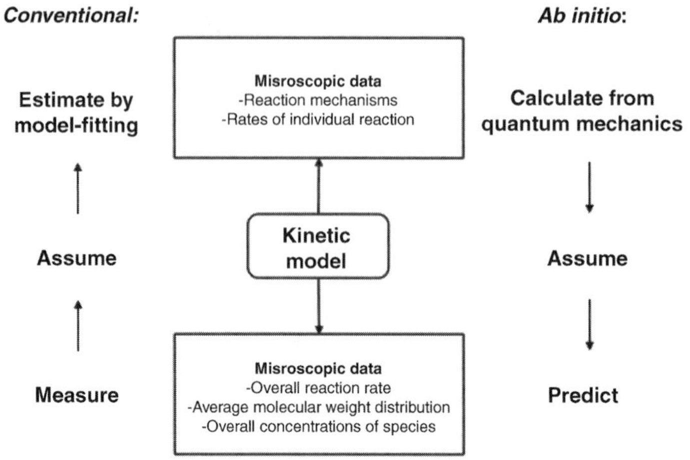

SCHEME 9.3 Schematic comparison of conventional and *ab initio* kinetic modeling.

individual reactions and their rate expressions and rate coefficients, one then solves the resulting coupled differential equations in the normal manner. However, as the values of the model parameters are already known, the model can be used to predict the macroscopic outcome of a process without the need for fitting to experimental data. These predictions can then be compared directly with experimental data so as to test the validity of the kinetic model.

Ab initio kinetic modeling essentially reverses the conventional modeling approach. Instead of predicting the microscopic information by fitting a kinetic model to the macroscopic experimental data, the model is instead used to predict macroscopic data from the microscopic information provided by quantum mechanics (see Scheme 9.3). As a result, in contrast to conventional modeling, *ab initio* kinetic model predictions are genuine predictions, totally independent of the experimental measurements used to test them. Hence, although a kinetic model is still assumed, it is much easier to test it and falsify it through direct comparison of the model predictions with experimental measurements. While it is easy to fit an invalid model to the experimental data, the probability that an invalid model can predict these same data correctly without the aid of fit parameters is very low.

Through the *ab initio* approach, it is also much easier to build a more complete kinetic model in the first place. Not only can individual reactions/side reactions be ruled in or out of the reaction scheme on the basis of the *ab initio* calculations, but also the practical limits to model complexity are less demanding than those of conventional kinetic models. Whereas with conventional modeling it is necessary to restrict the number of unknown parameters according to the degrees of freedom in the data, for *ab initio* kinetic modeling this is not necessary. Certainly each additional reaction/side reaction in a kinetic scheme represents an extra kinetic parameter that needs to be calculated. However, as each calculation is independent, the inclusion of additional reactions does not increase the overall degree of difficulty—provided computer time is available, one can consider as many reactions as necessary and the overall cost is linear scaling. This is important because, as we show below, complete kinetic models for many modern polymerization processes have many more unknown parameters than there are degrees of freedom in the accessible experimental data—studying such processes through conventional techniques is difficult, if not impossible.

9.3 QUANTUM CHEMICAL METHODOLOGY

The use of quantum chemistry to obtain the individual rate coefficients of a free-radical polymerization process frees them from errors due to kinetic model-based assumptions. However, this approach introduces a new source of error in the model predictions: the quantum chemical calculations themselves. As is well known, as there are no simple analytical solutions to a many-electron Schrödinger equation, numerical approximations are required. While accurate methods exist, they are generally very computationally intensive and their computational cost typically scales exponentially with the size of the system under study. The application of quantum chemical methods to radical polymerization processes necessarily involves a compromise in which small model systems are used to mimic the reactions of their polymeric counterparts so that high levels of theory may be used. This is then balanced by the need to make these models as realistic as possible; hence, lower cost theoretical procedures are frequently adopted, often to the detriment of the accuracy of the calculations. Nonetheless, aided by rapid and continuing increases to computer power, chemically accurate predictions are now possible, even for solvent-sensitive systems [8]. In this section we examine the "best-practice" methodology required to generate accurate gas- and solution-phase predictions of rate coefficients in free-radical polymerization.

9.3.1 Model Systems

Free-radical polymerization typically involves reactions of radicals having molecular weights of the order of 10^5 g/mol or more (i.e., thousands of non-hydrogen atoms). In contrast, even at relatively low levels of theory, quantum chemical calculations are typically only practical for systems with tens of non-hydrogen atoms. To study polymeric reactions, it is therefore necessary to model the polymeric species using related short-chain radicals. Thus, for example, one might model the propagating radical in vinyl chloride polymerization as the corresponding trimer radical $CH_3CH(Cl)CH_2CH(Cl)CH_2CH(Cl)^{\bullet}$. Such models are feasible because, provided there is no conjugation along the polymer chain, the influence of remote substituents diminishes rapidly with their distance from the reaction center. This is evident in experimental studies of the chain length dependence of propagation rate coefficients, which demonstrate convergence to within a factor of 2 or 3 by the trimer stage with complete convergence by the decamer stage or sooner [22]. It is also the basis of leading models for copolymerization kinetics, which ignore all substituent effects beyond the penultimate unit of the propagating radical [23].

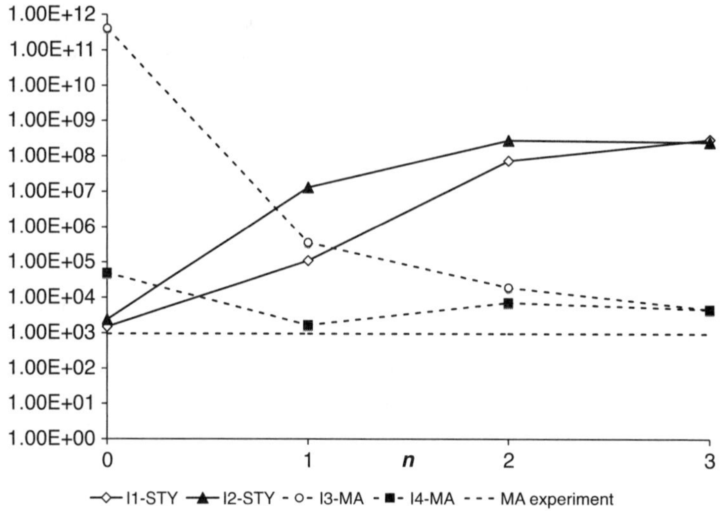

FIGURE 9.1 Chain length-dependent equilibrium constants for the RAFT-related reaction $I–M_n^{\cdot} + S{=}C(Z)SR \rightarrow I–M_n–SC^{\cdot}(Z)SR$. Owing to the size of the calculations, for the $n = 3$ case, the I-fragment is replaced with an H. The lines for I1-STY and I2-STY are for the gas-phase reaction at 60°C where Z = Ph, R = CH_3, M = $–CH_2–CHPh$ (i.e., a styrene-derived unit), I1 = $(CH_3)_2C(CN)$–, and I2 = $(CH_3)_2C(Ph)$ [11i]. The lines for I3-MA and I4-MA are for the reaction in toluene at −30°C where R = $CH(CH_3)COOCH_3$, Z = $SCH(CH_3)COOCH_3$, M = $CH_2CHCOOCH_3$ (i.e., a methyl acrylate-derived unit), and I3 and I4 are the acyl and alkyl fragments of the initiator 2-methyl-1-[4-(methylthio)phenyl]-2-morpholinopropan-1-one [11j]. The line "MA experiment" is the corresponding experimental value for these two reactions at the long-chain limit, as measured via laser flash photolysis [24].

To establish the minimal chain length necessary to model a polymerization reaction, one should, in principle, calculate the reaction as a function of chain length and thereby determine the point of acceptable convergence. However, such studies have now been performed for a variety of polymerization reactions and the emerging consensus is that, for species without conjugation along the backbone, a dimer or trimer radical is sufficient to model the corresponding polymeric reaction. As a couple of examples from the literature, Figure 9.1 shows the chain length-dependent equilibrium constants for the addition–fragmentation process in two very different RAFT polymerizations, each containing two different possible chain ends. For each polymerization considered, the equilibrium constant changes substantially from $n = 0$ to 1 and from $n = 1$ to 2; however, once the terminal and penultimate units are fixed, further changes beyond this are relatively small (less than a factor of 4). It is also seen that, where experimental data are available [24], the converged computational values agree well with the polymeric experimental data. Figure 9.2 shows the chain length dependence of the propagation rate coefficient for the monomers acrylonitrile, vinyl chloride, and ethylene. As in the case of RAFT polymerization, large changes can occur between the unimer and dimer; however, once these substituents are in place, further changes are fairly minimal (less than a factor of 1.5 in the three cases shown).

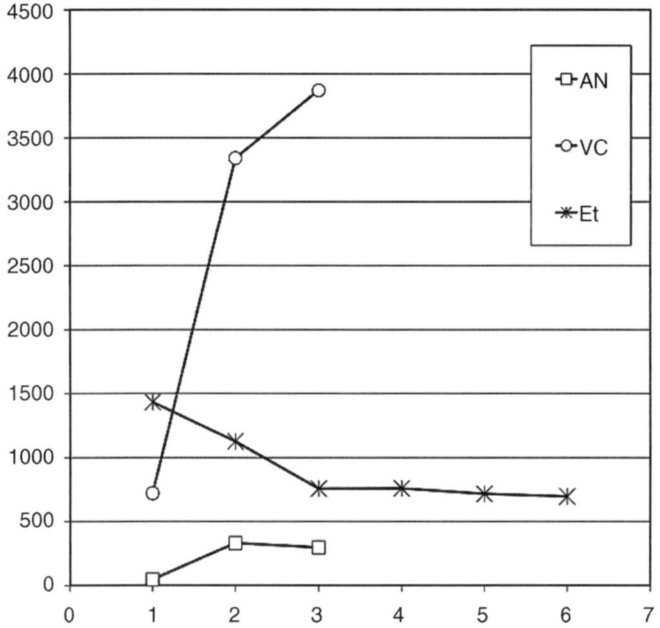

FIGURE 9.2 Plot of calculated gas-phase propagation rate coefficient (L/(mol s)) for vinyl chloride (VC), acrylonitrile (AN), and ethylene (Et) polymerization as a function of chain length. For VC and AN, results at 298.15K for the unimer, dimer, and trimer propagating radicals are shown [8a]; for Et, results at 300K are shown up to the hexamer [12k]. In all cases, the chains terminated with an H rather than an initiator fragment.

While a longer oligomer may be necessary for complete convergence of the rate coefficient, a dimer or trimer model can usually approximate the long-chain limit to within the limits of error.

9.3.2 Theoretical Procedures

Having established that (chemically controlled) radical polymerization processes can usually be studied using model reactions involving dimers or trimers in place of the corresponding polymeric species, the next question is whether there exist sufficiently low-cost theoretical procedures suitable for radical reactions of this size (about 30–40 non-hydrogen atoms). There is a large literature on the accuracy and applicability of computational procedures for radical chemistry in general [25], and radical polymerization in particular [7]. The following is a brief summary of the current "best practice" methodology, as used successfully to reproduce experimental values of the propagation rate coefficient of various monomers [8], and the equilibrium constant for the controlling reaction in various RAFT polymerization [11j], thioketone-mediated polymerization [16a], ATRP [9e,f], and NMP [10f] processes. For more details on these procedures and benchmarking studies, the reader is referred to the original references.

Low levels of theory such as small basis set DFT or HF calculations are usually suitable for geometry optimizations and frequency calculations, provided an IRCMax approach [26] is used to correct transition structures and provided frequencies are scaled by their appropriate scale factors. When combinatorial explosion of the conformational space becomes a problem for larger molecules, our new semisystematic algorithm, *energy-directed tree search*, can reliably locate the global minimum conformation at a fraction of the cost of full systematic conformation search [27].

Improved energies should be calculated using high-level *ab initio* molecular orbital theory procedures. For most classes of radical reactions, the most computationally inexpensive method that can still *reliably* deliver chemical accuracy (i.e., about 5 kJ/mol) is the composite procedure G3(MP2)-RAD [28]. Where computational expense renders even this procedure impractical, a good approximation is possible using a multilayer ONIOM [29] approach in which the reaction core (defined here as all forming and breaking bonds and any substituents attached to these bonds or conjugated with them) is studied at G3(MP2)-RAD and the remaining remote substituent effects are studied at ROMP2/6-311 + G(3df,2p). In a few cases, such as radical addition to thiocarbonyls, even G3(MP2)-RAD (or its ONIOM approximation) is not sufficient. For these reactions, one can define an additional ONIOM layer (i.e., an "inner core" that includes the forming and breaking bonds but not necessarily all of the α substituents), which is then studied at a higher level of theory such as W1 [30].

In the gas phase, it is usually sufficient to calculate the partition functions and associated thermal corrections to the enthalpy and entropy using the standard textbook formulae [31] for an ideal gas under the harmonic oscillator–rigid rotor approximation, provided one then makes explicit corrections for low-frequency torsional modes. These modes can be treated instead as one-dimensional hindered internal rotations using the torsional eigenvalue summation procedure described in Ref. [32]. Rate and equilibrium constants can then be obtained from the following standard textbook formulae [31]:

$$K(T) = (c^{\circ})^{\Delta n} e^{-\Delta G/RT} = (c^{\circ})^{\Delta n} \left(\frac{\prod_{\text{products}} Q_j}{\prod_{\text{reactants}} Q_i} \right) e^{-\Delta E/RT} \tag{9.1}$$

$$k(T) = \kappa(T) \frac{k_{\text{B}} T}{h} (c^{\circ})^{1-m} e^{-\Delta G^{\ddagger}/RT} = \kappa(T) \frac{k_{\text{B}} T}{h} (c^{\circ})^{1-m} \frac{Q_{\ddagger}}{\prod_{\text{reactants}} Q_i} e^{-\Delta E^{\ddagger}/RT} \tag{9.2}$$

In the above formulae, $\kappa(T)$ is the tunneling correction factor, T is the absolute temperature, k_B is the Boltzmann's constant, h is the Planck's constant, c° is the standard unit of concentration (mol/L), R is the universal gas constant, m is the molecularity of the reaction, Δn is the change in moles on reaction, $Q_{\ddagger}$, Q_i, and Q_j are the molecular partition functions of the transition structure, reactant i, and product j, respectively, $\Delta G^{\ddagger}$ is the Gibb's free energy of activation, ΔG is the Gibb's free energy of reaction, $\Delta E^{\ddagger}$ is the 0K, zero-point energy corrected energy barrier for the reaction, and ΔE is the 0K, zero-point energy corrected energy change for the reaction. The value of c° ($=P/RT$) depends on the standard-state concentration assumed in calculating the thermodynamic quantities (i.e., enthalpy, entropy, Gibb's free energy, and translational partition function). In essence, these are typically calculated for 1 mol of ideal gas at a specified temperature and pressure (P), whereas (bimolecular) rate coefficients are typically reported in units of L/(mol s). The tunneling coefficient $\kappa(T)$ corrects for quantum effects in motion along the reaction path, and is close to unity for radical addition reactions but is likely to be significant for hydrogen abstraction reactions. Ideally, tunneling coefficients should be calculated using multidimensional methods such as SCT or LCT [33]. However, for large systems these methods are expensive, and the Eckart method offers a low-cost alternative that is suitable for cases (such as those relevant to radical polymerization) with small-moderate levels of tunneling [34].

The development of accurate and inexpensive theoretical procedures to study reactions in the solution phase is an ongoing area of research. However, very promising results have recently been obtained for several solvent-sensitive polymerization reactions using simple dielectric continuum models, such as (C)PCM [35] to measure the free energy of solvation of the reactants and products [9e,f,10f,11j]. These solvation energies (ΔG_{solv}) are then combined with the accurate gas-phase energies (ΔG_{gas}) via a Hess cycle to obtain the solution free energies:

$$\Delta G_{\mathrm{soln}} = \Delta G_{\mathrm{gas}} + \Delta G_{\mathrm{solv}} + \Delta G^{1\ \mathrm{atm}\rightarrow 1\ \mathrm{M}} \tag{9.3}$$

In the above equation, $\Delta G^{1\ \mathrm{atm}\rightarrow 1\ \mathrm{M}}$ is required for converting from the gas-phase standard state for an ideal gas (typically 1 atm) to 1 M in solution:

$$\Delta G^{1\ \mathrm{atm}\rightarrow 1\ \mathrm{M}} = \Delta n\, RT \ln(V) = \Delta n\, RT \ln\left(\frac{RT}{P}\right) \tag{9.4}$$

where Δn is the number of moles of gas change from reactants to products. Having made this correction for the change in state, the standard unit of concentration in the rate and equilibrium constant expressions (Eqs (9.1) and (9.2)) becomes $c^\circ = 1$ mol/L, rather than its value for an ideal gas (e.g., 0.0408 mol/L at room temperature and pressure).

As continuum solvent models are heavily parameterized, they should be implemented using the same protocols (e.g., level of theory) for which they were parameterized. In this regard, it is worth noting that most simple continuum methods are currently only parameterized to reproduce experimental solvation free energies at a single temperature, usually 298K. To study reactions at other temperatures, and thereby isolate the constituent entropic and enthalpic contributions to the free energies in solution, one needs to use methods parameterized over a temperature range. One such method is COSMO-RS, which uses a scaled conductor boundary condition for the calculation of the polarization charges of a molecule in a continuum and further performs a statistical thermodynamic postprocessing of the results [36]. This method was recently successful in predicting the Arrhenius parameters of the propagation rate coefficients of methyl acrylate and vinyl acetate, systems for which the simple continuum models were less successful [8b]. This work also highlighted the importance of performing separate conformational searches for the gas- and solution-phase components of the calculations. While these results are very promising, further work is required to determine whether this good performance can be replicated for other solvent-sensitive polymerization reactions.

9.4 CASE STUDY: RAFT POLYMERIZATION

In this section, *ab initio* kinetic modeling is applied to the problem of modeling the RAFT process. As noted in Section 9.1, RAFT is one of several new techniques that have been developed over the last decade for controlling the molecular weight and architecture of the polymer produced from free-radical polymerization. In RAFT polymerization control is achieved through the reversible trapping of the propagating polymer radical as a dormant dithioester compound calls a RAFT agent via the following two-step chain transfer process:

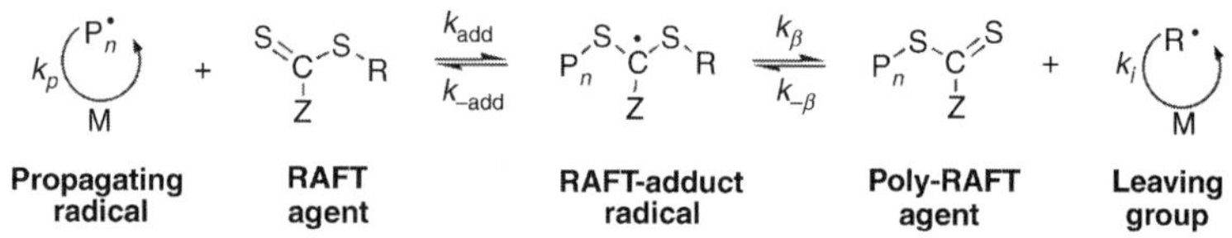

(9.5)

During the early stages of the RAFT process, the various low-molecular-weight radicals that are generated (e.g., initiating radical, unimer, dimer) may be transferred to the RAFT agent, giving rise to a variety of low-molecular-weight thiocarbonyl species. Accurately modeling the concentration profiles of these species is important as it is often desired to stop the reaction after a specific number of monomeric units have been added so as to generate a suitable precursor for block copolymer synthesis. Moreover, the rapidity and selectivity with which the propagating radical is transferred to initial RAFT agent is an important factor governing the polydispersity of the resulting polymer. Recently McLeary et al. [37,38] provided experimental measurements of the concentration profiles in a number of dithiobenzoate-mediated systems, but were unable to explain the results on a quantitative level in the context of the standard RAFT model. In this section, we show how the *ab initio* kinetic modeling approach can be used to predict these profiles from first principles. For full details, the reader is referred to our original published study [18].

The first step in both the standard and *ab initio* kinetic modeling methods is to assume a reaction scheme. For this purpose, we utilized the standard RAFT model [3] but augmented it to allow for chain length and other substituent effects in the addition–fragmentation equilibrium ($R^• + S{=}C(Z)SR' \rightarrow RSC^•(Z)SR'$); these are ignored in the standard model. In a recent *ab initio* study of simplified dithiobenzoate reactions, it was found that the equilibrium constant was affected by the chain length of the attacking radical ($R^•$) up to the trimer stage (see also Figure 9.1), and the RAFT agent substituent (R') up to the dimer, although in the latter case the most significant chain length effects converge at the unimer stage [11i]. We therefore included eight kinetically distinct addition–fragmentation reactions in our reaction scheme, corresponding to reactions of the initiating radical, the unimer, the dimer, and the trimer propagating radicals with the initial and unimeric RAFT agents. The remaining reactions comprise the standard initiation, propagation, and termination reactions, which are also common to the corresponding conventional radical polymerization process. The full reaction scheme is provided in Scheme 9.4.

The second step is to obtain values for the rate coefficients of the individual reactions. For this purpose we used high-level *ab initio* molecular orbital calculations to study all of the RAFT-related steps in Scheme 9.4, but standard experimental data for the initiation, propagation, and termination reactions of styrene (and related species) as these latter rate coefficients have been measured with a high degree of accuracy and precision in independent styrene homopolymerization experiments [39–43].[1] In cases where these latter data are not available, calculations could easily be done for any of the chemically controlled reactions, for example, chemically accurate predictions of rate coefficients for the propagation step in free-radical polymerization of several monomers have been recently published [8]. Only the diffusion-controlled termination processes would be difficult to calculate, requiring different theoretical approaches to those outlined here. This, however, does not pose a major disadvantage as, in contrast to the chemically controlled reactions, diffusion-controlled termination rate coefficients are relatively insensitive to chemical structure beyond very basic steric and chain length effects, and representative data are readily available in the literature.

To reduce the computational cost of the calculations further, we made the simplification of attributing the chain length dependence of the RAFT equilibrium entirely to the fragmentation reaction and assuming a standard value (consistent with both theory [11g][2] and experiment [44]) for the chain length-independent addition rate constant. The fragmentation rate coefficients could thus be obtained from the calculated equilibrium constants of these reactions using $K = k_{add}/k_{frag}$, and the more costly transition state calculations could then be avoided. This assumption is based on the evidence from previous experimental [44] and theoretical studies of these reactions [11g], which showed that the effect of changing the primary substituents (let alone remote substituents) of the attacking radical on the addition rate is negligible, even though the same substituents cause the fragmentation rate and equilibrium constants to vary over several orders of magnitude. This is because radical addition to the C=S bond is essentially barrierless for typical carbon-centered propagating radicals due to the poor overlap between 2p C and 3p S orbitals in the π-bond [11d,f,g,45]. There is a small free energy barrier due to the loss of translational

[1] A non-chain-length-dependent termination rate coefficient was implemented since during the initialization period the chain length does not change (i.e., increasing only from 1 to 3 U). It could be argued that a higher k_t (than the reported average k_t) should be employed for such small chain lengths. Therefore, the termination rate coefficient for the short chains (i.e., IMM, IM, and I terminating with each other) has been set to 1×10^9 L/(mol s). However, this parameter has a negligible effect on the predictions.

[2] In Ref. [11g], *ab initio* calculations for benzyl (2.76×10^6 L/(mol s)), cyanoisopropyl (8.29×10^5 L/(mol s)), and methyl (1.18×10^6 L/(mol s)) radicals adding to a common RAFT agent yield values of the same order of magnitude (i.e., $\log(k_{add}) = 6$).

I. Initiation

(Ia) $\text{Initiation} \xrightarrow[f]{k_d} I^{\bullet}$ (Ib) $I^{\bullet} + M \xrightarrow{k_i} I\text{-}M^{\bullet}$ $(I\text{-}M^{\bullet} = P_1^{\bullet})$

II. Raft equilibria

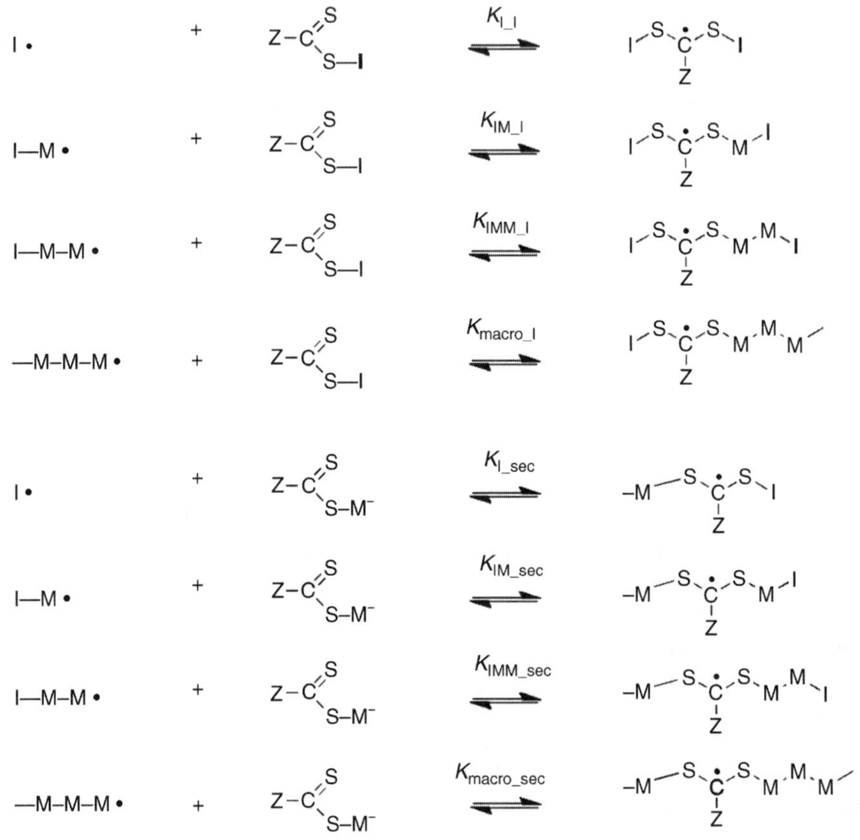

III. Propagation

(IIIa) $P_m^{\bullet} \xrightarrow[k_p]{\text{Monomer}} P_{m+1}^{\bullet}$ (*Note*: $P_m^{\bullet}$ includes $I\text{-}M^{\bullet}$, $I\text{-}MM^{\bullet}$, and $I\text{-}MMM^{\bullet}$)

V. Termination

(Va) $P_n^{\bullet} + P_m^{\bullet} \xrightarrow{k_t^{i,i}} P_{n+m}$ $n = m = i$

SCHEME 9.4 Chain length-dependent RAFT reaction scheme.

TABLE 9.1 Values of the Rate Parameters Used in the Kinetic Simulations

Coefficient	Value at 70°C	Source
k_{add}	1×10^6 L/(mol s)	Experiment = theory (see Refs [44])
$k_{frag\ I\_sec}$	5.8 s^{-1}	Theory using k_{add}/K_{I_sec} (see Ref. [18])
$k_{frag\ IM\_sec}$	0.3 s^{-1}	Theory using k_{add}/K_{IM_sec} (see Ref. [18])
$k_{frag\ IMM\_sec}$	3.7×10^{-4} s^{-1}	Theory using k_{add}/K_{IMM_sec} (see Ref. [18])
$k_{frag\ macro\_sec}$	2.5×10^{-4} s^{-1}	Theory using k_{add}/K_{macro_sec} (see Ref. [18])
$k_{frag\ I\_I}$	4.3 s^{-1}	Theory using k_{add}/K_{I_I} (see Ref. [18])
$k_{frag\ IM\_I}$	2.9×10^{-3} s^{-1}	Theory using k_{add}/K_{IM_I} (see Ref. [18])
$k_{frag\ IMM\_I}$	1.1×10^{-5}	Theory using k_{add}/K_{IMM_I} (see Ref. [18])
$k_{frag\ macro\_I}$	4.9×10^{-6}	Theory using k_{add}/K_{macro_I} (see Ref. [18])
k_p	480 L/(mol s)	Independent homopolymerization experiment [39]
$\langle k_t \rangle$	1.2×10^8 L/(mol s)	Independent homopolymerization experiment [40]
k_d	3.6×10^{-5} s^{-1}	Independent homopolymerization experiment [41]
f	0.7	Independent homopolymerization experiment [42]
$k_i = k_{p,rein}$	5200 L/(mol s)	Independent homopolymerization experiment [43]

and rotational entropy on forming the transition state in this bimolecular reaction but this contribution is relatively insensitive to structural changes in the reagents. In contrast, the reverse fragmentation barrier is large and governed by the reaction enthalpy; this can vary considerably with minor substitution changes according to the stabilities of the product and reactant radicals, and the strength of the C=S bond. It should also be noted that, unlike the fragmentation rate coefficient, there is general theoretical and experimental consensus as to the value of the addition constant, which would indicate that the value chosen for the study is reliable. The values of the rate coefficients used in the kinetic model are summarized in Table 9.1.

Finally, the model was implemented using PREDICI® so as to predict the concentration and conversion profiles under the reaction conditions used in the experimental study. The results obtained for the cyanoisopropyl dithiobenzoate-mediated polymerization of styrene at 70°C are shown in Figure 9.3 together with the corresponding experimental data from McLeary et al. [37]. From this figure, it is clear that the *ab initio* predictions match the (fully independent) experimental results almost perfectly, despite the fact that no model fitting of any kind was undertaken. While some minor discrepancies occur, these can be explained in terms of the simplifications that remain in the kinetic model. In particular, the chain length effects in the R′-position were terminated at the unimer stage (instead of the dimer stage) and this leads to minor errors in the high conversion region where reactions of the dimeric thiocarbonyl species become significant. Nonetheless, these predictions would be good enough for practical applications, for example, determining where the maximum yield of unimeric RAFT agent occurs, assessing how well controlled the process is, or studying the effect of changes to the reaction conditions. It is also worth noting that the predictions of the *ab initio* kinetic model were substantially better than those made with the standard RAFT model, reinforcing the importance of taking into account the chain length dependence of the fragmentation process [18].

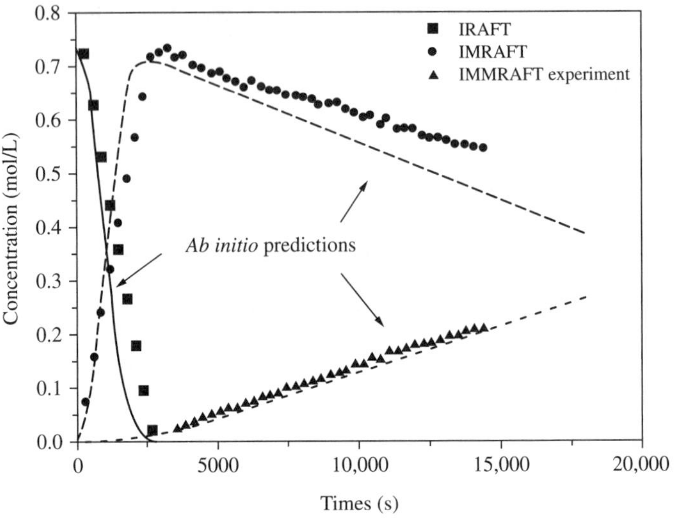

FIGURE 9.3 Experimental (symbols) [37] and theoretical (lines) [18] evolutions of the initial RAFT agent (cyanoisopropyl dithiobenzoate, IRAFT, full line) and the RAFT agents having one monomer unit (IMRAFT, dashed line) and two monomer units (IMMRAFT, dotted line) inserted. The concentrations of the reagents were $c^0_{\mathrm{IRAFT}} = 0.736\,\mathrm{mol/L}$, $c^0_{\mathrm{AIBN}} = 0.10\,\mathrm{mol/L}$, $c^0_{\mathrm{Styrene}} = 3.65\,\mathrm{mol/L}$, and $c^0_{\mathrm{C6D6}} = 5.4\,\mathrm{mol/L}$ at the temperature of 70°C.

9.5 OUTLOOK

The combination of quantum chemistry with kinetic modeling—*ab initio* kinetic modeling—is a mutually beneficial development that promises to improve the accuracy of kinetic models while at the same time expanding the scope of quantum chemical predictions from the microscopic to the macroscopic. As shown in this chapter, by providing access to *ab initio* values of the model parameters, quantum chemistry addresses a fundamental limitation of traditional kinetic modeling for complicated multistep processes, the need to "measure" the unknown model parameters by fitting the assumed model to experimental data. This begging of the question makes it difficult to test the validity of the assumed model and can place severe practical limitations on the model's complexity—problems that are avoided in the *ab initio* kinetic modeling approach. In this way, *ab initio* kinetic models can have fewer assumptions, and those that are made can be tested more thoroughly against experimental data.

While the accuracy of quantum chemical-derived parameters depends instead on the assumptions and numerical approximations used to solve the Schrödinger equation, in this chapter we have shown that, with appropriate care, these errors can be minimized, at least for the key steps in radical polymerization. Indeed, as shown in the case study provided, *ab initio* kinetic modeling can be used to make first-principles predictions of the concentration profiles of reagents, intermediates, and products in a complicated multistep process such as the RAFT process. In making these predictions, a pragmatic approach was adopted in which experimental values were used for those kinetic parameters that could be obtained with a high degree of accuracy and precision in independent experiments; theory was reserved for those remaining parameters that cannot be obtained without making model-based assumptions. In this way, one can avoid using theory to predict rate coefficients for diffusion-limited reactions such as termination, which are much more challenging.

As noted in Section 9.1, this powerful approach to kinetic modeling is by no means unique to the polymer field and is increasingly finding application in the study of other complicated multistep processes such as combustion and catalysis [19]. In fact, it is limited only by the ability of theory to predict accurate rate coefficients for the constituent reactions. While some processes, particularly those involving charged species in solution, remain computationally challenging, the scope of *ab initio* kinetic modeling is continually expanding thanks to ongoing developments in computational methods and algorithms, and increases in supercomputing power.

Within the radical polymer field, two broad areas of application for *ab initio* kinetic modeling are model development and process optimization. The RAFT case study presented in this chapter is an example of the former type of application; we are currently using the same *ab initio* approach to conventional radical polymerization to develop better models for predicting copolymer composition and microstructure, structural defect formation as a result of side reactions, and the effect of solvent and additives on the stereochemistry of the resulting polymer.

Having developed and established an accurate kinetic model for a particular process, it then becomes possible to use this model to optimize the process itself and evaluate new reagents prior to experimental testing. For example, we have used a hybrid *ab initio* kinetic model to evaluate whether or not a computer-designed RAFT reagent [15], *t*-butyl fluorodithioformate, would be the first reagent capable of controlling ethylene polymerization [46]. Our model was developed from the well-established conventional kinetic model of ethylene polymerization [47], which was then augmented to include reactions of the ethylene propagating radical with our proposed RAFT agent. Quantum chemical calculations were used to parameterize the RAFT reactions, as these were not accessible elsewhere; the remaining ethylene parameters were taken from experiment. Using this model, we predicted the molecular weight distribution of the resulting polymer as a function of the reaction conditions, finding that controlled polymerization should be possible provided that the correct reagent concentrations were used. In this way, *ab initio* kinetic modeling can be used to determine with a reasonable degree of confidence whether experimental investigations (which are typically more complicated and more expensive) are worth pursuing.

REFERENCES

1. (a) For more information on the chemistry and kinetics of free-radical polymerization, see, for example: K. Matyjaszewski and T. P. Davis, *Handbook of Radical Polymerization*, John Wiley and Sons, Inc., New York (2002); (b) G. Moad and D. H. Solomon, *The*

Chemistry of Free-Radical Polymerization, Pergamon, Oxford (1995); (c) G. Odian, *Principles of Polymerization*, Wiley Interscience, New York (1991).

2. M. Kamigaito and K. Satoh, *Macromolecules* **41**, 269 (2008).
3. G. Moad, E. Rizzardo, and S. H. Thang, *Aust. J. Chem.* **58**, 379 (2005).
4. K. Matyjaszewski, *Prog. Polym. Sci.* **30**, 858 (2005).
5. C. J. Hawker, A. W. Bosman, and E. Harth, *Chem. Rev.* **101**, 3661 (2001).
6. For a discussion of this problem in the context of the RAFT process, see, for example: C. Barner-Kowollik, M. L. Coote, T. P. Davis, L. Radom, and P. Vana, *J. Polym. Sci. A Polym. Chem.* **41**, 2828 (2003).
7. For a recent review, see, for example: M. L. Coote, *Macromol. Theory Simul.* **18**, 388 (2009).
8. (a) See, for example: E. I. Izgorodina and M. L. Coote, *Chem. Phys.* **324**, 96–110 (2006); (b) C. Y. Lin, E. I. Izgorodina, and M. L. Coote, *Macromolecules* **43**, 553 (2010).
9. (a) M. B. Gillies, K. Matyjaszewski, P.-O. Norrby, T. Pintauer, R. Poli, and P. Richard, *Macromolecules* **36** 8551 (2003); (b) D. A. Singleton, D. T. Nowlan, III, N. Jahed, and K. Matyjaszewski, *Macromolecules* **36** 8609 (2003); (c) K. Matyjaszewski and R. Poli, *Macromolecules* **38** 8093 (2005); (d) C. Y. Lin, M. L. Coote, A. Petit, P. Richard, R. Poli, and K. Matyjaszewski, *Macromolecules* **40** 5985 (2007); (e) W. Tang, Y. Kwak, W. Braunecker, N. V. Tsarevsky, M. L. Coote, and K. Matyjaszewski, *J. Am. Chem. Soc.* **130**, 10702 (2008); (f) C. Y. Lin, M. L. Coote, A. Gennaro, and K. Matyjaszewski, *J. Am. Chem. Soc.* **130**, 12762 (2008).
10. (a) P. Marsal, M. Roche, P. Tordo, and P. de Sainte Claire, *J. Phys. Chem. A* **103**, 2899 (1999); (b) D. Gigmes, A. Gaudel-Siri, S. R. A. Marque, D. Bertin, P. Tordo, P. Astolfi, L. Greci, and C. Rizzoli, *Helv. Chim. Acta* **89**, 2312 (2006); (c) A. Kaim and E. Megiel, *J. Polym. Sci. A* **44**, 914 (2005); (d) A. Kaim, *J. Polym. Sci. A* **45**, 232 (2006); (e) E. Megiel and A. Kaim, *J. Polym. Sci. A* **46**, 1165 (2008); (f) J. L. Hodgson, C. Y. Lin, M. L. Coote, S. R. A. Marque, and K. Matyjaszewski, *Macromolecules* **43**, 3728 (2010).
11. (a) S. C. Farmer and T. E. Patten, *J. Polym. Sci. A Polym. Chem.* **A40**, 555 (2002); (b) M. L. Coote and L. Radom, *J. Am. Chem. Soc.* **125**, 1490 (2003); (c) M. L. Coote and L. Radom, *Macromolecules* **37** 590 (2004); (d) M. L. Coote, *Macromolecules* **37** 5023 (2004); (e) A. Feldermann, M. L. Coote, M. H. Stenzel, T. P. Davis, and C. Barner-Kowollik, *J. Am. Chem. Soc.* **126**, 15915 (2004); (f) M. L. Coote and D. J. Henry, *Macromolecules* **38** 1415 (2005); (g) M. L. Coote, *J. Phys. Chem. A* **109**, 1230 (2005); (h) M. L. Coote, E. H. Krenske, and E. I. Izgorodina, *Macromol. Rapid Commun.* **27**, 473 (2006); (i) E. I. Izgorodina and M. L. Coote, *Macromol. Theory Simul.* **15**, 394 (2006); (j) C. Y. Lin and M. L. Coote, *Aust. J Chem.* **62**, 1479 (2009); (k) E. Chernikova, V. Golubev, A. Filippov, C. Y. Lin, and M. L. Coote, *Polym. Chem.* **1**, 1437–1440 (2010).
12. (a) G. Leroy, J.-P. Dewispelaere, H. Benkadour, and C. Wilante, *Macromol. Theory Simul.* **5**, 269 (1996); (b) J. P. A. Heuts, R. G. Gilbert, and L. Radom, *J. Phys. Chem.* **100**, 18997 (1996); (c) D. M. Huang, M. J. Monteiro, and R. G. Gilbert, *Macromolecules* **31** 5175 (1998); (d) J. S.-S. Toh, D. M. Huang, P. A. Lovell, and R. G. Gilbert, *Polymer* **42** 1915 (2001); (e) J. Filley, J. T. McKinnon, D. T. Wu, and G. H. Ko, *Macromolecules* **35** 3731 (2002); (f) C.-G. Zhan and D. A. Dixon, *J. Phys. Chem. A* **106**, 10311 (2002); (g) S. C. Thickett and R. G. Gilbert, *Polymer* **45** 6993 (2004); (h) K. Van Cauter, K. Hemelsoet, V. Van Speybroeck, M. F. Reyniers, and M. Waroquier, *Int. J. Quantum Chem.* **102**, 454 (2004); (i) S. Salman, A. Z. Albayrak, D. Avci, and V. Aviyente, *J. Polym. Sci. A* **43**, 2574 (2005); (j) H. Günaydin, S. Salman, N. S. Tüzün, D. Avci, and V. Aviyente, *Int. J. Quantum Chem.* **103**, 176 (2005); (k) K. Van Cauter, V. Van Speybroeck, P. Vansteenkiste, M.-F. Reyniers, and

M. Waroquier, *Chemphyschem* **7** 131 (2006); (l) I. Degirmenci, D. Avci, V. Aviyente, K. Van Cauter, V. Van Speybroeck, and M. Waroquier, *Macromolecules* **40**, 9590 (2007).

13. (a) J. Purmova, K. F. D. Pauwels, W. van Zoelen, E. J. Vorenkamp, A. J. Schouten, and M. L. Coote, *Macromolecules* **38** 6352 (2005); (b) K. Van Cauter, B. J. Van Den Bossche, V. Van Speybroeck, and M. Waroquier, *Macromolecules* **40** 1321 (2007); (c) J. Purmová, K. F. D. Pauwels, M. Agostini, M. Bruinsma, E. J. Vorenkamp, A. J. Schouten, and M. L. Coote, *Macromolecules* **41**, 5527 (2008).
14. (a) J. P. A. Heuts, Sudarko, and R. G. Gilbert, *Macromol. Symp.* **111**, 147–157 (1996); (b) J. P. A. Heuts, R. G. Gilbert, and I. A. Maxwell, *Macromolecules* **30** 726 (1997); (c) M. L. Coote, T. P. Davis, and L. Radom, *Theochem* **461-462** 91 (1999); (d) M. L. Coote, T. P. Davis, and L. Radom, *Macromolecules* **32** 5270 (1999); (e) M. L. Coote, T. P. Davis, and L. Radom, *Macromolecules* **32** 2935 (1999); (f) P. Cieplak and A. Kaim, *J. Polym. Sci. A* **42**, 1557 (2004).
15. (a) M. L. Coote and D. J. Henry, *Macromolecules* **38** 5774 (2005); (b) C. W. Barner-Kowollik, M. L. Coote, T. P. Davis, M. H. Stenzel, and A. Theis, *Polymerization Agent,* International Patent Number WO2006122344 A1 (2006), 54 pp.; (c) A. Theis, M. H. Stenzel, T. P. Davis, M. L. Coote, and C. Barner-Kowollik, *Aust. J. Chem.* **58**, 437 (2005).
16. (a) A. Ah Toy, H. Chaffey-Millar, T. P. Davis, M. H. Stenzel, E. I. Izgorodina, M. L. Coote, and C. Barner-Kowollik, *Chem. Commun.* **835** (2006); (b) H. Chaffey-Millar, E. I. Izgorodina, C. Barner-Kowollik, and M. L. Coote, *J. Chem. Theory Comput.* **2**, 1632 (2006).
17. (a) J. L. Hodgson and M. L. Coote, *Macromolecules* **38** 8902 (2005); (b) M. L. Coote, J. L. Hodgson, E. H. Krenske, M. Namazian, and S. B. Wild, *Aust. J. Chem.* **60**, 744 (2007).
18. M. L. Coote, E. I. Izgorodina, E. H. Krenske, M. Busch, and C. Barner-Kowollik, *Macromol. Rapid Commun.* **27**, 1015 (2006).
19. (a) See, for example: M. Martoprawiro, G. B. Bacskay, and J. C. Mackie, *J. Phys. Chem. A* **103**, 3923 (1999); (b) M. Saeys, J. W. Thybaut, M. Neurock, and G. B. Marin, *Mol. Phys.* **102**, 267 (2004); (c) L. K. Huynh, K. C. Lin, and A. Violi, *J. Phys. Chem. A* **112**, 13470 (2008); (d) S. Kozuch and S. Shaik, *J. Phys. Chem. A* **112**, 6032 (2008).
20. M. Wulkow, *Macromol. Theory Simul.* **5**, 393 (1996).
21. T. Fukuda, Y.-D. Ma, and H. Inagaki, *Macromolecules* **18**, 17 (1985).
22. (a) M. Deady, A. W. H. Mau, G. Moad, and T. H. Spurling, *Makromol. Chem.* **194**, 1691 (1993); (b) P. Zetterlund, W. Busfield, and I. Jenkins, *Macromolecules* **32**, 8041 (1999).
23. M. L. Coote and T. P. Davis, *Prog. Polym. Sci.* **24**, 1217 (1999).
24. M. Buback, P. Hesse, T. Junkers, and P. Vana, *Macromol. Rapid Commun.* **27**, 182 (2006).
25. For a review of the early work in this field, see: H. Fischer and L. Radom, *Angew. Chem. Int. Ed.* **40**, 1340–1371 (2001).
26. D. K. Malick, G. A. Petersson, and J. A. Montgomery, *J. Chem. Phys.* **108**, 5704 (1998).
27. E. I. Izgorodina, C. Y. Lin, and M. L. Coote, *Phys. Chem. Chem. Phys.* **9**, 2507 (2007).
28. D. J. Henry, M. B. Sullivan, and L. Radom, *J. Chem. Phys.* **118**, 4849 (2003).
29. (a) E. I. Izgorodina, D. R. B. Brittain, J. L. Hodgson, E. H. Krenske, C. Y. Lin, M. Namazian, and M. L. Coote, *J. Phys. Chem. A* **111**, 10754 (2007); (b) C. Y. Lin, J. L. Hodgson, M. Namazian, and M. L. Coote, *J. Phys. Chem. A* **113**, 3690 (2009).
30. J. M. L. Martin and G. J. De Oliveira, *J. Chem. Phys.* **111**, 1843 (1999).
31. (a) For textbooks on transition state theory and statistical thermodynamics, see, for example: (a) S. W. Benson, *Thermochemical Kinetics*, John Wiley & Sons, Inc., New

York (1976); (b) D. A. McQuarrie, *Statistical Mechanics*, Harper & Row, New York (1976); (c) R. G. Gilbert and S. C. Smith, *Theory of Unimolecular and Recombination Reactions*, Blackwell Scientific, Oxford (1990); (d) J. I. Steinfeld, J. S. Francisco, and W. L. Hase, *Chemical Kinetics and Dynamics*, 2nd edition, Prentice-Hall, Englewood Cliffs, NJ (1999); (e) P. W. Atkins, *Physical Chemistry*, 6th edition, W. H. Freeman and Company, San Francisco, CA (2000).

32. C. Y. Lin, E. I. Izgorodina, and M. L. Coote, *J. Phys. Chem. A* **112**, 1956 (2008).
33. For a recent review, see: A. Fernandez-Ramos, J. A. Miller, S. J. Klippenstein, and D. G. Truhlar, *Chem. Rev.* **106**, 4518 (2006).
34. M. L. Coote, M. A. Collins, and L. Radom, *Mol. Phys.* **101**, 1329 (2003).
35. (a) See, for example: A. Klamt and G. Schueuermann, *J. Chem. Soc. Perkin Trans.* **2** 799 (1993); (b) M. Cossi, N. Rega, G. Scalmani, and V. Barone, *J. Comput. Chem.* **24**, 669 (2003); (c) S. Miertus, E. Scrocco, and J. Tomasi, *J. Chem. Phys.* **55**, 117 (1981); (d) J. Tomasi, *Theor. Chem. Acc.* **112**, 184 (2004).
36. (a) A. Klamt, *J. Phys. Chem.* **99**, 2224 (1995); (b)A. Klamt, *COSMO-RS: From Quantum Chemistry to Fluid Phase Thermodynamics and Drug Design*, Elsevier Science Ltd. , Amsterdam, The Netherlands (2005); (c) A. Klamt, V. Jonas, T. Burger, and J. C. W. Lohrenz, *J. Phys. Chem. A* **102**, 5074 (1998).
37. J. B. McLeary, F. M. Calitz, J. M. McKenzie, M. P. Tonge, R. D. Sanderson, and B. Klumperman, *Macromolecules* **37**, 2383 (2004).
38. (a) See also: J. B. McLeary, J. M. McKenzie, M. P. Tonge, R. D. Sanderson, and B. Klumperman, *Chem. Commun.* **17**, 1950 (2004); (b) J. B. McLeary, F. M. Calitz, J. M. McKenzie, M. P. Tonge, R. D. Sanderson, and B. Klumperman, *Macromolecules* **38**, 3151 (2005).
39. M. Buback, R. G. Gilbert, R. A. Hutchinson, B. Klumperman, F.-D. Kuchta, B. G. Manders, K. F. O'Driscoll, G. T. Russell, and J. Schweer, *Macromol. Chem. Phys.* **196**, 3267 (1995).
40. M. Buback and F.-D. Kuchta, *Macromol. Chem. Phys.* **198**, 1455 (1997).
41. J. P. Van Hook and A. V. Tobolsky, *J. Am. Chem. Soc.* **80**, 779 (1958).
42. M. Buback, B. Huckestein, F.-D. Kuchta, G. T. Russell, and E. Schmid, *Macromol. Chem. Phys.* **195**, 2117 (1994).
43. K. Herberger and H. Fischer, *Int. J. Chem. Kinet.* **25**, 249–263 (1993).
44. (a) For example, compare the addition constants reported in: Y. Kwak, A. Goto, and T. Fukuda, *Macromolecules* **37**, 1219–1225 (2004); (b) A. Theis, T. P. Davis, M. H. Stenzel, and C. Barner-Kowollik, *Polymer* **47**, 999–1010 (2006); (c) A. Theis, A. Feldermann, N. Charton, M. H. Stenzel, T. P. Davis, and C. Barner-Kowollik, *Macromolecules* **38** 2595 (2005); (d) T. Junkers, A. Theis, T. P. Davis, M. Buback, M. H. Stenzel, P. Vana, and C. Barner-Kowollik, *Macromolecules* **38**, 9497–9508 (2005); (e) A. Theis, A. Feldermann, N. Charton, T. P. Davis, M. H. Stenzel, and C. Barner-Kowollik, *Polymer* **46**, 6797–6809 (2005).
45. D. J. Henry, M. L. Coote, R. Gómez-Balderas, and L. Radom, *J. Am. Chem. Soc.* **126**, 1732 (2004).
46. M. L. Coote, E. I. Izgorodina, G. E. Cavigliasso, M. Roth, M. Busch, and C. Barner-Kowollik, *Macromolecules* **39**, 4585 (2006).
47. M. Busch, *Macromol. Theory Simul.* **10**, 408 (2001).

CHAPTER 10

Intermolecular Electron Transfer Reactivity for Organic Compounds Studied Using Marcus Cross-Rate Theory

STEPHEN F. NELSEN

Department of Chemistry, University of Wisconsin, Madison, WI, USA

JACK R. PLADZIEWICZ

Department of Chemistry, University of Wisconsin, Eau Claire, WI, USA

10.1 INTRODUCTION

Relationships between structure and reactivity form the core of our understanding of chemistry, and are fundamental in allowing prediction of which compounds to employ for various functions. A basic determining factor of reactivity is the driving force for a reaction, ΔG°. The Hammett relationship examines structural changes that lead to predictable relative increments in ΔG° without affecting the barrier otherwise, and hence quantitatively correlates changes in rate constants when *meta* and *para* substituents on benzene rings located close enough to a reaction site change in charge during a reaction. However, most structural changes alter both the driving force and intrinsic reactivity—the reactivity at zero driving force. Especially well-studied series that cover very large ranges in reactivity include solvolyses in polar solvents and acid-catalyzed hydration of alkenes [1]. Detailed interpretation of most of these reactions is limited because the free energy change associated with the elementary steps is not known. Marcus pointed out that the natural comparison point for intrinsic reactivity for electron transfer (ET) reactions is the $\Delta G^\circ = 0$ self-ET rate constant, k_{ii}(self) [2]. Marcus described the relationship between intrinsic reactivity and the rate constant for outer sphere ET for a net chemical ET reaction, k_{ij}. Marcus

Rate Constant Calculation for Thermal Reactions: Methods and Applications, Edited by Herbert DaCosta and Maohong Fan.

used outer sphere in the Taube sense [3] of an ET reaction as one that does not involve simultaneous bridging by, and transfer of, ligands between electron donor and acceptor, and inner sphere for reactions in which ligand transfer is concerted with ET[1] [4]. For a reaction between a neutral species, $\mathbf{i}^0$, and its radical cation, $\mathbf{i}^+$ (Eq. (10.1), electrostatic factors are unimportant and the relationship

$$\mathbf{i}^0 + \mathbf{j}^+ \rightarrow \mathbf{i}^+ + \mathbf{j}^0 \tag{10.1}$$

obtained assuming that ET reactions are adiabatic is astonishingly simple (Eqs (10.2a) and (10.2b)) [2]. The self-exchange rate constants k_{ii} and k_{jj}:

$$k_{ij}(\text{calc}) = (k_{ii}k_{jj}K_{ij}f_{ij})^{1/2} \tag{10.2a}$$

$$\ln(f_{ij}) = [\ln(K_{ij})]^2/[4\ln(k_{ii}k_{jj}/Z^2)] \tag{10.2b}$$

and the equilibrium constant, K_{ij}, are the principal parameters determining k_{ij}. Z in Eq. (10.2b) is the pre-exponential factor for the rate constant of Eq. (10.1), and a collision number estimate, $10^{11}\ \text{M}^{-1}\ \text{s}^{-1}$, is often used as in Marcus's original work. For our systems the results are quite insensitive to the value employed, so we have retained use of $10^{11}\ \text{M}^{-1}\ \text{s}^{-1}$. The factor f_{ij} is 1 when $K_{ij} = 1$ and decreases as the reaction becomes more exoenergetic. No reactions in our data set have f_{ij} less than 0.1 because K_{ii} is always chosen to be reasonably small to keep the observed rate constant in the measurable range. Formal potentials were determined using cyclic voltammetry (CV) for all couples described here, and all the couples studied showed electrochemically reversible or nearly reversible CV curves, so K_{ij} used in Eqs (10.2a) and (10.2b) could be accurately determined from the differences in formal potential derived from them. A more general form of Eqs (10.2a) and (10.2b) that includes the work terms necessary for the multiply charged reactants has been successfully applied to a wide variety of inorganic, organic, organometallic, and biochemical reactions [5–7]. Implementation of these work terms has assumed that dielectric continuum theory properly describes how solvent reorganization energy, λ_s, changes with solvent. More recently it has been demonstrated that dielectric continuum theory does not work well for organic compounds in nonaqueous solvents [8]. Furthermore, most of the couples studied in the earlier work have rather similar k_{ii} values, which means that these studies principally test the dependence of k_{ij} on K_{ij}, which the success of the Hammett equation demonstrates works extremely well. Although Hammett used straight lines instead of parabolas in his analysis, large enough changes to observe the predicted curvature are essentially never caused by changing *meta* and *para* substituents.

[1] More recently others, who focused on organic systems instead of inorganic systems, especially Eberson and Kochi (see Ref. [4] for a few examples), have changed the meaning of inner sphere to indicate reactions for which a limited version of classical Marcus theory is perceived by them to greatly underestimate ET rate constants because Marcus originally assumed that electronic coupling could be ignored, which of course it cannot. However, Marcus soon included electronic coupling, calling it H_{ab}, and never used inner sphere to indicate anything about the size of the electronic coupling, making use of the altered meaning of inner sphere most unfortunate.

10.2 DETERMINATION OF $\Delta G^{\ddagger}_{ii}$ (fit) Values

We have measured k_{ij} using stopped-flow spectrophotometry for as wide a reactivity range as we can of 0/1 + couples for which $E^{\circ\prime}$ has been measured under the same conditions. Our principal contribution has been to include couples having a very large range of k_{ii} values. This has been achieved especially by including tetraalkylhydrazines of varied structural types in the data set. This causes a wide variation in the size of the geometry change between the oxidation states and hence the vibrational reorganization energy, λ_v [9–16]. Having both large and small k_{ii} couples available for study greatly expands the number of couples available for consideration compared to those that can be studied under self-ET conditions. Since no net chemical change occurs during self-ET, electron exchange is usually studied using magnetic resonance line broadening. However, the amount of broadening becomes too small to measure accurately by NMR in the slow-exchange region below about $7 \times 10^2\,M^{-1}\,s^{-1}$ [17]. In slow-exchange-limit NMR the theory for extracting the rate constant from signals that are singlets is exceptionally simple. The rate constant is given by Eq. (10.3), introduced over 50 years ago by McConnell and coworkers using copper NMR, and Weissman and coworkers using proton NMR [18]

$$k_{ii} = \pi\,\Delta\nu_{1/2}/[\mathbf{i}^+] \qquad (10.3)$$

where $\Delta\nu_{1/2}$ is the increase in line width caused by exchange. Few organic compound self-exchanges proceed slowly enough to make the slow-exchange region attainable and most organic radical cations are too unstable in the presence of the neutral compound, because of proton transfer decomposition of the radical cation caused by the neutral form, to do slow-exchange NMR experiments. The *N,N′*-bis(bicyclic) tetraalkylhydrazine$^{+/0}$ mixtures provide an exception to the usual instability of radical cations in the presence of the more basic neutral form because they benefit from Bredt's rule kinetic protection of the radical cation. The α-hydrogens of the radical cation, which are the ones removed by base, have a nearly 90° twist angle with the lone pair at nitrogen, so the deprotonated form would have a near 90° twist of the α-amino radical, destabilizing it so much that rapid deprotonation by the basic neutral hydrazine does not occur [19]. This allowed Nelsen's group to study several cases of kinetically protected *N,N′*-bis-bicyclic tetraalkylhydrazine radical cations, establishing that $\mathbf{R_4N_2}^{0/+}$ k_{ii} is exceptionally sensitive to the twist angle between the nitrogen atom lone pair axes *N,N*(θ) of the neutral form [17]. See structure **A** for an illustration of the electronically most stable

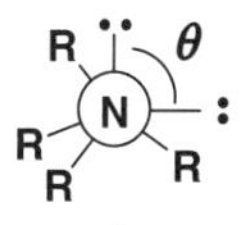

A

form, $\theta \sim 90°$, for acyclic hydrazines that have relatively small steric interactions between their alkyl groups [20]. It is easy to tell if one is in the slow-exchange NMR region because all peaks are broadened by the same amount, and the chemical shift is not affected by the concentration of $\mathbf{i}^+$. The theory is more complicated in the fast-exchange-limit NMR region, which has been especially used to measure k_{ii} values for substituted ferrocenes$^{0/+}$ [21]. In Ref. [10] we reported a significantly lower k_{11} value for $Cp^*CpFe^{0/+}$ than we had estimated in Ref. [9], from averaging the values reported for $Cp_2Fe^{0/+}$ and $Cp_2^*Fe^{0/+}$, and suggested that staying in the fast-exchange region was a problem for this compound. Much faster rate constants can be determined using EPR spectrometry, especially in the fast-exchange limit, where both the theory and the spectra being analyzed are more complicated [22]. There are also slow- and fast-exchange-limit regions for EPR. Although faster k_{ii} values have been determined, we have not seen k_{ii} values for any triarylamine derivatives determined by EPR, despite their ready availability, stability, and commercial interest in their use in devices. We presume that there are experimental problems in measuring them, and have studied several triarylamines in this work to provide k_{ii} values for them. Over half of the couples that we have studied by stopped flow (40 of 72) are triarylamines or have k_{ii} values that lie outside the rate constant range that could be studied under self-exchange conditions using magnetic resonance line broadening.

Stopped-flow kinetic data have often been used to establish relative reactivities by selecting a few isolable "oxidants" and determining k_{ij} values for their reaction with several reduced species. Extracting k_{ii} values using Eqs (10.2a) and (10.2b) then requires directly measuring k_{ii} values for some of the couples, using magnetic resonance methods. In contrast, we did not use any directly measured k_{ii} values in extracting intrinsic reactivities from our k_{ij} data. Instead, the k_{ij}(obs) and $E^{o\prime}$ values, both measured at 25°C in acetonitrile containing 0.1 M tetrabutylammonium perchlorate, were subjected to a least-squares fit of all the data [23] to Eqs (10.2a) and (10.2b) to extract k_{ii}(fit) values [12–16]. Our data set eventually included 206 reactions involving 72 couples. Information on each cross-reaction has been published: 1–47 [12], 48–91 [13], 92–141 [14], and 142–206 [16]. However, because of our global least-squares fit analysis, the calculated fit changes slightly as the data set is enlarged. The observed k_{ij} values are fit well using Eqs (10.2a) and (10.2b), as shown graphically in Figure 10.1. One hundred and ninety-one of the calculated k_{ij} values (93%) lie within a factor of 2 of the experimental value (the ratio $\boldsymbol{R} = k_{ij}(\text{obs})/k_{ij}(\text{calc})$ lying between 2.0 and 0.5, and $|RT\ln(R)| < 0.41$ kcal/mol) using the k_{ii}(fit) values and only three lie outside of the factor of 3. Since k_{ij}(calc) has a square root dependence on each k_{ii}(fit), a factor 2 discrepancy would correspond to a factor at most of 4 error in a particular k_{ii}(fit) and then only if all of the discrepancy were attributed to the uncertainty of one of the two relevant k_{ii}(fit) values. Using Eqs (10.2a) and (10.2b) to obtain reliable k_{ii}(fit) values requires measuring k_{ij} values between a wide range of the couples, which we have done. The total range of k_{ii}(fit) values obtained is very large, 3×10^{14}, but we have found no dependence of the scatter of the k_{ij}(fit) values on intrinsic reactivity, ΔG°, or on type of reaction partner.

The k_{ii}(fit) values have been converted to Eyring $\Delta G^{\ddagger}_{ii}$(fit) values to allow their linear comparison, and we shall refer to these $\Delta G^{\ddagger}_{ii}$(fit) values as intrinsic reactivities. For all of the reactions that we have been able to study, Eqs (10.2a) and (10.2b) work surprisingly well. Compounds of similar structure have similar intrinsic reactivity and individual $\Delta G^{\ddagger}_{ii}$(fit) values established from a few reactions do not change much as

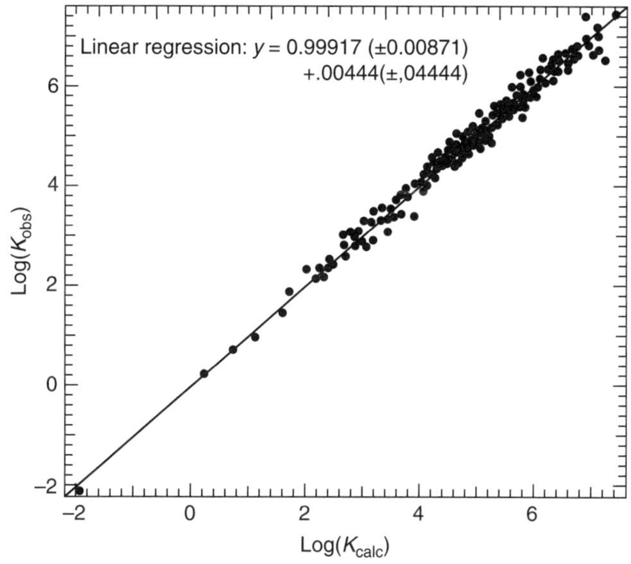

FIGURE 10.1 Plot of log(k_{obs}) versus log(k_{calc}) for the 206 reactions studied here. The slowest reaction is for tetraisopropyl hydrazine reacting with tetracyclohexyl hydrazine radical cation, which has zero driving force and is between two of the slower couples (entries 6 and 11 in Table 10.1).

more reactions are studied, suggesting that $\Delta G^{\ddagger}_{ii}$(fit) is indeed determined to a reliability of a few tenths of kcal/mol (the average $\Delta\Delta G^{\ddagger}_{ii}(\text{fit}) = |0.592 \ln[k_{ij}(\text{obs})/k_{ij}(\text{fit})]|$ for the entire data set is 0.22 kcal/mol). Consequently, we believe that these $\Delta G^{\ddagger}_{ii}$(fit) values are useful for understanding the relationship between structure and intrinsic reactivity for ET for the 72 compounds studied.

10.3 WHY IS THE SUCCESS OF CROSS-RATE THEORY SURPRISING?

The fundamental assumption giving rise to Eqs (10.2a) and (10.2b) is that the barrier for the cross-reaction will be the average of those for the self-reactions [2]. Marcus assumed that ET reactions are adiabatic in obtaining Eqs (10.2a) and (10.2b), and pre-exponential factors should be nearly constant for adiabatic reactions. The adiabatic pre-exponential factor is often approximated as $3 \times 10^{10} h\bar{\nu}_v$, where $h\bar{\nu}_v$ is the energy corresponding to the inherent barrier-crossing frequency. The $h\bar{\nu}_v$ values for all couples studied here are believed to vary between about 400 and 1600 cm^{-1}. Adiabaticity requires rather large H_{ab} values, which are not believed to occur for most intermolecular ET reactions, and certainly not for ones between compounds as hindered as for most of the cross-reactions reported here. In a recent review on ET, Bixon and Jortner concluded that although there was "lively discussion" in the 1960s about whether ET reactions were adiabatic or nonadiabatic, it has now been established that the great majority of ET reactions are nonadiabatic, which is no longer a topic for dispute [24]. However, one might expect Eqs (10.2a) and (10.2b) to not work well because nonadiabatic rate constants are controlled not only by activation barriers (or Franck–Condon factors using Bixon and Jortner's vibronic coupling theory) but also by widely varying pre-exponential factors. The nonadiabatic pre-exponential factor is proportional to the square of the electronic coupling, $(H_{ab})^2$, and is also directly proportional to e^{-S}, where S is the vibronic coupling constant (Huang–Rhys factor), which is the ratio of $h\bar{\nu}_v$ to the internal vibrational component of the reorganization energy, λ_v. Tetraalkylhydrazines have unusually large S values for the highest λ_v compounds, tetra-*n*-alkylhydrazines (about 20), while aromatic compounds and ferrocenes have much smaller S values (certainly under 6), so a factor of above a million-fold faster ET should occur for aromatic compounds and ferrocenes than for hydrazines simply from the e^{-S} term. However, we rapidly became convinced that Eqs (10.2a) and (10.2b) do work for a wide range of organic couples. We have not been able to observe the consequences of the predicted variation of the pre-exponential term in our experimental data. The same k_{ii} value suffices to calculate k_{ij} whether a hydrazine is reacted with a ferrocene, an aromatic compound, or another hydrazine. Equations (10.2a) and (10.2b) would not fit our data so well if the pre-exponential factor had the predicted sensitivity to H_{ab} and S. Our conclusion is that while one certainly expects variations in H_{ab} and S for the self-exchange reactions studied here, these variations are effectively averaged by Eqs (10.2a) and (10.2b) allowing the extraction of accurate and consistent self-exchange rate constants from the cross-reaction rate constants.

10.4 MAJOR FACTORS DETERMINING INTRINSIC REACTIVITIES OF HYDRAZINE COUPLES

Abbreviations are used for the structures studied, as indicated in Chart 10.1. *N,N*-monocyclic rings are abbreviated **r** (ring size), *N,N*-bicyclic rings by (ring size)(ring size)*N*, a γ,γ-keto group substitution in 9-azabicyclo[3.3.1]nonyl groups by a **k**, the ring size in monocyclic *N,N′*-hydrazine rings is enclosed in brackets, a γ,γ-unsaturation in an *N,N′*-cyclic or -bicyclic system by a **u**, 2,4-di-*t*-butylphenyl by **b$_2$Ph**, and the nitrogens in *N,N′*-bicyclic systems are separated from the alkyl or aryl substituents by a slash. Two bishydrazines were included, both monoaryl 2,3-diazabicyclo[2.2.2]octyl-substituted, abbreviated **$Hy_2{}^{14}Ph$** and **$(BuPh)_2 6\sigma$**.

Thirty-nine tetrasubstituted hydrazine couples were used in this work (see Table 10.1, where they are listed in order of decreasing $\Delta G^{\ddagger}_{ii}$(fit), i.e., increasing ET reactivity).

The couple numbers assigned are completely arbitrary, depending principally on the order in which we studied them. Hydrazine couples span an astonishing range of intrinsic reactivity, 16.6 kcal/mol, which is 83% of the total $\Delta G^{\ddagger}_{ii}$(fit) range observed for all compounds studied. The very unreactive, high $\Delta G^{\ddagger}_{ii}$(fit) hydrazines are essential to study the intrinsically most reactive couples. Only by having slow couples can the cross-reaction ET rate constant be kept small relative to diffusion rates, so the

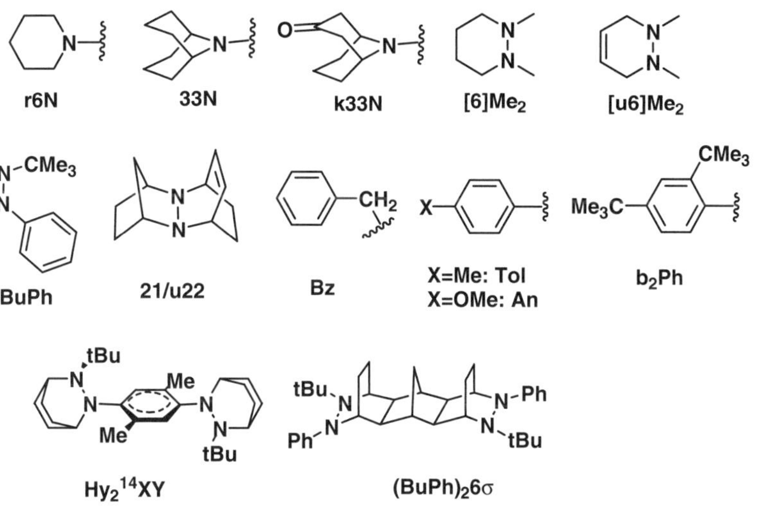

CHART 10.1 Examples of the abbreviations used for hydrazine couples.

reactions being studied are not diffusion limited when very fast couples are included. It also becomes more difficult to measure k_{ij} using stopped flow when it exceeds about $3 \times 10^7\,M^{-1}\,s^{-1}$ for technical reasons. Because Eqs (10.2a) and (10.2b) are for activation barrier-limited reactions, we also do not wish to study faster reactions, which ought to be corrected for diffusion effects, so we limited our studies to reactions having k_{ij} values below 3×10^7. A ladder of couples with different formal potentials (E°) is essential to expanding the range of compounds that can be studied because ΔE° of 0.4 V provides a factor of 2.4×10^3 in Eq. (10.2a), so one could only study systems having $(k_{ii}k_{jj})^{1/2}$ below about 10^4 with such a large ΔE°. The abbreviations for the 16 couples used in the E° ladder for determining k_{ij} values for reactions involving triarylamines are shown in Figure 10.2. The need for relatively small ΔE° makes the ability to tune E° by structural changes important for being able to carry out these studies. The sixth column of Table 10.1 shows the number of cross-reactions studied, and in parentheses the root mean square average $|\Delta\Delta G^\ddagger|$ corresponding to 0.592 $\ln[k_{ij}(\text{obs})/k_{ij}(\text{fit})]$ (when several reactions are available, or the simple average deviation if under three reactions were studied), which is a measure of how well Eqs (10.2a) and (10.2b) fit the data. Obviously no estimate of fit to Eqs (10.2a) and (10.2b) can be made when only a single cross-reaction rate constant has been measured for a particular couple. Consequently, several cross-reactions were studied for each couple whenever possible. Nonetheless, for 10 of the 72 compounds it was possible to measure only a single cross-reaction rate constant because of difficulty in matching a reaction partner of suitable E° and intrinsic reactivity, or because of low stability of one of the reactants. The fact that there is little variance in $k_{ii}(\text{fit})$ as more cross-reactions are studied for the other 62 compounds that were more extensively studied gives us confidence that even when a single cross-reaction was possible for a couple we have determined a useful estimate of its intrinsic reactivity.

TABLE 10.1 Intrinsic Reactivities, $\Delta G^{\ddagger}_{ii}$(fit), Number of Cross-Reactions Studied, Their Average Deviation, and NN Twist Angles of Neutral Hydrazines (θ_0) and Their Radical Cations (θ_+)

Entry No.	Couple No.	Couple Abbreviation	E° (V) Versus SCE	$\Delta G^{\ddagger}_{ii}$(fit) (kcal/mol)	No. of Reactions (rms Average Deviation)	θ_0 (°)	θ_+ (°)[a]
1	20	**$(nPr_2N)_2$**$^{0/+}$	0.29	22.0	10 (0.12)	~90 (PE)	(~0)
2	31	**iPr_2NNMe_2**$^{0/+}$	0.24	22.0	3 (0.15)	~90 (PE)	(~0)
3	19	**$(Et_2N)_2$**$^{0/+}$	0.29	21.9	3 (0.16)	~90 (PE)	(~0)
4	21	**$(nHx_2N)_2$**$^{0/+}$	0.29	21.4	5 (0.19)	(~90)	(~0)
5	30	**$(iPrMeN)_2$**$^{0/+}$	0.24	21.4	3 (0.28)	~90 (PE)	(~0)
6	1	**$(iPr_2N)_2$**$^{0/+}$	0.26	21.0	29 (0.23)	**89** [29e]	**8**
7	47	**$(Bz_2N)_2$**$^{0/+}$	0.60	20.8	15 (0.18)	(~90)	(~0)
8	43	**$(nPrMeN)_2$**$^{0/+}$	0.25	20.6	3 (0.15)	~90 (PE)	(~0)
9	42	**nPr_2NNMe_2**$^{0/+}$	0.25	20.5	4 (0.34)	~90 (PE)	(~0)
10	38	**$(nBuMeN)_2$**$^{0/+}$	0.25	20.5	4 (0.18)	~90 (PE)	(~0)
11	11	**$(cHx_2N)_2$**$^{0/+}$	0.26	19.6	27 (0.29)	**88** [29e]	**0, ~44**
12	41	**$r7NNMe_2$**$^{0/+}$	0.18	19.4	2 (0.25)	~90 (PE)	(~0)
13	37	**$33NNiPr_2$**	0.34	19.2	5 (0.29)		(0)
14	22	**$(Me_2N)_2$**$^{0/+}$	0.28	18.5	3 (0.29)	~90 (PE)	(~0)
15	40	**$r6NNMe_2$**$^{0/+}$	0.31	18.0	2 (0.23)	~90 (PE)	(~0)
16	39	**$r5NNMe_2$**$^{0/+}$	0.12	17.8	3 (0.44)	~90 (PE)	(~0)
17	37	**r6NNr5**$^{0/+}$	0.34	17.5	2 (0.33)	~90 (PE)	(0)
18	45	**$[u6]Me_2$**$^{0/+}$	0.28	17.4	7 (0.40)	Only *gauche* (PE)	(~0)
19	46	**$21/Me_2$**$^{0/+}$	0.15	17.3	5 (0.23)	~120 (PE)	(~0)
20	44	**$[6]Me_2$**$^{0/+}$	0.18	16.5	4 (0.24)	*anti* > *gauche*	(~0)
21	24	**22/tBuiPr**$^{0/+}$	−0.10	15.8	3 (0.22)	(~120)	
22	14	**$(k33N)_2$**$^{0/+}$	0.45	15.5	20 (0.22)	(~180)	(~180)

23	23	**22/tBuMe$^{0/+}$**	0.11	15.2	3 (0.10)	~120 (PE)	(~0)
24	25	**22/iPr$_2$$^{0/+}$**	0.08	14.9	1 (–)	(~120)	(~0)
25	2	**22/22$^{0/+}$**	−0.53	14.7	2 (0.16)	**15**	**4**
26	13	**k33NN33$^{0/+}$**	0.22	14.3	13 (0.15)	(~180)	(~180)
27	36	**33NNMe$_2$$^{0/+}$**	0.11	13.6	4 (0.17)	*anti* + *gauche* (PE)	(~180)
28	12	**(33N)$_2$$^{0/+}$**	−0.01	13.6	11 (0.20)	**180**	**180**
29	3	**21/u22$^{0/+}$**	0.058	13.3	9 (0.18)	**3**	(~0)
30	26	**22/tBuPh$^{0/+}$**	0.26	13.3	6 (0.06)	(~120)	**146**
31	5	**22/u22$^{0/+}$**	−0.241	13.2	5 (0.21)	**0**	**0**
32	4	**22/u23$^{0/+}$**	−0.298	12.8	3 (0.11)	(~0)	(~0)
33	6	**21/21$^{0/+}$**	0.01	12.7	4 (0.12)	(~0)	(~0)
34	49	**Hy$_2$^{14}XY$^{0/+}$**	0.03	12.4	2 (0.08)	(~120)	(~0)
35	50	**(BuPh)$_2$6$\sigma^{0/+}$**	0.185	11.8	1 (–)	(~120)	(~0)
36	62	**(iPrPhN)$_2$$^{0/+}$**	0.70	11.1	1 (–)	(~90)	(0)
37	34	**22/Ph$_2$$^{0/+}$**	0.48	11.0	13 (0.33)	**114**	**155**
38	33	**(b$_2$Ph)$_2$N)$_2$$^{0/+}$**	0.61	8.0	7 (0.42)[b]	**113**[c]	**125**[c]
39	32	**(tol$_2$N)$_2$$^{0/+}$**	0.65	5.4	4 (0.20)	**111**[c]	**124**[c]

[a] Photoelectron spectroscopic data, marked "(PE)", are from Ref. [20, 25]. The numbers in boldface type are from X-ray crystallography [12, 26]. For discussion of the conformations of the six-membered ring hydrazines, see Ref. [27]. Entries in parentheses are for data taken on very close structural analogues to compounds that have experimental data, for which we believe the θ values given also apply.

[b] Three of the reactions studied for **(b$_2$Ph)$_2$N)$_2$** have k_{ii}(fit)/k_{ii}(obs) values lying outside the range 3 or 0.33, so half of the six reactions studied that fit Eqs (10.2a) and (10.2b) most poorly involve this couple.

[c] Obtained by DFT calculation with the (U)B3LYP/6-31G(d) basis set.

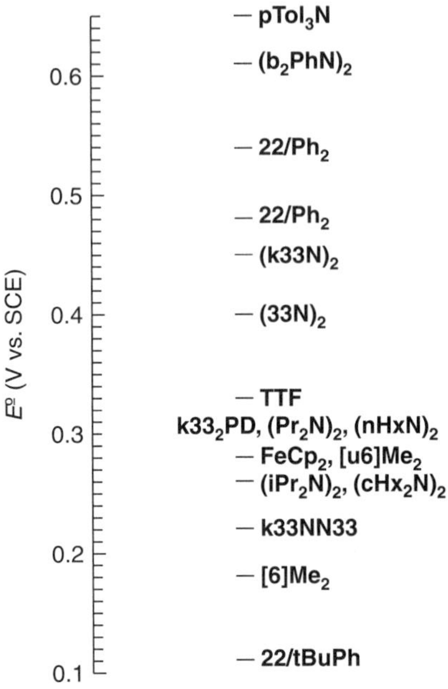

FIGURE 10.2 E° ladder used to measure cross-reactions involving triarylamines [16].

The ET reactivities of these hydrazines clearly fall into groups based on the connectivity of their substituents, and hence their λ_v values. The hydrazines having the largest geometry changes on electron loss have the largest λ_v changes and are the least reactive for ET. The lone pair twist angles in the neutral form (θ_0) and radical cation (θ_+), shown in the last two columns in Table 10.1, depend on the connectivity of the substituents, and the difference in these twist angles is the most important structural feature influencing the size of $\Delta G^{\ddagger}_{ii}$(fit). The spectroscopic information most directly related to θ_0 is the photoelectron spectroscopic lone pair vertical ionization potential difference (ΔvIP), which is approximately proportional to cos θ_0 [20, 25]. X-ray crystallographic structures are available for several of these hydrazines in either or both oxidation states [12, 26], and X-ray θ values, where available, are shown in boldface type in Table 10.1. Neutral hydrazines electronically prefer to be twisted at the NN bond in $\theta_0 \sim 90°$ conformations because this allows maximum lone pair, σ^* stabilizing interactions (i.e., avoiding overlap of the lone pair orbitals). Their nitrogens are distinctly pyramidalized [13]. Hydrazine radical cations are "3e–π systems" that have one antibonding (π^*) and two bonding (π) electrons, and therefore have very different geometries from the neutral forms. Their NN bonds have a strong electronic preference for being untwisted (θ_+ near 0 or 180°), and their nitrogens are considerably flattened (0 and 180° twists become the same when the nitrogens are planar and the lone pair orbitals become pure p orbitals). The largest geometry changes and hence largest reorganization energy and ET barriers occur for $\theta_0 = \sim 90°$; $\theta_+ \sim 0°$ compounds, which include the first 17 entries of Table 10.1

($\Delta G^{\ddagger}_{ii}$(fit), 22.0–17.5), include compounds that have neutral forms with θ values near 90°, but those with *N,N*-cyclic rings are more reactive than those without them. We will return to why this occurs below. Including *N,N′*-rings in the structures precludes the 90° conformations that give rise to the largest λ_v values and hence the highest $\Delta G^{\ddagger}_{ii}$(fit) values. Consequently, the six-membered ring monocyclic and bicyclic compounds that have *trans* alkyl groups have similar and somewhat greater reactivities. Entries 18–21, 23, and 24 in Table 10.1 having $\Delta G^{\ddagger}_{ii}$(fit) values between 17.4 and 14.9 kcal/mol are of this sort. When the substituents at nitrogen are α-branched and also are unable to rotate away from N–C_α twist angles that give large nonbonded steric interactions between α-substituents, as is the case for the (bis)*N,N*-bicyclononyl derivatives studied, $\theta_0 = 180°$ and $\theta_+ = 180°$ conformations occur. Examples are entries 22, 26, and 28, having $\Delta G^{\ddagger}_{ii}$(fit) values of 15.5, 14.3, and 13.6 kcal/mol for the (bis)bicyclononyl derivatives with two, one, and zero γ-keto groups, respectively, demonstrating that the presence of γ-keto group systematically lowers ET reactivity for these compounds. (Bis)*N,N′*-bicyclic compounds with five- to seven-membered rings force the CNNC angles to be near 0°, which makes them assume the electronically least favorable *syn* neutral conformations having $\theta_0 \sim$ near 0°. They therefore have relatively small geometry changes, and are the lowest $\Delta G^{\ddagger}_{ii}$(fit) group of compounds that lack aryl substitution. The highest barrier couple of this group at 14.7 kcal/mol, **22/22**$^{0/+}$ (entry 25), was shown by crystallography to be twisted 15° in the neutral form [26b]. This NN twist is expected to increase its vertical reorganization energy because twist of the +1 oxidation state is rather difficult. The other couples in this series, **21/u22**$^{0/+}$, **22/u22**$^{0/+}$, **23/u23**$^{0/+}$, and **21/21**$^{0/+}$ (entries 29 and 31–33), are significantly more reactive, with ET barriers of 13.3, 13.2, 12.8, and 12.7 kcal/mol. This order is that caused by the second major effect on the size of the geometry change on electron loss: the pyramidality change. Decreasing ring size makes the nitrogens more pyramidal in both oxidation states, and the nitrogens are sufficiently pyramidal for **21/21**$^{+}$ that the barrier to their becoming planar was measured by EPR [28]. There is a substantial ET barrier-lowering effect of replacing alkyl by aryl substituents, caused principally by flattening at nitrogen in the neutral form in the aryl-substituted compounds. For example, replacing the isopropyl of **22/tBuiPr** (entry 21) by phenyl to give **22/tBuPh** (entry 30) lowers $\Delta G^{\ddagger}_{ii}$(fit) by 2.5 kcal/mol and a second aryl substitution in going to **22/Ph$_2$** (entry 37) lowers $\Delta G^{\ddagger}_{ii}$(fit) an additional 2.3 kcal/mol. The lowest barrier hydrazine couples are the two tetraarylhydrazines studied, entries 38 and 39. The $\Delta G^{\ddagger}_{ii}$(fit) for **(tol$_2$N)$_2$** (entry 39) lies between the unhindered nearly planar aminoaromatic couples, **TMPD** (entry 11, Table 10.2) and **DMP** (entry 18, Table 10.2).

10.5 NONHYDRAZINE COUPLES

Structural abbreviations for examples of the nonhydrazine couples studied appear in Chart 10.2. The only two trialkylamines with radical cation lifetimes long enough to study are 1,4-diazabicyclo[2.2.2]octane, **N[222]N**, and Alder's tris-trimethylene bridged diamine **N[333]N**. Three tetraalkyl-2-tetrazenes (abbreviated **N$_4$**, preceded

TABLE 10.2 Nonhydrazine Couples Studied

Entry No.	Couple No.	Couple Abbreviation	$E°$ (V) Versus SCE	$\Delta G^{\ddagger}_{ii}$(fit) (kcal/mol)	No. of Reactions (rms Average Deviation)
1	28	**N[333]N**$^{0/+}$	−0.165	15.1	3 (0.24)
2	54	**N[222]N**$^{0/+}$	0.58	12.2	1 (–)
3	55	**k33$_2$N$_4$**$^{0/+}$	0.75	10.8	4 (0.14)
4	35	**33$_2$N$_4$**$^{0/+}$	0.40	9.4	10 (0.48)
5	58	**nPr$_4$N$_4$**$^{0/+}$	0.35	8.9	3 (0.72)
6	7	**FeCp$_2'$**$^{0/+}$	0.28	8.3	4 (0.16)
7	8	**FeCp*Cp**$^{0/+}$	0.124	8.0	11 (0.17)
8	9	**FeCp$_2^*$**$^{0/+}$	−0.109	7.9	9 (0.22)
9	17	**FeCp$_2$**$^{0/+}$	0.395	7.7	4 (0.13)
10	16	**(k33)$_2$PD**$^{0/+}$	0.29	7.1	21 (0.24)
11	10	**TMPD**$^{0/+}$	0.12	6.5	8 (0.15)
12	63	**Xy$_2$pBrN**$^{0/+}$	0.847	6.5	3 (0.49)
13	68	**pTol$_2$PhN**$^{0/+}$	0.812	6.5	1 (–)
14	53	**Z$_4$PD**$^{0/+}$	0.51	6.4	3 (0.08)
15	15	**(33)$_2$PD**$^{0/+}$	0.02	6.3	3 (0.33)
16	59	**(pBr)$_3$N**$^{0/+}$	1.10	5.9	1 (–)
17	71	**Xy$_2$pOPhN**$^{0/+}$	0.723	5.3	2 (0)
18	27	**DMP**$^{0/+}$	0.138	5.3	7 (0.17)
19	72	**XyAn$_2$N**$^{0/+}$	0.602	5.1	1 (–)
20	69	**Xy$_2$ptBuN**$^{0/+}$	0.739	5.1	2 (0.07)
21	61	**(pTol)$_3$N**$^{0/+}$	0.775	5.1	4 (0.20)
22	48	**iPrPT**$^{0/+}$	0.737	5.0	1 (–)
23	51	**An$_4$PD**$^{0/+}$	0.35	5.0	6 (0.26)
24	67	**XyptBu$_2$N**$^{0/+}$	0.766	5.0	2 (0.07)
25	52	**Z$_2$An$_2$PD**$^{0/+}$	0.49	5.0	4 (0.24)
26	64	**Xy$_2$pTolN**$^{0/+}$	0.733	5.0	1 (–)
27	70	**Xy$_2$AnN**$^{0/+}$	0.653	4.8	3 (0.07)
28	65	**Xy$_2$pBiN**$^{0/+}$	0.770	4.4	3 (0.12)
29	66	**XypBi$_2$N**$^{0/+}$	0.823	4.3	1 (–)
30	18	**TTF**$^{0/+}$	0.33	3.5	15 (0.27)
31	60	**An$_4$ND**$^{0/+}$	0.456	3.5	4 (0.18)
32	56	**pAn$_3$N**$^{0/+}$	0.560	3.3	2 (0.61)
33	29	**TMTSF**$^{0/+}$	0.423	2.2	13 (0.18)

by abbreviations of the alkyl groups) were studied. Methylated ferrocenes are abbreviated **Fe** (followed by abbreviations for the two ferrocene rings), using **Cp** for cyclopentadienylidine, **Cp*** for pentamethyl-**Cp**, and **Cp**$'$ for monomethyl-**Cp**. Several 1,4-*p*-phenylenediamine derivatives and one 2,6-naphthalene-diamine derivative were studied, abbreviated **PD** and **ND**, preceded by abbreviations for the other four groups attached, but we used the familiar **TMPD** abbreviation for the tetramethyl example. *N,N'*-Dimethylphenazine is abbreviated **DMP**, *N*-isopropyl

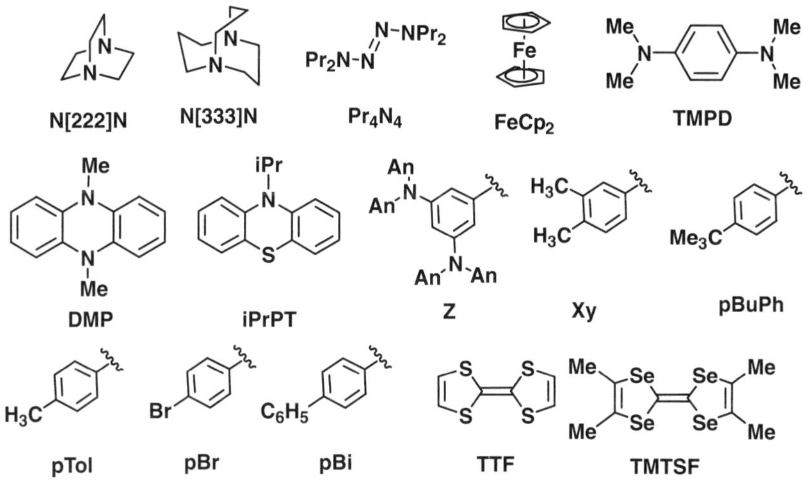

CHART 10.2 Abbreviations used for nonhydrazine couples.

phenothiazine by **iPrPT**, the 3,5-bis-dianisyl phenyl group appearing in some of the couples studied has been abbreviated as **Z**, several of the triarylamines obtained from H. Bruce Goodbrand (Xerox Research Centre of Canada) having 3,4-dimethylphenyl substituents have been abbreviated as **Xy**, aryl groups with *para* substituents are prefixed by **p**, tetrathiafulvalene has been abbreviated as **TTF**, and its tetramethyl tetraseleno analogue as **TMTSF**.

Table 10.2 contains reaction data for the nonhydrazine couples studied, also listed in order of decreasing ET reactivity. The least reactive compounds of Table 10.2 are the bicyclic trialkyldiamines, which gave $\Delta G^{\ddagger}_{ii}$(fit) 2.9 kcal/mol larger for **N[333]N** than for **N[222]N** (corresponding to k_{ii}(fit) about 150 times higher than for **N[222]N**). The 2-tetrazenes are much more reactive than their hydrazine counterparts, and the acyclic compound difference, $\Delta\Delta G^{\ddagger}_{ii}(\text{fit}) = 13.1\ \text{kcal/mol}$ for the **nPr$_2$N**-subtituted compounds, is far larger than for the bicyclic-substituted ones, 4.7 kcal/mol for the **k33**-substituted systems and 4.2 kcal/mol for the **33**-substituted ones. The next four more reactive compounds are the four ferrocene derivatives, which have $\Delta G^{\ddagger}_{ii}$(fit) values lying in a rather narrow range of 0.6 kcal/mol, which are somewhat less reactive than any of the heteroatom-substituted aromatic compounds. The remaining third of the couples are more reactive and all give formally heteroatom-centered radical cations and are aromatic or have heteroaromatic π systems. The k_{ii}(fit) values for the three most reactive couples are near or above the diffusion limit—**TMTSF**, $1.6 \times 10^{11}\ \text{M}^{-1}\,\text{s}^{-1}$; **An$_3$N**, $2.3 \times 10^{10}\ \text{M}^{-1}\,\text{s}^{-1}$; and **TTF**, $1.6 \times 10^{10}\ \text{M}^{-1}\,\text{s}^{-1}$—so their intrinsic reactivities can only be properly determined using cross-reactions. The k_{ii}(fit) obtained for **An$_4$PD**, $1.2 \times 10^{9}\ \text{M}^{-1}\,\text{s}^{-1}$, is over a factor of 100 less than that obtained for **An$_3$N**, but faster than that obtained under self-exchange conditions by NMR in $CDCl_3$, 3.5 (3) $\times 10^{8}\ \text{M}^{-1}\,\text{s}^{-1}$ [29].

10.6 COMPARISON OF $\Delta G^{\ddagger}_{ii}$ (fit) with $\Delta G^{\ddagger}_{ii}$ (self) Values

Because the $\Delta G^{\ddagger}_{ii}(\text{fit})$ values are entirely determined from experimental rate constants for cross-reactions, they must include the effect of averaged pre-exponential factors for the cross-reactions studied, while $\Delta G^{\ddagger}_{ii}(\text{self})$ values only depend on the pre-exponential factor for the self-reaction of the couple. In part, this appears to rationalize the observed pattern of deviations between $\Delta G^{\ddagger}_{ii}(\text{fit})$ and $\Delta G^{\ddagger}_{ii}(\text{self})$ for cases where directly measured values are available, although there is also clearly scatter in the values observed. Directly measured self-ET rate constants have been reported for 11 of the couples studied, and are summarized in Table 10.3, where they are compared with the values estimated from stopped-flow data that are discussed here.

Figure 10.3 is a plot of the data in Table 10.3, and shows that there is a rather good correlation between $\Delta G^{\ddagger}_{ii}(\text{fit})$ and $\Delta G^{\ddagger}_{ii}(\text{self})$ values, but the $\Delta G^{\ddagger}_{ii}(\text{fit})$ values are systematically larger than $\Delta G^{\ddagger}_{ii}(\text{self})$ values by an average of $\Delta\Delta G^{\ddagger}_{ii} = \Delta G^{\ddagger}_{ii}(\text{fit}) - \Delta G^{\ddagger}_{ii}(\text{self}) = 0.68\ \text{kcal/mol}$ or a factor of about 3 in the self-exchange rate constant. The largest $\Delta\Delta G^{\ddagger}_{ii}$ determined is for the lowest barrier couple for which data are available, **TMPD**$^{0/+}$, at 1.5 kcal/mol. The $\Delta G^{\ddagger}_{ii}(\text{fit})$ value for **TMPD**$^{0/+}$ was determined from reactions with eight hydrazines, of which six are tetra-α-branched and thus quite hindered, and the other two are **22/tBuMe** and **33NNMe**$_2$. It seems unlikely that any of the cross-reactions can have as good overlap between the reaction partners at the transition state as can **TMPD**^{0} with **TMPD**$^+$,

TABLE 10.3 Comparison of Self-ET Rate Constants ($M^{-1}\,s^{-1}$) and Barriers (kcal/mol) for Self-ET Reactions Measured Under Self-Exchange Conditions with $\Delta G^{\ddagger}_{ii}$(fit) Values Determined from Stopped-Flow Studies

Entry	Couple	$k_{ii}(\text{self})^a$	$\Delta G^{\ddagger}_{ii}(\text{self})$	$\Delta G^{\ddagger}_{ii}(\text{fit})$	$\Delta\Delta G^{\ddagger}_{ii}$
1	**(iPr$_2$N)$_2$**$^{0/+}$	$3.0\ (3) \times 10^{-3}$	20.9	21.0	0.1
2	**22/22**$^{0/+}$	$\sim$700	13.6	14.7	1.2
3	**21/u22**$^{0/+}$	$2.29\ (9) \times 10^3$	12.9	13.6	0.7
4	**22/u23**$^{0/+}$	$3.84\ (11) \times 10^3$	12.6	13.3	0.5
5	**(33N)$_2$**$^{0/+}$	$\sim 2.2 \times 10^3$	11.9	13.6	0.7
6	**22/u22**$^{0/+}$	$1.21\ (10) \times 10^4$	11.9	13.2	1.3
7	**21/21**$^{0/+}$	$1.85\ (6) \times 10^4$	11.6	12.7	1.1
8	**FeCp$_2$**$^{0/+}$	8.5×10^6	8.0	8.0	0.0
9	**FeCp*Cp**$^{0/+}$	8.1×10^6	8.0	7.7	−0.3
10	**FeCp$_2^{*0/+}$**	2.9×10^7	7.3	7.9	0.6
11	**TMPD**$^{0/+}$	1.47×10^9	5.0	6.5	1.5

^{a}Measured under self-exchange conditions by magnetic resonance line broadening, except for entry 1, which used the rate constant for exchange measured for partially deuterium-labeled material determined by NMR [11,12]. The hydrazines in entries 2–7 used slow-exchange region NMR: 2 [30], 3, 4, and 7 [17c], 5 [10] (measured in CD_2Cl_2 and estimated in CD_3CN from the rate constant obtained), and 6 [17b]. The ferrocenes in entries 8–10 used fast-exchange region NMR: 8 and 10 [21b, 31] and 9 [10]. Entry 11 used fast-exchange region EPR [22].

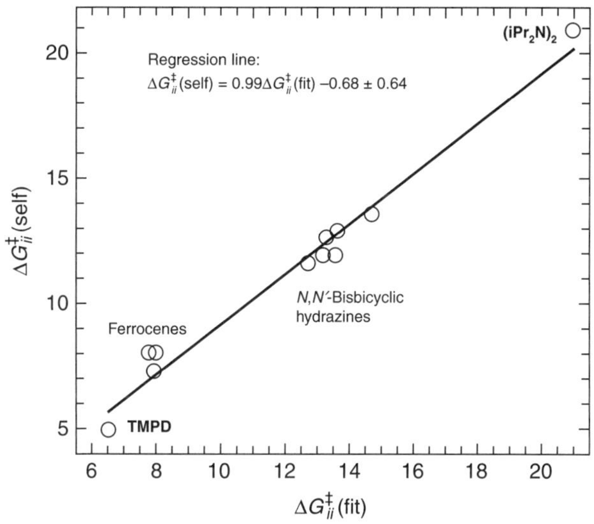

FIGURE 10.3 Plot of the data of Table 10.3.

which ought to lower H_{ab} for the cross-reactions, and might be the principal factor causing the 13-fold larger k_{ii}(self) than k_{ii}(fit). **TMPD** is the least hindered couple for which we know both $\Delta G^\ddagger_{ii}$(self) and $\Delta G^\ddagger_{ii}$(fit), so it should show the largest $\Delta\Delta G^\ddagger_{ii}$ if the H_{ab} change in going from self-exchange to cross-reactions is responsible. At the other reactivity extreme for which both barriers are available, the very hindered couple **(iPr$_2$N)$_2$** has $\Delta\Delta G^\ddagger_{ij}$ within experimental error of zero. In sum, the agreement between $\Delta G^\ddagger_{ii}$(self) and $\Delta G^\ddagger_{ii}$(fit) is truly remarkable and gives confidence in the approach used for obtaining useful self-exchange rate constants and activation energies. The average $\Delta\Delta G^\ddagger$ of 0.68 kcal/mol is nearly within the fitted uncertainty of ±0.64 kcal/mol. This in itself is an interesting result because a much larger systematic discrepancy [32] between experimental cross-reaction rate constants and those calculated from Eqs (10.2a) and (10.2b) from independently measured self-exchange rate constants has been observed for reactions of inorganic coordination compounds. For those systems the discrepancy is typically a factor of 10 or more with the observed cross-reaction rate constants being lower than calculated [32]. This discrepancy could result if H_{ab} for cross-reactions is systematically lower than the geometric average of the self-exchange values. In a similar fashion, since our k_{ii} values are determined by fitting Eqs (10.2a) and (10.2b) to the set of cross-reactions, we might anticipate that the values derived from the cross-reactions would be systematically lower than those determined by direct exchange measurement. While we observe such a systematic discrepancy, it is considerably less than reported for the other systems and the apparent discrepancy is very close to the uncertainty in the fit of the data. Better agreement in our systems may in part reflect the absence of work terms related to more highly charged donors and acceptors involved in the previous reported studies as well.

10.7 ESTIMATION OF H_{AB} FROM EXPERIMENTAL EXCHANGE RATE CONSTANTS AND DFT-COMPUTED λ

The semiclassical treatment for nonadiabatic cross-ET in the high temperature limit developed by Marcus [33] and Levich and Dogonadze [34] is based on Fermi's golden rule, and given in Eq. (10.4):

$$k = \frac{2\pi}{\hbar}\frac{(H_{ab})^2}{\sqrt{4\pi\lambda RT}}\exp\left[-\frac{(\lambda+\Delta G^\circ)^2}{4\lambda RT}\right] \tag{10.4}$$

where k is the first-order rate constant for intramolecular ET (or for ET from within a donor–acceptor precursor complex), H_{ab} is the electronic coupling matrix element between donor and acceptor, λ is the total reorganizational energy associated with the ET, and ΔG° is the standard reaction free energy. Solving Eq. (10.4) for H_{ab}, in units of kcal/mol, at 25°C results in Eq. (10.5):

$$H_{ab} = \sqrt{6.58\times 10^{-15} k\lambda^{1/2}\exp\left[\frac{(\lambda+\Delta G^\circ)^2}{4\lambda RT}\right]} \tag{10.5}$$

For reactions treated in this work, where one observes second-order intermolecular ET between freely diffusing small donor and acceptor molecules in homogeneous solution, ET is assumed to be preceded by the diffusion-controlled formation of a donor–acceptor collision complex. When formation and disassociation occur at or near diffusion control and are rapid relative to ET, the observed second-order ET rate constant, $k_{(\text{obsvd})}$, is the product of the equilibrium constant for collision complex formation, K_c, and the unimolecular rate constant from Eq. (10.4) for ET from within the precursor complex (Eq. (10.6) [32, 33b]:

$$k_{(\text{obsvd})} = kK_c, \qquad k = k_{(\text{obsvd})}/K_c \tag{10.6}$$

Although we suggest that it is unlikely that K_c is actually 1.0 for the reactions that we have studied, we doubt that it is very much larger in most cases. ΔG° for association of anthracene with its radical cation at 25°C was measured as −8.7 kcal/mol by Meot-Ner in the gas phase [35]. Masnovi and Kochi determined ΔG° in CH_2Cl_2 at 25°C as −2.3 kcal/mol in methylene chloride [36]. Although there is a decrease of a factor of 3.8 on going from the vapor phase to solution, the K_c value remains 50 M^{-1} in solution for this unhindered example of a collision complex formation constant. However, most of our cross-reactions have at least one hindered compound, which will cause K_c to decrease significantly.

Substitution of $k_{(\text{obsvd})}/K_c$ for k in Eq. (10.5) allows the calculation of H_{ab} directly from the experimentally determined second-order ET rate constant, provided λ and K_c can be estimated independently. Since no effects were observable—kinetically or spectroscopically—for stable complex formation between any of the donor–acceptor couples studied here, we assume that both formation and disassociation of the precursor complex are essentially under diffusion control and make the approximation that K_c is not far from 1 in most cases studied here. This approximation introduces uncertainty into the calculated H_{ab} values and an error in this approximation by a factor of 10 in K_c would result in a difference of approximately a factor of 3 in the calculated H_{ab} since H_{ab} has an inverse square root dependence on K_c.

When self-exchange reactions, for which ΔG° is zero, are considered, Eqs (10.4) and (10.5) simplify to Eqs (10.7) and (10.8), respectively:

$$k = \frac{2\pi}{\hbar}\frac{(H_{ab})^2}{\sqrt{4\pi\lambda RT}}\exp\left[-\frac{\lambda}{4RT}\right] \tag{10.7}$$

$$H_{ab} = \sqrt{6.58 \times 10^{-15} k\lambda^{1/2} \exp[0.424\lambda]} \tag{10.8}$$

Equation (10.8) allows the direct calculation of H_{ab} provided that k and λ can either be measured or accurately estimated by computation. This is an important result because values for H_{ab} have been notoriously difficult to evaluate computationally because H_{ab} for many reactions is exponentially dependent on the donor–acceptor distance, which itself is very difficult to estimate for small molecules in solution undergoing outer sphere ET. As described above, we have accurately measured the second-order rate constants for self-exchange, k_{ii}(fit), for 72 compounds, which leaves an independent determination of λ as the principal barrier to obtaining H_{ab}. We have calculated the vibrational reorganization energies λ_v for the 46 couples shown in Table 10.4 using the "four-point" method developed by Nelsen and coworkers using (U)B3LYP/6-31 + G(d) calculations for the smaller systems and 6-31G(d) basis sets for the larger ones to obtain the energies of each of the four structures required to apply this method [17b]. Application to larger basis set calculations has been discussed [16]. These values are summarized in Table 10.4. Difficulty in estimation of λ_v increases rapidly as the number of energy minima increases because each minimum in each oxidation level usually results in a different value. We discussed this effect at length for **$(Et_2N)_2$** and **$(iPr_2N)_2$** [16], and suggested that averaged values of 67.7 and 61.1 kcal/mol, respectively, might be more appropriate than the entries in Table 10.4, 67.9 and 61.7, but these differences are too small to change interpretation for these compounds. Our general conclusion was that one only needs to consider minimum energy conformations in such calculations because tetra-*n*-alkylhydrazines have much higher barriers than those of cyclic hydrazines, and rotation about their NN bonds is rather low energy. For 19 of 21 couples calculated both ways, the + G(d) calculation using diffuse functions gave a smaller result than the 6-31G(d) calculation lacking diffuse functions, but the differences are generally small.

Using our experimental k_{ii}(fit) values, λ_v values obtained by DFT calculated as just described, and λ_s (solvent reorganization) approximated at 8.0 kcal/mol for each system, we have calculated self-exchange H_{ab} for each of the 46 couples for which λ_v was calculated. These values are collected in Table 10.4. Only H_{ab} values derived from the first, larger basis set, λ_v value cited below are tabulated.

TABLE 10.4 Observed k_{ii}(fit), (U)B3LYP λ_v, and H_{ab} Resulting from Using λ_s of 8.0 kcal/mol

Entry No.	Couple	Couple No.	k_{ii}(fit) ($M^{-1}s^{-1}$)	λ_v (DFT)[a]	H_{ab} ($\lambda_s = 8.0$)
1	**$(Et_2N)_2{}^{0/+}$**	19	5.30×10^{-4}	67.92, 69.71√	0.05
2	**$(iPrMeN)_2{}^{0/+}$**	30	1.40×10^{-3}	63.89, 64.69√	0.04
3	**$r5NNMe_2{}^{0/+}$**	39	5.8×10^{-1}	63.67, 64.49√	0.71
4	**$(Me_2N)_2{}^{0/+}$**	22	1.71×10^{-1}	62.55, 64.95√	0.30
5	**$(iPr_2N)_2{}^{0/+}$**	1	2.70×10^{-3}	61.68√	0.03
6	**$r6NNMe_2{}^{0/+}$**	40	4.00×10^{-1}	61.20, 62.85√	0.35
7	**$(Bz_2N)_2{}^{0/+}$**	47	3.60×10^{-3}	57.83√	0.02
8	**$[u6]Me_2{}^{0/+}$ (ae)**	45	8.00×10^{-2}	55.01, 58.08	0.04
9	**$21/Me_2{}^{0/+}$**	46	1.40	50.94, 51.97√	0.07
10	**$[6]Me_2{}^{0/+}$ (ee)**	44	4.90	43.88, 46.91√	0.04
11	**$22/tBuiPr^{0/+}$**	24	1.50×10	42.01, 42.34√	0.03
12	**$22/tBuMe^{0/+}$**	23	4.60×10	41.25, 42.23√	0.05
13	**$(k33N)_2{}^{0/+}$**	14	2.60×10	40.03, 40.20√	0.03
14	**$(33N)_2{}^{0/+}$**	12	7.10×10^2	37.58, 38.15√	0.09
15	**$21/21^{0/+}$**	6	3.20×10^3	36.49, 37.54√	0.15
16	**$22/22^{0/+}$**	2	9.60×10	37.09√	0.03
17	**$21/u22^{0/+}$**	3	1.00×10^3	34.40, 35.18√	0.05
18	**$22/u23^{0/+}$**	4	2.40×10^3	33.84, 34.37√	0.07
19	**$22/u22^{0/+}$**	5	1.20×10^3	32.10, 32.88√	0.04
20	**$22/tBuPh^{0/+}$**	26	1.10×10^3	30.00, 31.19√	0.02
21	**$(k33)_2N_4{}^{0/+}$**	55	7.20×10^4	30.93√	0.21
22	**$(33)_2N_4{}^{0/+}$**	39	8.00×10^5	28.72√	0.43
23	**$22/Ph_2{}^{0/+}$**	34	5.60×10^4	26.67, 27.53√	0.07
24	**$N[333]N^{0/+}$**	28	4.90×10	23.70	0.001
25	**$(iPrPhN)_2{}^{0/+}$**	62	4.40×10^4	20.79	0.02
26	**$(k33)_2PD^{0/+}$**	16	4.00×10^7	17.79√	0.27
27	**$N[222]N^{0/+}$**	54	7.30×10^3	16.07	0.003
28	**$(33)_2PD^{0/+}$**	15	1.60×10^8	15.98√	0.37
29	**$TMPD^{0/+}$** (syn)	10	1.10×10^8	12.92, 12.50√	0.15
30	**$DMP^{0/+}$**√	27	8.20×10^8	11.34, 8.91√	0.29
31	**$iPrPT^{0/+}$**	48	1.30×10^9	10.64√	0.32
32	**$(b_2Ph)_2N)_2{}^{0/+}$**	33	8.00×10^6	7.70√	0.01
33	**$(An)_3N^{0/+}$**	56	2.30×10^{10}	6.28√	0.49[b]
34	**$TTF^{0/+}$**	18	1.60×10^{10}	6.12, 7.58√	0.40
35	**$An_4PD^{0/+}$**	51	1.20×10^9	5.66√	0.10
36	**$An_4ND^{0/+}$**	60	1.60×10^{10}	5.45√	0.34
37	**$(pTol_2N)_2{}^{0/+}$**	32	6.40×10^8	5.18√	0.06
38	**$Xy_2AnN^{0/+}$**	70	1.70×10^9	4.63√	0.09
39	**$FeCp_2{}^{0/+}$**	17	1.30×10^7	4.51[c]	0.008
40	**$TMTSF^{0/+}$**	29	1.60×10^{11}	3.40, [9.79√][d]	0.67
41	**$Xy_2pBrN^{0/+}$**	67	1.00×10^8	3.21√	0.02
42	**$(pBr)_3N^{0/+}$**	59	2.70×10^8	3.17√	0.03

43	$\mathbf{Xy_2ptBuN^{0/+}}$	69	1.20×10^9	2.99√	0.05
44	$\mathbf{Xy_2pTolN^{0/+}}$	64	1.40×10^9	2.98√	0.06
45	$\mathbf{Xy_2pBiN^{0/+}}$	65	3.30×10^9	2.94√	0.09
46	$\mathbf{pTol_3N^{0/+}}$	61	1.10×10^9	2.92√	0.05

Energies are in kcal/mol and H_{ab} in kcal/(mol $M^{1/2}$); couples are listed in order of decreasing λ_v.
[a] From (U)B3LYP/6-31 + G(d) calculations if not marked, and from (U)B3LYP/6-31G(d) calculations for cases marked "√.".
[b] The $\Delta G^{\ddagger}_{ii}$(fit) value, which is based on two reactions (see Table 10.2), is probably too low because the H_{ab} it gives is anomalously high (see text).
[c] From (U)B3LYP/LACV3P** calculations using Jaguar.
[d] See text for discussion of why these numbers are anomalous.

What H_{ab} values are appropriate for intermolecular ET reactions has not been obvious. For one of the most studied examples, Grampp and Jaenicke suggested that the $\mathbf{TMPD^{0/+}}$ transition state should have a geometry having the aryl rings pressed together, close to their orientation in crystalline material, and that a rather low value of $H_{ab} = 0.1$ kcal/mol is necessary to be consistent with the observed k_{ii}(self) value [22], while Rauhut and Clark have calculated values for transition states having a wider range of geometries for $\mathbf{TMPD^{0/+}}$, estimating about 0.65 kcal/mol using *ab initio* theory [37].

Estimation of λ_s is also necessary to obtain the total reorganizational energy leading to the determination of the H_{ab} values just described. Although it has been traditional to use dielectric continuum theory to estimate λ_s for ET reactions, we have pointed out that dielectric continuum theory estimates have produced numbers that are systematically too high [16], and that there is little evidence in our data that λ_s increases in acetonitrile for $\mathbf{(nR_2N)_2}$ for changing alkyl groups from two to six carbons, although use of dielectric continuum theory predicts a significant change. Formosinho et al. have concluded independently and for different reasons that dielectric continuum theory overestimates λ_s [38]. We therefore used our most reactive couple, $\mathbf{TMTSF^{0/+}}$, to estimate what a typical λ_s value is for these experiments because here the contribution from λ_v is at a minimum. The observed $\Delta G^{\ddagger}_{ii}$(fit) value is only 2.2 kcal/mol, which produces the plot shown in Figure 10.4 for its ΔG^* versus log H_{ab} values. We note that curvature is quite evident in this plot because of the low $\Delta G^{\ddagger}_{ii}$(fit) value. While λ_v is clearly quite small for TMTSF, the entire activation energy of 2.2 kcal/mol cannot be assigned to solvent reorganization. Because ΔG_v is 1/4 of λ_v, its calculated value is 0.85 kcal/mol based on the DFT calculation result in Table 10.4. This makes it very unlikely that ΔG_s can be more than 1.5–2 kcal/mol for this compound, consequently making the upper bound for λ_s for TMTSF 8 kcal/mol. Although this is considerably less than people have estimated for reactions in acetonitrile using dielectric continuum theory, it is consistent with our data, and we have used $\lambda_s = 8$ kcal/mol in calculating the H_{ab} values shown in the last column of Table 10.4. We have used the same λ_s for all compounds without considering the effect of molecular size, although larger molecules should have smaller λ_s. While this is somewhat arbitrary, we do not believe that λ_s varies greatly for the systems studied and that it can be much higher than this. Our data show that the

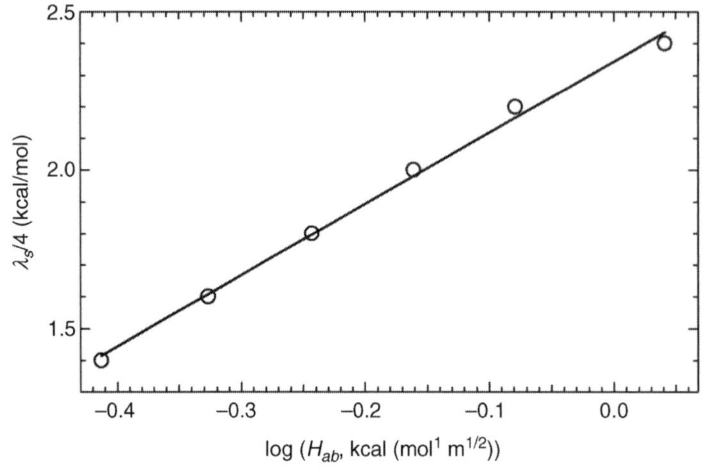

FIGURE 10.4 Plot of ΔG_s versus log H_{ab} for **TMTSF**$^{0/+}$.

effect is smaller than that calculated using dielectric continuum theory, and we do not know how to estimate what the size effect actually is, but note that $\Delta G^{\ddagger}_{ii}$(fit) only drops 0.5 kcal/mol between $\mathbf{(Et_2N)_2}^{0/+}$ and $\mathbf{(nHx_2N)_2}^{0/+}$. Because we expect the λ_v values to be the same for these compounds, this suggests that at least increasing the molecular size from $C_8H_{20}N_2$ to $C_{24}H_{54}N_2$ has little effect on λ_s. Based on continuum theory, one would have expected to see a significant decrease in λ_s and consequently a lower $\Delta G^{\ddagger}_{ii}$(fit) as one changes from $\mathbf{(Et_2N)_2}^{0/+}$ to $\mathbf{(nHx_2N)_2}^{0/+}$. The fact that this does not happen reinforces the confidence in our estimate that λ_s cannot be much more than 8 kcal/mol and that it does not change much over the series of reactions we have studied.

We were initially confused by our failure to find a steric effect on these intermolecular ET reactions, but finally realized that although there is a large steric effect, it does not change for replacing ethyl by longer *n*-alkyl groups, which we had not anticipated. However, replacing methyl by *n*-alkyl in $\mathbf{(R_2N)_2}$ shows a large steric effect; $\Delta G^{\ddagger}_{ii}$(fit) for $\mathbf{(Et_2N)_2}$ is 3.4 kcal/mol larger than for $\mathbf{(Me_2N)_2}$. However, λ_v should be similar for *n*-alkylhydrazines to that for tetramethylhydrazine (we calculate $\mathbf{(Et_2N)_2}$ to have a 5.4 kcal/mol higher λ_v, corresponding to a 1.35 increment in ΔG), so most of the increase in reactivity for $\mathbf{(Me_2N)_2}$ compared to $\mathbf{(Et_2N)_2}$ presumably results from a decrease in H_{ab} for going from methyl to *n*-alkyl substituents at nitrogen. If anything, one expects λ_s to be less for the larger volume $\mathbf{(Et_2N)_2}$ as compared to $\mathbf{(Me_2N)_2}$.

The increment in $\Delta G^{\ddagger}_{ii}$(fit) for changing four methyl to *n*-alkyl groups corresponds to −1.5 to −1.7 in $\log_{10}H_{ab}$, and for two groups an increment of −0.74 to −0.77 in $\log_{10}H_{ab}$. How could such changes result? It is known from EPR splitting constants and calculations that the methyl groups of $\mathbf{(Et_2N)_2}^+$ alternate up and down around its nearly planar R_4N_2 structure, which we suggested precludes direct overlap of the two-atom π system at nitrogen with another molecule [16]. However, there is no further increase in $\Delta G^{\ddagger}_{ii}$(fit) for further lengthening the alkyl groups, which we believe is because the ET has to proceed through nonbonded alkyl–alkyl interactions when hydrazines with *n*-alkyl groups approach. When molecules with *n*-alkyl groups touch, they have a smaller value of H_{ab} that is nearly constant and significantly smaller than occurs for **(Me$_2$N)$_2$**, 0.30, using $\lambda_s = 8$ kcal/mol (Table 10.4, entry 4). The couples **r5NNMe$_2$** (entry 3) and **r6NNMe$_2$** (entry 6) also show large H_{ab} values (**r5NNMe$_2$** anomalously large, perhaps suggesting a systematic error of some sort), and their structures preclude the alternating methyl group conformations adopted by **(Et$_2$N)$_2$**$^+$ and larger *n*-alkyl compounds. The H_{ab} values obtained for the 17 hydrazines (Table 10.4) that are *n*-alkyl or more hindered in their substitution are in the range 0.02–0.09, suggesting a scatter due to our assumptions in obtaining H_{ab} of about $\pm 0.03\,M^{-1/2}\,kcal^{-1}$, but **21/21**$^{0/+}$ (entry 15), the most pyramidal and hence least hindered of these compounds for approach of a partner to its NN π system, produces a larger H_{ab} of 0.15 kcal/mol that may indicate partial direct π contact for this couple. The extremely hindered **(b$_2$Ph)$_2$N)$_2$** (entry 32) produces H_{ab} of 0.01, one-fifth that of the less hindered **(pTol$_2$N)$_2$** (entry 37).

Figure 10.5 compares color-coded calculated electron density at the van der Waals surfaces projected onto the van der Waals surfaces for four hydrazine radical cations, the H_{ab} values obtained from our analysis of the cross-rates for these couples (from Table 10.4), and how these H_{ab} value changes would affect the rate constants if only differences in H_{ab} affected the rate constants (labeled k'_{rel}). It is seen that although λ_v changes are by far the most important factor in determining the 2×10^{14} range in k_{ii}(fit) obtained for the couples studied, changes in H_{ab} also lead to clearly detectable effects on intrinsic rate constants.

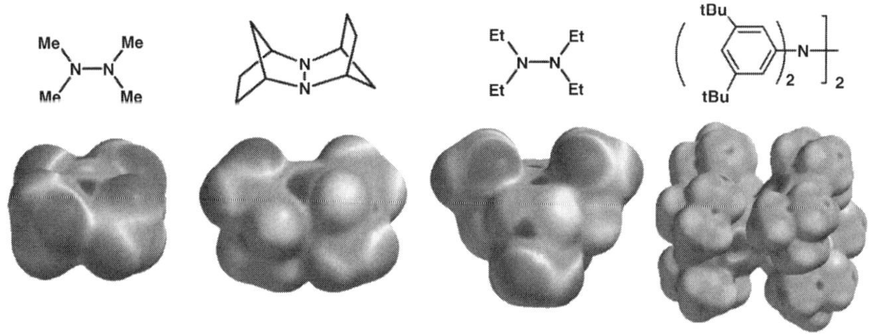

H_{ab}	0.30	0.15	0.05	0.01
k'_{rel}	900	225	25	1

FIGURE 10.5 Projections of color-coded charge (red, most positive charge; blue, most negative) on the van der Waals surfaces for **(Me$_2$N)$_2$**$^+$, **21/21**$^+$, **(Et$_2$N)$_2$**$^+$, and **(b$_2$Ph)$_2$N)$_2$**$^+$, along with the H_{ab} values obtained and the relative rate constants that would be observed if only H_{ab} differences affected the rate constants. Drawn using Spartan'02 (Spartan'02, Wavefunction, Inc., Irvine, CA).

The smaller H_{ab} for **(k33N)$_2$** (0.03, Table 10.4, entry 13) than for **(33N)$_2$** (0.09, entry 14) and for **(k33)$_2$PD** (0.27, entry 26) than for **(33)$_2$PD** (0.37, entry 28) still seem to us to be consistent with the regions near the electron-withdrawing carbonyl groups being ineffective for ET because of low spin density in these regions for these radical cations [14,15]. This effect might prove to be a useful design feature for decreasing ET rate constant without changing λ.

For the nonhydrazine couples, larger H_{ab} values are obtained for the unhindered heteroaromatic π systems having nearly planar radical cations, **TMPD** (0.15, Table 10.4, entry 29), **DMP** (0.29, entry 30), **iPrPT** (0.32, entry 31), **TTF** (0.40, entry 34), and **TMTSF** (0.67, entry 40) increase in this order, we suggested because of better overlap for the larger S and Se heteroatoms [16]. Larger basis sets are required for heavier atoms, and the neutral form of **TMTSF** was incorrectly calculated to be nonplanar using a 6-31G(d) basis set, leading to an anomalous λ_v estimate, as shown in Table 10.4. Considering the lowest H_{ab} values obtained, **N[333]N** (0.001, entry 24) has an exceptionally large geometry change and has its nitrogens pyramidalized inward as the radical cation [39], and **N[222]N** (0.003, entry 27) has a large geometry change in the very stiff CH_2–CH_2 bonds [40]. Ferrocene, **FeCp$_2$** (entry 39), produced the next lowest H_{ab} value of 0.008, which is much lower than the value of 0.1 kcal/mol obtained by Weaver and coworkers from considering solvent friction effects [21b,c]. As Weaver and coworkers point out, the charge is mainly centered on iron in **FeCp$_2$$^+$**, which appears to us to be even more important in lowering H_{ab} than they expected. Ferrocene and dimethylferrocene have similar ΔG_{ii}(fit) values of 8.0 and 8.3, respectively (Table 10.2). The range in ΔG_{ii}(fit) values observed including the other two methylated ferrocenes studied, **FeCp$_2^*$** and **FeCp*Cp**, is smaller than that expected from the reported self-exchange measurements [21], although we suggested from our measurements on pentamethylferrocene that the 10-fold larger rate constant reported for **FeCp$_2^{*+}$** than for **FeCp$_2$** self-ET might be caused by a problem with the couple not being in the slow-exchange limit [10].

Turning to the triarylamine couples, **An$_3$N** (Table 10.4, entry 33) gave an anomalously large intrinsic reactivity that is based on two reactions, those with **(iPr$_2$N)$_2$$^+$** and **(cHx$_2$N)$_2$$^+$**. They produce k_{calcd}/k_{obsd} values with $\Delta G^\ddagger$ differing by the anomalously large amount of 1.22 kcal/mol. If the lower rate constant is taken as the correct one (they obviously both cannot be correct), it would make $\Delta G^\ddagger_{ii}$(fit) be 4.57 kcal/mol, and produce an H_{ab} value of 0.17, which, although larger than any of the other triarylamines, is more reasonable. Anisyl substitution on triarylamines consistently leads to larger H_{ab} values than other substitution. There is good evidence that electron-releasing substituents raise H_{ab} and electron-withdrawing ones lower it. Both **Xy$_2$AnN** (entry 38) and **Xy$_2$pBiN** (entry 45), which is also calculated to raise the positive charge at the *para* position relative to the **pTol**-substituted compounds [16], give $H_{ab} \sim 0.09$, the largest of the other monotriarylamines. Both **Xy$_2$pTolN** (entry 44) and **pTol$_3$N** (entry 46) give intermediate H_{ab} values of 0.05–0.06, and **Xy$_2$pBrN** (0.02, entry 41) and **(pBr)$_3$N** ($H_{ab} = 0.03$, entry 42) give smaller values. This may be an echo of the clearer H_{ab}-lowering effect of the electron-withdrawing 3-carbonyl groups on bicyclononyl-substituted systems. An electron-withdrawing group may contract the acceptor LUMO in the region of the molecule critical to overlap for ET and since it is presumably the ability of electrons in the neutral component of the reaction to donate electrons to the hole of the cation that enhances H_{ab}, the lack of hole character in the vicinity of electron-withdrawing groups might produce "dead spots" for ET approaches to such regions that effectively lower H_{ab}. The effect may be principally at the *para* positions because approach to the *ortho* positions is partially blocked by the adjacent rings. An indication of the importance of steric lowering of H_{ab} in triarylamines is given by comparing the value for the bis-dianisylamino-*p*-phenylene **An$_4$PD** (entry 35, 0.10) with that for its naphthalene analogue, **An$_4$ND** (entry 36, 0.34). The 1,4-substitution on a benzene ring for the former places the large **An$_2$N** substituents closer than does the 2,6-naphthalene ring substitution for the latter, which apparently has the effect of blocking access to the aryl ring between the nitrogens. For couples for which we did not calculate λ_v because of their size, **Z$_2$An$_2$PD** and **Z$_4$PD** successively replace two and all four of the *p*-methoxy-phenyl substituents of **An$_4$PD** with the much bulkier *m,m'*-dianisylamin-phenyl groups (**Z**). Although the intrinsic barrier for **Z$_2$An$_2$PD** was indistinguishable from that of **An$_4$PD**, the intrinsic barrier for **Z$_4$PD** increased by about 1.4 kcal/mol, possibly reflecting the redox inversion between groups on the periphery and the interior suggested by the makers of these compounds, Selby and Blackstock [29].

In a similar fashion we have used Eq. (10.5) to calculate H_{ab} for each of the cross-ET reactions for which the necessary data are available. The cross-reaction ΔG° values were obtained from the standard reduction potentials for the couples obtained under the same solvent, temperature, and electrolyte conditions as for the measurement of the cross-reaction ET rate constants; the second-order cross-reaction ET rate constants were converted to unimolecular rate constants by dividing by K_c, approximated to be $1\,M^{-1}$; cross-reaction λ (total) values were obtained by taking the arithmetic average of the two relevant self-exchange λ_v values found in Table 10.4 and again approximating λ_s at 8 kcal/mol for each reaction. The values used and the resulting H_{ab} are summarized in Table 10.5.

Sutin [41] has shown that for outer sphere reactions that satisfy the assumptions implicit in Eq. (10.4), a good approximation for the cross-reaction H_{ab} can be obtained from the geometric mean of the H_{ab} values of the two related self-exchange reactions (Eq. (10.9)):

$$(H_{ab}\text{Geo}) = [(H_{ii})(H_{jj})]^{1/2} \tag{10.9}$$

For comparison we have computed the "geometric mean" H_{ab} for each cross-reaction and listed them in Table 10.5.

The H_{ab} values estimated by these two methods for all but 13 of the 92 reactions listed agree within 30% and none differ by more 62%, a factor of 1.6. Considering the approximations and assumptions that underlie Eqs (10.5) and (10.9), the agreement is quite good, as illustrated in Figure 10.6.

We note that entries 63, 66, and 67 of Table 10.5, which have noticeably larger positive deviations than most entries in Figure 10.6 and occur at the upper right end of the plot, are between the planar heteroaromatics **TTF$^+$** and **TMTSF$^+$** with high H_{ab} hydrazine couples **r5NNMe$_2$^{0}**, **(Me$_2$N)$_2$^{0}**, and **r6NNMe$_2$^{0}**, respectively. These are

TABLE 10.5 Observed Cross-Reaction Rate Constants, k_{ij} ($M^{-1}\,s^{-1}$), Reaction Free Energies, ΔG° (kcal/mol), Reorganization Energies, λ°_{ij} (kcal/mol), and Electronic Coupling Matrix Element, H_{ab} and H_{ab} Geo (kcal/(mol $M^{1/2}$)), Calculated Using Eqs (10.5) and (10.9), Respectively

Entry	RxnReaction: $i^+ + j^o$	k_{ij}	ΔG°	λ_{ij} DFT	H_{ab}	H_{ab} Geo	% Difference[a]
1	$(iPr_2N)_2^+ + TMPD^0$	1.20×10^4	−3.22	45.3	0.09	0.07	24
2	$(iPr_2N)_2^+ + 21/21^0$	2.60×10^2	−5.75	57.085	0.06	0.07	−8
3	$(iPr_2N)_2^+ + 21/u22^0$	5.50×10	−4.65	56.04	0.04	0.04	−14
4	$(iPr_2N)_2^+ + 22/22^0$	3.20×10^5	−18.2	57.385	0.04	0.03	16
5	$(iPr_2N)_2^+ + 22/u22^0$	1.10×10^4	−11.5	54.89	0.03	0.03	0
6	$(iPr_2N)_2^+ + 33N)_2^0$	1.10×10^2	−6.21	57.63	0.04	0.05	−35
7	$(iPr_2N)_2^+ + 22/u23^0$	3.10×10^4	−12.8	55.76	0.04	0.05	−12
8	$(iPr_2N)_2^+ + (33)_2PD^0$	4.90×10^4	−5.52	46.83	0.11	0.11	−2
9	$(iPr_2N)_2^+ + DMP^0$	2.00×10^4	−2.81	44.51	0.12	0.10	17
10	$(k33N)_2^+ + iPr_2N)_2^0$	1.35×10	−4.37	58.86	0.04	0.03	18
11	$(k33)PD_2^+ + (iPr_2N)_2^0$	7.70×10^2	−0.690	47.74	0.11	0.09	14
12	$TTF^+ + (iPr_2N)_2^0$	1.43×10^4	−1.61	41.9	0.09	0.11	−24
13	$TMTSF^+ + (iPr_2N)_2^0$	5.20×10^5	−3.75	40.54	0.18	0.15	17
14	$(b_2Ph)_2N)_2^+ + (iPr_2N)_2^0$	1.90×10^5	−8.05	42.69	0.04	0.02	43
15	$22/Ph_2^+ + (iPr_2N)_2^0$	1.02×10^3	−5.06	52.18	0.06	0.05	17
16	$(33)_2N_4^+ + (iPr_2N)_2^0$	4.70×10^2	−3.22	53.2	0.10	0.12	−17
17	$(pTol_2N)_2^+ + (iPr_2N)_2^0$	1.60×10^6	−8.97	41.43	0.06	0.05	21
18	$An_4PD^+ + (iPr_2N)_2^0$	8.10×10^3	−2.07	41.67	0.05	0.06	−4
19	$(An)_3N^+ + (iPr_2N)_2^0$	5.70×10^5	−6.90	41.98	0.08	0.13	−62
20	$(k33)_2N_4^+ + (iPr_2N)_2^0$	8.80×10^4	−1.13	54.31	0.09	0.08	9
21	$An_4ND^+ + (iPr_2N)_2^0$	2.30×10^5	−4.51	41.57	0.11	0.10	4
22	$pTol_3N^+ + (iPr_2N)_2^0$	1.40×10^7	−1.19	40.3	0.05	0.04	26
23	$Xy_2AnN^+ + (iPr_2N)_2^0$	2.30×10^6	−9.04	41.16	0.06	0.05	15
24	$(k33N)_2^+ + 21/u22^0$	2.20×10^5	−9.02	45.22	0.05	0.04	15

25	**$(k33N)_2^+ + 21/21^0$**	8.30×10^5	−1.01	46.26	0.08	0.07	14
26	**$(k33N)_2^+ + FeCp_2^0$**	4.80×10^4	−1.27	30.27	0.02	0.02	1
27	**$(k33N)_2^+ + TTF^0$**	5.40×10^6	−2.76	31.08	0.11	0.11	−1
28	**$(k33N)_2^+ + (k33)_2PD^0$**	9.80×10^5	−3.68	36.91	0.11	0.09	21
29	**$(k33N)_2^+ + (33)_2N_4^0$**	8.30×10^3	−1.15	42.38	0.09	0.11	−21
30	**$(k33N)_2^+ + TMTSF^0$**	2.30×10^6	−0.62	29.72	0.12	0.14	−15
31	**$(k33N)_2^+ + An_4PD^0$**	1.19×10^6	−2.30	30.85	0.06	0.05	6
32	**$(pTol_2N)_2^+ + (k33N)_2^0$**	4.40×10^6	−4.60	30.61	0.04	0.04	1
33	**$(b_2Ph)_2N)_2^+ + (k33N)_2^0$**	1.98×10^5	−3.68	31.87	0.02	0.02	−12
34	**$22/Ph_2^+ + (k33N)_2^0$**	2.80×10^3	−0.69	41.35	0.05	0.05	12
35	**$An_4PD^+ + (k33N)_2^0$**	4.40×10^4	2.30	30.85	0.08	0.05	31
36	**$An_4ND^+ + (k33N)_2^0$**	1.03×10^6	−0.134	30.74	0.12	0.10	20
37	**$(pTol_2N)_2^+ + (Bz_2N)_2^0$**	2.30×10^3	−1.15	39.51	0.03	0.03	−22
38	**$iPrPT^+ + (Bz_2N)_2^0$**	2.80×10^4	−3.15	42.24	0.07	0.07	4
39	**$(k33)_2N_4^+ + (Bz_2N)_2^0$**	3.00×10^2	−3.45	52.38	0.06	0.06	5
40	**$(pBr)_3N^+ + (Bz_2N)_2^0$**	4.60×10^6	−11.5	38.5	0.02	0.02	17
41	**$pTol_3N^+ + (Bz_2N)_2^0$**	4.40×10^4	−4.03	38.38	0.03	0.03	2
42	**$Xy_2pBrN^+ + (Bz_2N)_2^0$**	1.15×10^5	−5.68	38.52	0.03	0.02	38
43	**$(pTol_2N)_2^+ + (Bz_2N)_2^0$**	2.60×10^4	−3.06	38.41	0.03	0.03	7
44	**$Xy_2pBiN^+ + (Bz_2N)_2^0$**	6.90×10^4	−3.91	38.39	0.04	0.04	2
45	**$Xy_2ptBuN^+ + (Bz_2N)_2^0$**	3.10×10^4	−3.20	38.41	0.03	0.03	13
46	**$Xy_2AnN^+ + (Bz_2N)_2^0$**	5.90×10^3	−1.22	39.23	0.04	0.04	1
47	**$(k33)PD_2^+ + (Me_2N)_2^0$**	1.70×10^3	−0.23	48.17	0.22	0.29	−32
48	**$(k33)PD_2^+ + (Et_2N)_2^0$**	2.10×10^2	0.000	50.86	0.15	0.12	19
49	**$(k33)PD_2^+ + (iPrMeN)_2^0$**	1.20×10^3	−1.15	48.84	0.14	0.10	30
50	**$(k33)PD_2^+ + r5NNMe_2^0$**	1.02×10^5	−3.91	48.73	0.43	0.44	−3
51	**$(k33)PD_2^+ + (Me_2N)_2^0$**	1.49×10^5	−2.53	38.84	0.10	0.09	15
52	**$(k33)PD_2^+ + 21/Me_2^0$**	1.50×10^5	−3.22	42.37	0.17	0.14	19
53	**$(33N)_2^+ + 22/u22^0$**	6.30×10^4	−5.31	42.84	0.06	0.06	1

(Continued)

TABLE 10.5 (*Continued*)

Entry	RxnReaction: $\mathbf{i}^{+} + \mathbf{j}^{o}$	k_{ij}	ΔG°	λ_{ij} DFT	H_{ab}	H_{ab} Geo	% Difference[a]
54	$\mathbf{(33N)_2^{+} + N[333]N^0}$	3.20×10^{3}	−3.57	38.64	0.01	0.01	4
55	$\mathbf{TMPD^{+} + (33N)_2^0}$	2.00×10^{6}	−2.99	33.25	0.10	0.12	−23
56	$\mathbf{(33)_2PD^{+} + (33N)_2^0}$	1.20×10^{6}	0.690	34.78	0.26	0.18	30
57	$\mathbf{21/21^{+} + (33N)_2^0}$	2.09×10^{3}	−0.460	45.04	0.11	0.11	−3
58	$\mathbf{22/tBuPh^{+} + (33N)_2^0}$	1.02×10^{5}	−6.21	41.79	0.04	0.04	−6
59	$\mathbf{DMP^{+} + (33N)_2^0}$	1.30×10^{7}	−3.40	32.46	0.17	0.16	7
60	$\mathbf{TTF^{+} + (Me_2N)_2^0}$	1.40×10^{5}	−1.15	42.34	0.38	0.35	9
61	$\mathbf{TTF^{+} + (Et_2N)_2^0}$	4.50×10^{3}	−0.920	45.02	0.13	0.15	−9
62	$\mathbf{TTF^{+} + (iPrMeN)_2^0}$	1.97×10^{4}	−2.07	43.01	0.11	0.12	−7
63	$\mathbf{TTF^{+} + r5NNMe_2^0}$	1.10×10^{7}	−4.83	42.90	0.89	0.53	40
64	$\mathbf{TTF^{+} + 21Me_2^0}$	6.50×10^{6}	−4.14	36.53	0.224	0.17	25
65	$\mathbf{TMTSF^{+} + (Et_2N)_2^0}$	1.05×10^{5}	−3.06	43.66	0.202	0.19	6
66	$\mathbf{TMTSF^{+} + (Me_2N)_2^0}$	4.10×10^{6}	−3.29	40.975	0.645	0.45	30
67	$\mathbf{TMTSF^{+} + r6NNMe_2^0}$	3.00×10^{6}	−2.60	40.3	0.63	0.48	23
68	$\mathbf{TMTSF^{+} + (iPrMeN)_2^0}$	2.80×10^{5}	−4.21	41.645	0.14	0.16	−15
69	$\mathbf{TMPD^{+} + 22/tBuMe^0}$	1.10×10^{5}	−0.230	35.09	0.10	0.09	14
70	$\mathbf{22/tBuPh^{+} + TMPD^0}$	4.30×10^{6}	−3.22	29.46	0.06	0.06	−3
71	$\mathbf{(33)_2N_4^{+} + FeCp_2^0}$	3.30×10^{6}	−0.115	24.62	0.06	0.06	−2
72	$\mathbf{(33)_2N_4^{+} + [u6]Me_2^0}$	8.70×10^{3}	−2.76	49.87	0.25	0.13	47
73	$\mathbf{(33)_2N_4^{+} + [6]Me_2^0}$	1.40×10^{5}	−5.06	44.3	0.12	0.11	11
74	$\mathbf{(33)_2N_4^{+} + 21/Me_2^0}$	5.80×10^{4}	−5.75	47.83	0.13	0.18	−33
75	$\mathbf{22/Ph_2^{+} + (33)_2N_4^0}$	4.30×10^{5}	−1.84	35.70	0.12	0.18	−50
76	$\mathbf{22/Ph_2^{+} + FeCp_2^0}$	3.50×10^{6}	−1.96	23.59	0.02	0.02	−7
77	$\mathbf{22/Ph_2^{+} + r5NNMe_2^0}$	4.60×10^{4}	−8.28	53.17	0.15	0.23	−57
78	$\mathbf{22/Ph_2^{+} + [6]Me_2^0}$	1.35×10^{5}	−6.90	43.28	0.05	0.05	8
79	$\mathbf{22/Ph_2^{+} + [u6]Me_2^0}$	4.90×10^{3}	−4.60	48.84	0.07	0.05	26

80	**$pTol_3N^+$ + $(iPrPhN)_2^0$**	2.80×10^7	−1.73	19.86	0.03	0.03	1
81	**$pTol_3N^+$ + $(k33)_2N_4^0$**	1.41×10^7	−0.575	24.93	0.11	0.10	3
82	**$(b_2Ph)_2N)_2^+$ + $r6NNMe_2^0$**	2.60×10^5	−6.90	42.45	0.06	0.07	−15
83	**$(b_2Ph)_2N)_2^+$ + $[u6]Me_2^0$**	1.07×10^5	−7.59	39.36	0.02	0.02	−51
84	**$(b_2Ph)_2N)_2^+$ + $(iPr_2N)_2^0$**	4.30×10^5	−0.690	19.89	0.01	0.01	−9
85	**$22/Ph_2^+$ + $21/Me_2^0$**	9.40×10^4	−7.59	46.81	0.07	0.07	−3
86	**$22/tBuPh^+$ + $21/u22^0$**	4.60×10^4	−4.65	40.2	0.03	0.03	4
87	**$22/tBuPh^+$ + DMP^0**	8.00×10^6	−2.806	28.67	0.08	0.08	−6
88	**DMP^+ + $21/u22^0$**	2.80×10^6	−1.84	30.87	0.10	0.12	−19
89	**$22/tBuiPr^+$ + $22/u22^0$**	1.14×10^3	−2.622	45.06	0.03	0.03	0
90	**An_4PD^+ + $[u6]Me_2^0$**	3.20×10^4	−1.61	38.34	0.06	0.06	−1
91	**An_4PD^+ + $[6]Me_2^0$**	8.80×10^5	−3.91	32.77	0.040	0.05	−33
92	**An_4ND^+ + $[u6]Me_2^0$**	5.70×10^5	−4.05	38.23	0.10	0.12	−19

[a] More digits than reported for H_{ab} and H_{ab} Geo were used in the calculation of the % difference.

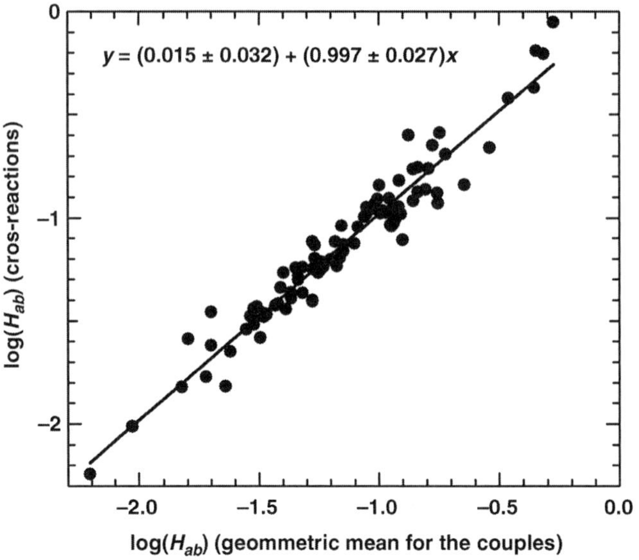

FIGURE 10.6 Plot of log (H_{ab}) determined from Eq. (10.5) versus log (H_{ab}) determined from Eq. (10.9).

cases for which H_{ab} Geo is significantly smaller than H_{ab} for the cross-reaction. This is the direction that one would expect deviation if K_c were significantly larger than 1. As noted previously [14–16], lack of steric hindrance will raise both K_c and H_{ab}.

The H_{ab} values obtained are small enough that the assumption that the reactions are nonadiabatic seems reasonable. While the values calculated from experimental data are obviously preferred, the geometric average is clearly a useful approach when limited information in available.

Various values have been offered for the approximate magnitude of H_{ab} defining the demarcation from adiabatic to nonadiabatic reactions and it is clearly expected to be system dependent [32,33b,c]. Generally one expects adiabatic behavior when H_{ab} is large enough to insure that the reaction passes smoothly on the lower energy surface from reactants to products through the transition state. Newton and Sutin [32,33b,c] suggest that a value above ~0.6 kcal/mol would insure an adiabatic reaction for outer sphere ET in typical small transition metal complexes. Given that almost all of the reactions, both self-exchange reactions and cross-reactions, reported here have H_{ab} considerably less than this suggests that these reactions are appropriately considered to be nonadiabatic. As mentioned in Section 10.3, the highly successful Bixon and Jortner treatment of nonadiabatic photo-ET reactions cannot be successfully applied to our data, despite the fact that electron tunneling is clearly involved in intramolecular ET in charge-localized mixed valence radical ions, because the appropriate H_{ab} for them is that obtained from Hush theory, and thus that at the ground state of the system, and not that at the transition state that is calculated for ET. H_{ab} at the transition state would be far larger than that at the ground state, making which value is appropriate unambiguous [8].

10.8 COMPARISON WITH GAS-PHASE REACTIONS

Equations (10.2a) and (10.2b) also have been successfully applied to 40 gas-phase ET reactions between some of the hydrazines described above in acetonitrile, demonstrating not only that these reactions are activation limited but also that solvent reorganization energy is not especially important when vibrational reorganization energy is large [42]. At least one high $\Delta G^{\ddagger}_{ii}$(fit) acyclic hydrazine must be included and the reaction cannot be too exoergonic or the ET activation barrier becomes too small to exceed the ion–neutral stabilization energy of the precursor complex, which has been measured to be −13 kcal/mol for $\mathbf{Me_2N)_2}^{0/+}$ itself. The same approach to extracting intrinsic reactivity for gas-phase reactions was used and the internal agreement was similar to the reactions in acetonitrile. Moreover, the order of reactivity was similar to that found in solution, but with a notable reversal of the reactivity of **[u6]Me$_2$** and **[6] Me$_2$**, which we have discussed [42].

10.9 CONCLUSIONS

The principal approximation involved in the treatment described here was the need to estimate the typical solvent reorganization energy, λ_o, for the reactions studied in order to complete the calculations of H_{ab}. While we are confident that the approximation used—λ_s is about 8 kcal/mol—is a reasonable one based on our experimental results, it points to the need for better theoretical and experimental methods for measuring solvent reorganization, as none of the existing approaches appear to provide accurate estimates of this quantity.

The classical Marcus cross-rate expression (Eqs (10.2a) and (10.2b)) correlates all experimental data that we have obtained extraordinarily well. This is surprising since Eqs (10.2a) and (10.2b) were derived for adiabatic reactions and the reactions studied here seem best to be characterized as nonadiabatic. These experiments generated an internally consistent intrinsic reactivity series for couples ranging in structure from **TMTSF** to **(nPr$_2$N)$_2$**, whose self-exchange rate constants differ by a factor of 2×10^{14}, and include many very hindered compounds that appear limited to transferring an electron through nonbonded contact of saturated alkyl groups. The correlation observed demonstrates that pre-exponential factors and barriers must effectively average for the cross-reactions studied, and that the same intrinsic rate constants apply whether a couple is reacted with a hydrazine, a ferrocene, or an aromatic compound. One facet of these studies that we believe contributed to exceptional internal agreement was the ability to obtain the relevant information, such as redox potentials and rate constants, measured under identical experimental conditions of solvent, ionic strength, and temperature. Moreover, restricting the study to reactions involving a neutral molecule obviated the need to estimate or account for electrostatic work terms, both simplifying the treatment and raising confidence in the result. The close agreement of the k_{ii}(fit) obtained by this approach with self-exchange rate constants measured directly by magnetic resonance line broadening, where available, gives us confidence that this approach provides accurate self-exchange rate constants. Moreover, this treatment has allowed the evaluation of a large number of couples with insufficiently stable cations or too high intrinsic reactivity to evaluate by any other method. Of particular note in this regard is the fact that several of the couples such as **TMTSF** are so reactive that their self-exchange rates approach diffusion control. Yet by pairing them with much less reactive reaction partners and studying cross-reactions, we have been able to extract accurate estimates of their intrinsic reactivity uncomplicated by corrections for diffusion.

The combination of cross-reaction studies to deduce self-exchange rate constants [23] as just described with modern DFT calculations using Nelsen's four-point energy analysis [30] to determine the relevant inner shell reorganizational energies, λ_v, for the related self-exchange ET reactions, has allowed the direct calculation of the electronic coupling matrix element between donor and acceptor, H_{ab}, for 46 self-exchange reactions and 92 of their cross-reactions. Historically, the estimation of H_{ab}, either computationally or experimentally, has been a challenge. These results show the value of combining cross-reaction studies with modern computations and applying current theory to extract these quantities and gain deeper insight to outer sphere ET dynamics of small organic molecules.

REFERENCES

1. See, for example, T. H. Lowry and K. S. Richardson, *Mechanism and Theory in Organic Chemistry*, 3rd edition, Harper and Row, New York (1987), for discussion of Hammett correlations (p. 143), solvolysis reactions (p. 425), and alkene hydration (p. 569).
2. (a) R. A. Marcus, *J. Chem. Phys.* **24**, 966 (1956);(b) R. A. Marcus, *Discuss. Faraday Soc.* **29**, 21 (1960); (c) R. A. Marcus, *J. Phys. Chem.* **67**, 853, 2889 (1963); (d) R. A. Marcus and N. Sutin, *Inorg. Chem.* **14**, 213 (1975).
3. H. Taube and E. S. Gould, *Acc. Chem. Res.* **2**, 321 (1969).
4. (a) L. Eberson and S. S. Shaik, *J. Am. Chem. Soc.* **112**, 4484–4489 (1990); (b) S. Fukuzumi, C. L. Wong, and J. K. Kochi, *J. Am. Chem. Soc.* **102**, 2928 (1980); (c) S. Fukuzumi and J. K. Kochi, *Bull. Chem. Soc. Jpn.* **56**, 969 (1983); (d) J. K. Kochi, *Acc. Chem. Res.* **25**, 39–47 (1992); (e) S. M. Hubig, T. M. Bockman, and J. K. Kochi, *J. Am. Chem. Soc.* **118**, 3842–3851 (1996).
5. (a) R. A. Marcus and N. Sutin, *Biochim. Biophys. Acta* **811**, 265 (1985); (b) N. Sutin, *Prog. Inorg. Chem.* **30**, 441 (1983).
6. S. Wherland, *Coord. Chem. Rev.* **123**, 169–199 (1993).
7. L. Eberson, *Electron Transfer Reactions in Organic Chemistry*, Springer-Verlag, Heidelberg (1987).
8. S. F. Nelsen, in J. P. Richard (ed.), *Advances in Physical Organic Chemistry*, Vol. 41, Academic Press Ltd., London (2006), pp. 183–215.
9. S. F. Nelsen, Y. Wang, M. T. Ramm, M. A. Accola, and J. R. Pladziewicz, *J. Phys. Chem.* **96**, 10654 (1992).
10. S. F. Nelsen, L.-J Chen, M. T. Ramm, G. T. Voy, D. R. Powell, M. A. Accola, T. Seehafer, J. Sabelko, and J. R. Pladziewicz, *J. Org. Chem.* **61**, 1405 (1996).

11. S. F. Nelsen, R. F. Ismagilov, L.-J. Chen, J. L. Brandt, X. Chen, and J. R. Pladziewicz, *J. Am. Chem. Soc.* **118**, 1555 (1996).
12. S. F. Nelsen, M. T. Ramm, R. F. Ismagilov, M. A. Nagy, D. A. Trieber II, D. R. Powell, X. Chen, J. J. Gengler, Q. Qu, J. L. Brandt, and J. R. Pladziewicz, J. R. *J. Am. Chem. Soc.* **119**, 5900–5907 (1997).
13. S. F. Nelsen, R. F. Ismagilov, K. E. Gentile, M. A. Nagy, H. Q. Tran, Q. Qu, D. T. Halfen, A. L. Oldegard, and J. R. Pladziewicz, *J. Am. Chem. Soc.* **120**, 8230–8240 (1998).
14. S. F. Nelsen, D. A. Trieber II, M. A. Nagy, A. Konradsson, D. T. Halfen, K. A. Splan, and J. R. Pladziewicz, *J. Am. Chem. Soc.* **122**, 5940–5946 (2000).
15. S. F. Nelsen, and J. R. Pladziewicz, *Acc. Chem. Res.* **35**, 247–254 (2002).
16. S. F. Nelsen, M. N. Weaver, Y. Luo, J. R. Pladziewicz, L. Ausman, T. L. Jentzsch, and J. J. O'Konneck, *J. Phys. Chem. A* **110**, 11665–11676 (2006).
17. (a) S. F. Nelsen and S. C. Blackstock, *J. Am. Chem. Soc.* **107**, 7189 (1985); (b) S. F. Nelsen, Y. Kim, and S. C. Blackstock, *J. Am. Chem. Soc.* **111**, 2045 (1989); (c) S. F. Nelsen and Y. Wang, *J. Org. Chem.* **59**, 1655 (1994); (d) Ref. [10].
18. (a) H. M. McConnell and H. E. Weaver, *J. Chem. Phys.* **25**, 307 (1956); (b) H. M. McConnell and S. B. Berger, *J. Chem. Phys.* **27**, 230 (1957); (c) C. R. Bruce, R. E. Norberg, and S. I. Weissman, *J. Chem. Phys.* **24**, 473 (1956).
19. S. F. Nelsen, C. R. Kessel, and D. J. Brien, *J. Am. Chem. Soc.* **102**, 702–711 (1980).
20. (a) S. F. Nelsen, *Acc. Chem. Res.* **14**, 131–138 (1981); (b) S. F. Nelsen, Hydrazine, hydrazine cation radical electron transfer reactions, in J. F. Liebmanand A. Greenberg (eds), *Molecular Structures and Energetics*, Chapter 1, Vol. 3, VCH Publishers, Inc., Deerfield Beach, FL (1986), pp. 1–56.
21. (a) E. S. Yang, M.-S. Chan, and A. C. Wahl, *J. Phys. Chem.* **84**, 3094 (1980); (b) G. E. McManis, R. M. Nielson, A. Gochev, and M. J. Weaver, *J. Am. Chem. Soc.* **111**, 5533 (1989); (c) R. M. Nielson, G. E. McManis, L. K. Safford, and M. J. Weaver, *J. Phys. Chem.* **93**, 2152 (1989); (d) K. Kirchner, K.-S. Dang, M. Stebler, H. W. Dodgcn, S. Wherland, and J. P. Hunt, *Inorg. Chem.* **28**, 3604 (1989); (e) Ref. [10].
22. (a) G. Grampp and W. Jaenicke, *Ber. Bunsenges. Phys. Chem.* **88**, 325 (1984); (b) G. Grampp and W. Jaenicke, *Ber. Bunsenges. Phys. Chem.* **88**, 335 (1984); (c) G. Grampp, *J. Magn. Reson.* **70**, 1 (1986); (d) G. Grampp and W. Jaenicke, *Ber. Bunsenges. Phys. Chem.* **95**, 904 (1991).
23. J. R. Pladziewicz and J. H. Espenson, *J. Am. Chem. Soc.* **95**, 56–63 (1973).
24. M. Bixon and J. Jortner, *Adv. Chem. Phys.* **106**, 35–202 (1999).
25. (a) S. F. Nelsen and J. M. Buschek, *J. Am. Chem. Soc.* **95**, 2011–2013 (1973); (b) S. F. Nelsen, J. M. Buschek, and P. J. Hintz, *J. Am. Chem. Soc.* **95**, 2013–2014 (1973); (c) S. F. Nelsen and J. M. Buschek, *J. Am. Chem. Soc.* **96**, 2392–2397 (1974); (d) S. F. Nelsen and J. M. Buschek, *J. Am. Chem. Soc.* **96**, 6982–6987 (1974); (e) S. F. Nelsen and J. M. Buschek, *J. Am. Chem. Soc.* **96**, 6987–6993 (1974); (f) S. F. Nelsen, V. E. Peacock, and G. R. Weisman, *J. Am. Chem. Soc.* **98**, 5269–5277 (1976).
26. (a) S. F. Nelsen, W. C. Hollinsed, C. R. Kessel, and J. C. Calabrese, *J. Am. Chem. Soc.* **100**, 7876 (1978); (b) S. F. Nelsen, S. C. Blackstock, and K. J. Haller, *Tetrahedron* **42** 6101 (1986); (c) S. F. Nelsen, Y. Wang, D. R. Powell, and R. K. Hiyashi, *J. Am. Chem. Soc.* **115**, 5246 (1993); (d) S. F. Nelsen, L.-J. Chen, D. R. Powell, and F. A. Neugebauer, *J. Am. Chem. Soc.* **117**, 11434 (1995); (e) S. F. Nelsen, R. F. Ismagilov, and D. R. Powell, *J. Am. Chem. Soc.* **119**, 10213 (1997); (f) S. F. Nelsen, H. Q. Tran, R. F. Ismagilov, L.-J. Chen, and D. R.

Powell, *J. Org. Chem.* **63**, 2536 (1998); (g) J. V. Lockard, J. I. Zink, D. A. Trieber, II, A. E. Konradsson, M. N. Weaver, and S. F. Nelsen, *J. Phys. Chem. A* **109**, 1205–1215 (2005).

27. S. F. Nelsen, *Acc. Chem. Res.* **11**, 14–20 (1978).
28. S. F. Nelsen, T. B. Frigo, Y. Kim, and S. C. Blackstock, *J. Am. Chem. Soc.* **111**, 5387 (1989).
29. T. D. Selby and S. C. Blackstock, *J. Am. Chem. Soc.* **120**, 12155 (1998).
30. S. F. Nelsen, S. C. Blackstock, and Y. Kim, *J. Am. Chem. Soc.* **109**, 677–682 (1987).
31. G. E. McManis, A. Gochev, R. M. Nielson, and M. J. Weaver, *J. Phys. Chem.* **93**, 7733–7739 (1989).
32. M. D. Newton and N. Sutin, *Annu. Rev. Phys. Chem.* **35**, 437–480 (1984).
33. (a) R. A. Marcus, *Annu. Rev. Phys. Chem.* **15**, 155–196 (1964);(b) R. A. Marcus, in H. Gerischerand J. J. Katz (eds), *Light-Induced Charge Separation in Biology and Chemistry*, Verlag Chemie, Berlin (1979), pp. 15–43; (c) J. R. Bolton, and M. D. Archer, in J. R. Bolton, N. Mataga, and G. McLendon (eds), *Electron Transfer in Inorganic, Organic, and Biological Systems*, Vol. 228, ACS, Washington (1991), pp. 7–23; (d) N. Sutin,in J. R. Bolton, N. Mataga, and G. McLendon (eds), *Electron Transfer in Inorganic, Organic, and Biological Systems*, Vol. 228, ACS, Washington (1991), pp. 26–43.
34. (a) V. G. Levich and R. R. Dogonadze, *Dokl. Phys. Chem.* **124**, 9 (1959) (English translation); (b) V. G. Levich, *Adv. Electrochem. Electrochem. Eng.* **4**, 249 (1966); (c) V. G. Levich and R. R. Dogonadze, *Coll. Czech. Chem. Commun.* **26**, 193–214 (1961) (O. Boshko, Trans., University of Ottawa, Ontario, Canada).
35. M. Meot-Ner (Mautner), *J. Phys. Chem.* **84**, 2724–2728 (1980).
36. J. M. Masnovi and J. K. Kochi, *J. Phys. Chem.* **91**, 1878–1881 (1987).
37. G. Rauhut and T. Clark, *J. Chem. Soc. Faraday Trans.* **90**, 1783 (1994).
38. J. J. Formosinho, L. G. Arnault, and R. Fausto, *Prog. React. Kinet.* **23**, 1–90 (1998).
39. R. W. Alder and R. B. Sessions, Special properties of di- and polyamines, in S. Patai (ed.), *The Chemistry of Amino Nitroso and Nitro Compounds and their Derivatives*, Chapter 18, John Wiley and Sons, Inc. , Chichester, NY (1982), pp. 763–803.
40. G. Balakrishnan, T. Keszthelyi, R. Wilbrandt, J. M. Zwier, A. M. Brouwer, and W. J. Buma, *J. Phys. Chem. A* **104**, 1834–1841 (2000).
41. (a) N. Sutin, *Acc. Chem. Res.* **1**, 225–231 (1968); (b) N. Sutin, *Acc. Chem. Res.* 15, 275–282 (1982).
42. S. F. Nelsen, A. Konradsson, T. L. Jentzsch, J. J. O'Konek, and J. R. Pladziewicz, *J. Chem. Soc. Perkin Trans. 2* 1552–1556 (2001).

INDEX

Rate Constant Calculation for Thermal Reactions: Methods and Applications, Edited by Herbert DaCosta and Maohong Fan.